AN INTRODUCTION TO COMPUTATIONAL BIOCHEMISTRY

AN INTRODUCTION TO COMPUTATIONAL BIOCHEMISTRY

C. Stan Tsai, Ph.D.

Department of Chemistry
and Institute of Biochemistry
Carleton University
Ottawa, Ontario, Canada

A JOHN WILEY & SONS, INC., PUBLICATION

Copyright © 2002 by Wiley-Liss, Inc., New York. All rights reserved.

Published simultaneously in Canada.

For ordering and customer service information please call 1-800-CALL-WILEY.

Library of Congress Cataloging-in-Publication Data:

Tsai, C. Stan.
     An introduction to computational biochemistry / C. Stan Tsai.
        p. cm.
     Includes bibliographical references and index.
     ISBN 0-471-40120-X (pbk. : alk. paper)
        1. Biochemistry--Data processing. 2. Biochemistry--Computer simulation. 3.
     Biochemistry--Mathematics. I. Title.

     QP517.M3 T733 2002
     572′.0285--dc21   2001057366

Printed in the United States of America.

10 9 8 7 6 5 4 3 2 1

CONTENTS

APPENDIX 343

INDEX 357

PREFACE

Since the arrival of information technology, biochemistry has evolved from an interdisciplinary role to becoming a core program for a new generation of interdisciplinary courses such as bioinformatics and computational biochemistry. A demand exists for an introductory text presenting a unified approach for the combined subjects that meets the need of undergraduate science and biomedical students.

This textbook is the introductory courseware at an entry level to teach students biochemical principles as well as the skill of using application programs for acquisition, analysis, and management of biochemical data with microcomputers. The book is written for end users, not for programmers. The objective is to raise the students' awareness of the applicability of microcomputers in biochemistry and to increase their interest in the subject matter. The target audiences are undergraduate chemistry, biochemistry, biomedical sciences, molecular biology, and biotechnology students or new graduate students of the above-mentioned fields.

Every field of computational sciences including computational biochemistry is evolving at such a rate that any book can seem obsolete if it has to discuss the technology. For this reason, this text focuses on a conceptual and introductory description of computational biochemistry. The book is neither a collection of presentations of important computational software packages in biochemistry nor the exaltation of some specific programs described in more detail than others. The author has focused on the description of specific software programs that have been used in his classroom. This does not mean that these programs are superior to others. Rather, this text merely attempts to introduce the undergraduate students in biochemistry, molecular biology, biotechnology, or chemistry to the realm of computer methods in biochemical teaching and research. The methods are not alternatives to the current methodologies, but are complementary.

This text is not intended as a technical handbook. In an area where the speed of change and growth is unusually high, a book in print cannot be either comprehensive or entirely current. This book is conceived as a textbook for students who have taken biochemistry and are familiar with the general topics. However, the book aims to reinforce subject matter by first reviewing the fundamental concepts of biochemistry briefly. These are followed by overviews on computational approaches to solve biochemical problems of general and special topics.

This book delves into practical solutions to biochemical problems with software programs and interactive bioinformatics found on the World Wide Web. After the introduction in Chapter 1, the concept of biochemical data analysis and management is described in Chapter 2. The interactions between biochemists and computers are the topics of Chapter 3 (Internet resources) and Chapter 4 (computer graphics). Computational applications in structural biochemistry are described in Chapter 5 (biochemical compounds) and then in Chapters 14 and 15 (molecular modeling). Dynamic biochemistry is treated in Chapter 6 (biomolecular interactions), Chapter 7 (enzyme kinetics), and Chapter 8 (metabolic simulation). Information biochemistry that overlaps bioinformatics and utilizes the Internet resources extensively is discussed in Chapters 9 and 10 (genomics), Chapters 11 and 12 (proteomics), and Chapter 13 (phylogenetic analysis).

I would like to thank all the authors who elucidate sequences and 3D structures of nucleic acids as well as proteins, and they kindly place such valuable information in the public domain. The contributions of all the authors who develop algorithms for free access on the Web sites and who provide highly useful software programs for free distribution are gratefully acknowledged. I thank them for granting me the permissions to reproduce their web pages, online and e-mail returns. I am grateful to Drs. Athel Cornish-Bowden (Leonora), Tom Hall (BioEdit), Petr Kuzmic (DynaFit), and Pedro Mendes (Gepasi) for the consents to use their software programs. The effort of all the developers and managers of the many outstanding Web sites are most appreciated. The development of this text would not have been possible without the contribution and generosity of these investigators, authors, and developers. I am thankful to Dr. D. R. Wiles for reading parts of this manuscript. It is my pleasure to state that the writing of this text has been a family effort. My wife, Alice, has been most instrumental in helping me complete this text by introducing and continuously coaching me on the wonderful world of microcomputers. My son, Willis, and my daughter, Ellie, have assisted me in various stages of this endeavor. The credit for the realization of this textbook goes to Luna Han, Editor, and Danielle Lacourciere, Associate Managing Editor, of John Wiley & Sons. This book is dedicated to Alice.

C. Stan Tsai
Ottawa, Ontario, Canada

1

INTRODUCTION

The use of microcomputers will certainly become an integral part of the biochemistry curriculum. *Computational biochemistry* is the new interdisciplinary subject that applies computer technology to solve biochemical problems and to manage and analyze biochemical information.

1.1. BIOCHEMISTRY: STUDIES OF LIFE AT THE MOLECULAR LEVEL

All the living organisms share many common attributes, such as the capability to extract energy from nutrients, the power to respond to changes in their environments, and the ability to grow, to differentiate, and to reproduce. Biochemistry is the study of life at the molecular level (Garrett and Grisham, 1999; Mathews and van Holde, 1996; Voet and Voet, 1995; Stryer, 1995; Zubay, 1998). It investigates the phenomena of life by using physical and chemical methods dealing with (a) the structures of biological compounds (biomolecules), (b) biomolecular transformations and functions, (c) changes accompanying these transformations, (d) their control mechanisms, and (e) impacts arising from these activities.

The distinct feature of biochemistry is that it uses the principles and language of one science, chemistry, to explain the other science, biology at the molecular level. Biochemistry can be divided into three principal areas: (1) Structural biochemistry focuses on the structural chemistry of the components of living matter and the relationship between chemical structure and biological function. (2) Dynamic biochemistry deals with the totality of chemical reactions known as metabolic processes that occur in living systems and their regulations. (3) Information biochemistry is

Structure: organization in space (length/radius in m.)

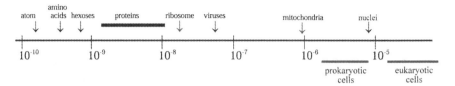

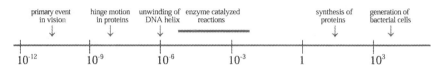

Reaction: organization in time (sec.)

Information: organization in number (number of nucleotides)

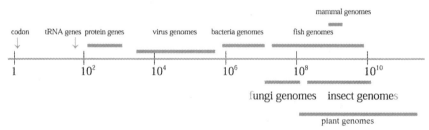

Figure 1.1. Representative organizations of biochemical components. Three component areas of biochemistry—structural, dynamic, and information biochemistry—are represented as organizations in space (dimensions of biomolecules and assemblies), time (rates of typical biochemical processes), and number (number of nucleotides in bioinformatic materials).

concerned with the chemistry of processes and substances that store and transmit biological information (Figure 1.1). The third area is also the province of molecular genetics, a field that seeks to understand heredity and the expression of genetic information in molecular terms.

Among biomolecules, water is the most common compound in living organisms, accounting for at least 70% of the weight of most cells, because water is both the major solvent of organisms and a reagent in many biochemical reactions. Most complex biomolecules are composed of only a few chemical elements. In fact, over 97% of the weight of most organisms is due to six elements (% in human): oxygen (62.81%), carbon (19.37%), hydrogen (9.31%), nitrogen (5.14%), phosphorus (0.63%), and sulfur (0.64%). In addition to covalent bonds (3000 ± 150 kJ/mol for single bonds) that hold molecules together, a number of weaker chemical forces (ranging from 4 to 30 kJ/mol) acting between molecules are responsible for many of the important properties of biomolecules. Among these noncovalent interactions (Table 1.1) are van der Waals forces, hydrogen bonds, ionic bonds/electrostatic interactions, and hydrophobic interactions.

TABLE 1.1. Energy Contribution and Distance of Noncovalent Interactions in Biomolecules

Chemical Force	Description	Energy (kJ/mol)	Distance (nm)	Remark
Van der Waals interactions	Induced electronic interactions between closely approaching atoms/molecules.	0.4–4.0	0.2	The limit of approach is determined by the sum of their vdW radii and related to the separation (r) of the two atoms by r^{-6}.
Hydrogen bonds	Formed between a covalently bonded hydrogen atom and an electronegative atom that serves as the hydrogen bond acceptor.	12–38	0.15–0.30	Proportional to the polarity of the donor and acceptor, stable enough to provide significant binding energy, but sufficiently weak to allow rapid dissociation.
Ionic bonds	Attractive forces between oppositely charged groups in aqueous solutions.	~ 20	0.25	Depending on the polarity of the interacting charged species and related to $q_i q_j / D r_{ij}$.
Hydrophobic interactions	Tendency of nonpolar groups or molecules to stick together in aqueous solutions.	~ 25	—	Proportional to buried surface area for the transfer of small molecules to hydrophobic solvents, the energy of transfer is 80–100 kJ/mol/Å^2 that becomes buried.

All biomolecules are ultimately derived from very simple, low-molecular-weight *precursors* (M.W. = 30 ± 15), such as CO_2, H_2O, and NH_3, obtained from the environment. These precursors are converted by living matter via series of metabolic *intermediates* (M.W. = 150 ± 100), such as acetate, α-keto acids, carbamyl phospahate etc., into the building-block biomoleucles (M.W. = 300 ± 150) such as glucose, amino acids, fatty acids and mononucleotides. They are then linked to each other covalently in a specific manner to form *biomacromolecules* (M.W. = $10^7 \pm 10^4$) or biopolymers. The unique chemistry of living systems results in large part from the remarkable and diverse properties of biomacromolecules. Macromolecules from each of the four major classes (e.g., polysaccharides, lipid bilayers, proteins, nucleic acids) may act individually in a specific cellular process, whereas others associate with one another to form supramolecular structures (particle weight $> 10^6$) such as proteosome, ribosomes, and chromosomes. All of these structures are involved in important cellular processes. The supramolecular complexes/systems are further assembled into organelles of eukaryotic cells and other types of structures. These organelles and substructures are enveloped by cell membrane into intracellular structures to form cells that are the fundamental units of living organisms. Viruses are supramolecular complexes of nucleic acids (either DNA or RNA) encapsulated in a protein coat and, in some instances, surrounded by a membrane envelope. Viruses infecting bacteria are called bacteriophages.

The cell is the basic unit of life and is the setting for most biochemical phenomena. The two classes of cell, eukaryotic and prokaryotic, differ in several respects but most fundamentally in that a eukaryotic cell has a nucleus and a prokaryotic cell has no nucleus. Two prokaryotic groups are the *eubacteria* and the *archaebacteria* (*archaea*). Archaea, which include thermoacidophiles (heat- and acid-tolerant bacteria), halophiles (salt-tolerant bacteria), and methanogens (bacteria that generate methane), are found only in unusual environments where other cells cannot survive. Prokaryotic cells have only a single membrane (plasma membrane or cell membrane), though they possess a distinct nuclear area where a single circular DNA is localized. Eukaryotic cells are generally larger than prokaryotic cells and more complex in their structures and functions. They possess a discrete, membrane-bounded nucleus (repository of the cell's genetic material) that is distributed among a few or many chromosomes. In addition, eukaryotic cells are rich in internal membranes that are differentiated into specialized structures such as the endoplasmic reticulum and the Golgi apparatus. Internal membranes also surround certain organelles such as mitochondria, chloroplasts (in plants), vacuoles, lysosomes, and peroxisomes. The common purpose of these membranous partitions is the creation of cellular compartments that have specific, organized metabolic functions. All complex multicellular organisms, including animals (Metazoa) and plants (Metaphyta), are eukaryotes.

Most biochemical reactions are not as complex as they may at first appear when considered individually. Biochemical reactions are enzyme-catalyzed, and they fall into one of six general categories: (1) oxidation and reduction, (2) functional group transfer, (3) hydrolysis, (4) reaction that forms or breaks carbon–carbon bond, (5) reaction that rearranges the bond structure around one or more carbons, and (6) reaction in which two molecules condense with an elimination of water. These enzymatic reactions are organized into many interconnected sequences of consecutive reactions known as metabolic pathways, which together constitute the metabolism of cells. Metabolic pathways can be regarded as sequences of the reactions organized to accomplish specific chemical goals. To maintain homeostatic conditions (a constant internal environment) of the cell, the enzyme-catalyzed reactions of metabolism are intricately regulated. The metabolic regulation is achieved through controls on enzyme quantity (synthesis and degradation), availability (solubility and compartmentation), and activity (modifications, association/dissociation, allosteric effectors, inhibitors, and activators) so that the rates of cellular reactions and metabolic fluxes are appropriate to cellular requirements.

An inquiry into the continuity and evolution of living organisms has provided great impetus to the progress of information biochemistry. Double-stranded DNA molecules are duplicated semiconservatively with high fidelity. The triplet-code words of genetic information encoded in DNA sequence are transcribed into codons of messenger RNA (mRNA) which in turn are translated into an amino acid sequence of polypeptide chains. The semantic switch from nucleotides to amino acids is aided by a 64-membered family of transfer RNA (tRNA). The ensuing folding process of polypeptide chains produces functional protein molecules. The processes of information transmission involve the coordinated actions of numerous enzymes, factors, and regulatory elements. One of the exciting areas of studies in information biochemistry is the development of recombinant DNA technology (Watson et al., 1992) which makes possible the cloning of tailored made protein molecules. Its impact on our life and society has been most dramatic.

1.2. COMPUTER SCIENCE AND COMPUTATIONAL SCIENCES

A computer is a machine that has the ability to store internally sequenced instructions that will guide it automatically through a series of operations leading to a completion of the task (Goldstein, 1986; Morley, 1997; Parker, 1988). A microcomputer, then, is regarded as a small stand-alone desktop computer (strictly speaking, a microcomputer is a computer system built around a microprocessor) that consists of three basic units:

1. The central processor unit (CPU) including the control logic that coordinates the whole system and manipulates data.
2. The memory consisting of random access memory (RAM) and read-only memory (ROM).
3. The buses and input/output interfaces (I/O) that connect the CPU to the other parts of the microcomputer and to the external world.

Computer science (Brookshear, 1997; Forsythe et al., 1975; Palmer and Morris, 1980) is concerned with four elements of computer problem solving namely problem solver, algorithm, language and machine. An algorithm is a list of instructions for carrying out some process step by step. An instruction manual for an assay kit is a good example of an algorithm. The procedure is broken down into multiple steps such as preparation of reagents, successive addition of reagents, and time duration for the reaction and measurement of an increase in the product or a decrease in the reactant. In the same way, an algorithm executed by a computer can combine a large number of elementary steps into a complicated mathematical calculation. Getting an algorithm into a form that a computer can execute involves several translations into different languages — for example,

English $\rightarrow$ Flowchart language $\rightarrow$ Procedural language $\rightarrow$ machine language

A flowchart is a diagram representing an algorithm. It describes the task to be executed. A procedure language such as FORTRAN and C enables a programmer to communicate with many different machines in the same language, and it is easier to comprehend than machine language. The programmer prepares a procedure language program, and the computer compiles it into a sequence of machine language instructions.

To solve a problem, a computer must be given a clear set of instructions and the data to be operated on. This set of instructions is called a *program*. The program directs the computer to perform various tasks in a predetermined sequence.

It is well known at a very basic level that computers are only capable of processing quantities expressed in binary form — that is, in machine code. In general, the computational scientist uses a high-level language to program the computer. This allows the scientist to express his/her algorithms in a concise and understood form. FORTRAN and C++ are the most commonly used high-level programming languages in scientific computations.

Recent years have seen considerable progress in computer technology, in computer science, and in the computational sciences. To a large extent, developments in these fields have been mutually dependent. Progress in computer technology has led to (a) increasingly larger and faster computing machines, (b) the supercomputers, and (c) powerful microcomputers. At the same time, research in computer science has explored new methods for the optimal use of these resources, such as the formulation of new algorithms that allow for the maximum amount of parallel computations. Developments in computer technology and computer science have had a very significant effect on the computational sciences (Wilson and Diercksen, 1997), including computational biology (Clote and Backofen, 2000; Pevzner, 2000; Setubai and Meidanis, 1997; Waterman, 1995), computational chemistry (Fraga, 1992; Jensen, 1999; Rogers, 1994), and computational biochemistry (Voit, 2000).

The main tasks of a computer scientist are to develop new programs and to improve efficiency of existing programs, whereas computational scientists strive to apply available software intelligently on real scientific problems.

1.3. COMPUTATIONAL BIOCHEMISTRY: APPLICATION OF COMPUTER TECHNOLOGY TO BIOCHEMISTRY

There is a general trend in biochemistry toward more quantitative and sophisticated interpretations of experimental data. As a result, demand for accurate, complex, and elaborate calculation increases. Recent progress in computer technology, along with the synergy of increased need for complex biochemical models coupled with an improvement in software programs capable of meeting this need, has led to the birth of computational biochemistry (Bryce, 1992; Tsai, 2000).

Computational biochemistry can be considered as a second-generation interdisciplinary subject derived from the interaction between biochemistry and computer science (Figure 1.2). It is a discipline of computational sciences dealing with all of the three aspects of biochemistry, namely, structure, reaction, and information. Computational biochemistry is used when biochemical models are sufficiently well developed that they can be implemented to solve related problems with computers. It may encompass bioinformatics. Bioinformatics (Baxevanis and Ouellete, 1998; Higgins and Taylor, 2000; Letovsky, 1999; Misener and Krawetz, 2000) is information technology applied to the management and analysis of biological data with the aid of computers. Computational biochemistry then applies computer technology to solve biochemical problems, including sequence data, brought about by the wealth of information now becoming available. The two subjects are highly intertwined and extensively overlapped.

Computational biochemistry is an emerging field. The contribution of "computational" has contributed initially to its development; however, as the field broadens and grows in its importance, the involvement of "biochemistry" increases prominently. In its early stage, computational biochemistry has been exclusively the domain of those who are knowledgeable in programming. This hindered the appreciation of computational biochemistry in the early days. The wide availability of inexpensive microcomputers and application programs in biochemistry has helped to relieve these restrictions. It is now possible for biochemists to rely on existing software programs and Internet resources to appreciate computational biochemistry in biochemical research and biochemical curriculum (Tsai, 2000). Well-established

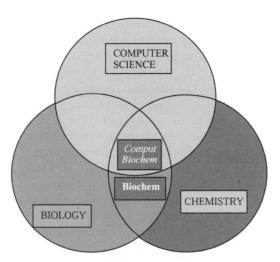

Figure 1.2. Relationship showing computational biochemistry as an interdisciplinary subject. Biochemistry is represented by the overlap (interaction) between biology and chemistry. A further overlap (interaction) between biochemistry and computer science represents computational biochemistry.

techniques have been reformulated to make more efficient use of the new computer technology. New and powerful algorithms have been successfully implemented.

Furthermore, it is becoming increasingly important that biochemists are exposed to databases and database management systems due to exponential increase in information of biochemical relevance. Visual modeling of biochemical structures and phenomena can provide a more intuitive understanding of the process being evaluated. Simulation of biochemical systems gives the biochemist control over the behavior of the model. Molecular modeling of biomolecules enables biochemists not only to predict and refine three-dimensional structures but also to correlate structures with their properties and functions.

The field has matured from the management and analysis of sequence data, *albeit* still the most important areas, into other areas of biochemistry. This text is an attempt to capture that spirit by introducing computational biochemistry from the biochemists' prospect. The material content deals primarily with the applications of computer technology to solve biochemical problems. The subject is relatively new and perhaps a brief description of the text may benefit the students.

After brief introduction to biostatistics, Chapter 2 focuses on the use of spreadsheet (Microsoft Excel) to analyze biochemical data, and of database (Microsoft Access) to organize and retrieve useful information. In the way, a conceptual introduction to desktop informatics is presented. Chapter 3 introduces Internet resources that will be utilized extensively throughout the book. Some important biochemical sites are listed. Molecular visualization is an important and effective method of chemical communication. Therefore, computer molecular graphics are treated in Chapter 4. Several drawing and graphics programs such as ISIS Draw, RasMol, Cn3D, and KineMage are described. Chapter 5 reviews biochemical compounds with an emphasis on their structural information and characterizations. Dynamic biochemistry is described in the next three chapters. Chapter 6 deals with ligand–receptor interaction and therefore receptor biochemistry including signal transductions. DynaFit, which permits free access for academic users, is employed to analyze interacting systems. Chapter 7 discusses quasi-equilibrium versus steady-state kinetics of enzyme reactions. Simplified derivations of kinetic equations as well as Cleland's nomenclature for enzyme kinetics are described. Leonora is used to evaluate kinetic parameters. Kinetic analysis of an isolated enzyme system is extended to metabolic pathways and simulation (using Gepasi) in Chapter 8. Topics on metabolic control analysis, secondary metabolism, and xenometabolism are presented in this chapter. The next two chapters split the subject of genomic analysis. Chapter 9 discusses acquisition (both experimental and computational) and analysis of nucleotide sequence data and recombinant DNA technology. The application of BioEdit is described here, though it can be used in Chapter 11 as well. Chapter 10 describes theory and practice of gene identifications. The following two chapters likewise share the subject of proteomic analysis. Chapter 11 deals with protein sequence acquisition and analysis. Chapter 12 is concerned with structural predic-tions from amino acid sequences. Internet resources are extensively used for genomic as well as proteomic analyses in Chapters 9 to 12. Since there are many outstanding Web sites that provide genomic and proteomic analyses, only few readily accessible sites are included. The phylogenetic analysis of nucleic acid and protein sequences is introduced in Chapter 13. The software package Phylip is used both locally and online. Chapter 14 describes general concepts of molecular modeling in biochemistry. The application of molecular mechanics in energy calculation, geometry optimiz-ation, and molecular dynamics are described. Chapter 15 discusses special aspect of molecular modeling as applied to protein structures. Freeware programs KineMage and Swiss-Pdb Viewer are used in conjunction with WWW resources. For a comprehensive modeling, two commercial modeling packages for PC (Chem3D and HyperChem) are described in Chapter 14 and they are also applicable in Chapter 15.

Each chapter is divided into four sections (except Chapter 1). From Chapters 5 to 15, biochemical principles are reviewed/introduced in the first section. The general topics covered in most introductory biochemistry texts are mentioned for the purpose of continuity. Some topics not discussed in general biochemistry are also introduced. References are provided so that the students may consult them for better understanding of these topics. The second section describes practices of the computa-tional biochemistry. Some backgrounds to the application programs or Internet resources are presented. Descriptions of software algorithms are not the intent of this introductory text and mathematical formulas are kept to the minimum. The third section deals with the application programs and/or Internet resources to perform computations. Aside from economic reasons, the use of suitable PC-based freeware programs and WWW services have the distinct appeal of portability, so that the students are able to continue and complete their assignments after the regular workshop period. There has been no attempt to exhaustively search for the many outstanding software programs and Web sites or to provide in-depth coverage of the functionalities of the selected application programs or Web sites. The focus is on their uses to solve pertinent biochemical problems. By these initial exposures, it is hoped that interest in these programs or resources may serve as catalysts for the students to delve deeper into the full functionalities of these programs or resources. Arrows ($\rightarrow$) are used to indicate a series of operations; for example, Select $\rightarrow$ Secondary Structure $\rightarrow$ Helix indicates that from the Select menu, choose Secondary Structure Pop-up Submenu (or Command) and then go to Helix Tool (or Option). For submission of amino acid/nucleotide sequences to the WWW servers for genomic/proteomic analyses, fasta format is generally preferred. The query sequence can be uploaded from the local file via browsing the directories/files or entering the path and the filename directly (e.g., [drive]:\[directory]\[file]). The copy-and-paste procedure (copying the sequence into the clipboard and pasting it onto the query box) is recommended for the online submission of the query sequence if the browser mechanism is unavailable. The requested executions by the Web servers appeared in capital letter(s), in italics or with underlines and are duplicated as they are on the Web pages. It is also helpful to know that the right mouse button is useful to bring up context sensitive commands that shortcut going to the menu bar for selection. Workshops in the last section are not merely exercises. They are designed to review familiar biochemical knowledge and to introduce some new biochemical concepts. Most of them are simple for a practical reason to minimize human and computer time.

REFERENCES

Baxevanis, A. D., and Ouellete, B. F. F., Eds. (1998) *Bioinformatics: A Practical Guide to the Analysis of Genes and Proteins.* Wiley-Interscience, New York.

Brookshear, J. G. (1997) *Computer Science: An Overview.* 5th edition, Addison-Wesley, Reading, MA.

Bryce, C. F. A. (1992) *Microcomputers in Biochemistry: A Practical Approach.* IRL Press, Oxford.

Clote, P., and Backofen, R. (2000) *Computational Molecular Biology: An Introduction.* John Wiley & Sons, New York.

Forsythe, A. I., Keenan, T. A., Organick, E. I., and Stenberg, W. (1975). *Computer Science, A First Course.* 2nd edition. John Wiley & Sons, New York.

Fraga, S., Ed. (1992) *Computational Chemistry: Structure, Interactions and Reactivity.* Elsevier, New York.

Garrett, R. H., and Grisham, C. M. (1999) *Biochemistry*, 2nd edition. Saunders College Publishing, San Diego.

Goldstein, L. J. (1986) *Computers and Their Applications.* Prentice-Hall, Englewood Cliffs, NJ.

Higgins, D., and Taylor, W., Eds. (2000) *Bioinformatics: Sequence, Structure and Databank.* Oxford University Press, Oxford.

Jensen, F. (1999) *Introduction to Computational Chemistry.* John Wiley & Sons, New York.

Letovsky, S. (1999) *Bioinformatics: Databases and Systems.* Kluwer Academic Publishers, Boston, MA.

Mathews, C. K., and van Holde, K. E. (1996) *Biochemistry*, 2nd edition. Benjamin/Cummings, New York.

Misener, S., and Krawetz, S. A. (2000) *Bioinformatics: Methods and Protocols.* Humana Press, Totowa, NJ.

Morley, D. (1997) *Getting Started with Computers.* Dryden Press/Harcourt Brace, FL.

Parker, C. S. (1988) *Computers and Their Applications.* Holt, Rinehart and Winston, New York.

Palmer, D. C., and Morris, B. D. (1980) *Computer Science.* Arnold, London.

Pevzner, P. A. (2000) *Computational Molecular Biology: An Algorithmic Approach.* MIT Press, Cambridge, MA.

Rogers, D. W. (1994) *Computational Chemistry Using PC*, 2nd edition. VCH, New York.

Setubai, J. C., and Meidanis, J. (1997) *Introduction to Computational Molecular Biology*. PWS Publishing Company, Boston, MA.

Stryer, L. (1995) *Biochemistry*, 4th edition. W. H. Freeman, New York.

Tsai, C. S. (2000) *J. Chem. Ed.* **77**:219–221.

Voet, D., and Voet, J. G. (1995) *Biochemistry*, 2nd edition. John Wiley & Sons, New York.

Voit, E. O. (2000) *Computational Analysis of Biochemical Systems: A Practical Guide for Biochemists and Molecular Biologists*. Cambridge University Press, New York.

Waterman, M. S. (1995) *Introduction to Computational Biology: Maps, Sequences and Genomes*. Chapman and Hall, New York.

Watson, J. D., Gilman, M., Witkowski, J., and Zoller, M. (1992) *Recombinant DNA*, 2nd edition, W. H. Freeman, New York.

Wilson, S., and Diercksen, G. H. F. (1997) *Problem Solving in Computational Molecular Science: Molecules in Different Environments*. Kluwer Academic, Boston, MA.

Zubay, G. L. (1998) *Biochemistry*, 4th edition. W. C. Brown, Chicago.

2

BIOCHEMICAL DATA:
ANALYSIS AND MANAGEMENT

This chapter is aimed at introducing the concepts of *biostatistics* and *informatics*. Statistical analysis that evaluates the reliability of biochemical data objectively is presented. Statistical programs are introduced. The applications of spreadsheet (Excel) and database (Access) software packages to analyze and organize biochemical data are described.

2.1. STATISTICAL ANALYSIS OF BIOCHEMICAL DATA

Many investigations in biochemistry are quantitative. Thus, some objective methods are necessary to aid the investigators in presenting and analyzing research data (Fry, 1993). *Statistics* refers to the analysis and interpretation of data with a view toward objective hypothesis testing (Anderson et al., 1994; Milton et al., 1997; Williams, 1993; Zar, 1999). *Descriptive statistics* refers to the process of organizing and summarizing the data in a way as to arrive at an orderly and informative presentation. However, it might be desirable to make some generalizations from these data. *Inferential statistics* is concerned with inferring characteristics of the whole from characteristics of its parts in order to make generalized conclusions.

2.1.1. The Quality of Data

All numerical data are subject to uncertainty for a variety of reasons; but because decisions will be made on the basis of analytical data, it is important that this uncertainty be quantified in some way. Variation between replicate measurements may be due to a variety of causes, the most predictable being random error that occurs as a cumulative result of a series of simple, indeterminate variations. Such error gives rise to results that will show a normal distribution about the mean. The number (n) of measurements (x_i or x_i) falling within the range of a particular group is known as *frequency*. The measurement occurring with the greatest frequency is known as the *mode*. The middle measurement in an ordered set of data is typically defined as *median*. That is, there are just as many observations larger than the median as there are smaller. The average of all measurements is known as the *mean*, and in theory, to determine this value (μ), many replicates are required. In practice, when the number of replicates is limited, the calculated mean ($\bar{x}$ or $\bar{x}$) is an acceptable approximation of the true value.

The sum of all deviations from the mean — that is, $\Sigma(x_i - \bar{x})$ — will always equal zero. Summing the absolute values of the deviations from the mean results in a quantity that expresses dispersion about the mean. This quantity is divided by n to yield a measure known as the *mean deviation* or the *standard error of the mean* (SEM), which expresses the confidence in the resulting mean value:

$$\text{SEM} = \Sigma|(x_i - \bar{x})|/n$$

An approach to eliminate the sign of the deviations from the mean is to square the deviations. The sum of the squares of the deviations from the mean is called the *sum of squares* (SS) and is defined as

$$\text{Population SS} = \Sigma(X_i - \mu)^2$$
$$\text{Sample SS} \quad = \Sigma(X_i - \bar{X})^2$$

As a measure of variability or dispersion, the sum of squares considers how far the X_i's deviate from the mean. The mean sum of squares is called the *variance* (or *mean squared deviation*), and it is denoted by σ^2 for a population:

$$\sigma^2 = \Sigma(X_i - \mu)^2/N$$

The best estimate of the population variance is the sample variance, s^2:

$$s^2 = \Sigma(X_i - \bar{X})^2/(n - 1) = [\Sigma X_i^2 - \{\Sigma X_i\}^2/n]/(n - 1)$$

Dividing the sample sum of squares by the degree of freedom ($n - 1$) yields an unbiased estimate. If all observations are equal, then there is no variability and $s^2 = 0$. The sample variance becomes increasingly large as the amount of variability or dispersion increases.

The most acceptable way of expressing the variation between replicate measurements is by calculating the *standard deviation* (s) of the data:

$$s = [\Sigma(\bar{x} - x)^2/n - 1]^{1/2}$$

where x is an individual measurement and n is the number of individual measurements. An alternative, more convenient formula to use is

$$s = [\{\Sigma\, x^2 - (\Sigma\, x)^2/n\}/(n-1)]^{1/2}$$

The calculation of standard deviation requires a large number of replicates. For any number of replicates less than 30, the value for s is only an approximate value and the function $(n-1)$ known as the degrees of freedom (DF) is used rather than (n).

In addition to random errors derived from samplings, systematic errors are peculiar to each particular method or system. They cannot be assessed statistically. A major effect of systematic error known as *bias* is a shift in the position of the mean of a set of readings relative to the original mean.

Analytical methods should be precise, accurate, sensitive, and specific. The precision or reproducibility of a method is the extent to which a number of replicate measurements of a sample agree with one another and is expressed numerically in terms of the standard deviation of a large number of replicate determinations. Statistical comparison of the relative precision of two methods uses the variance ratio (F_ϕ) or the F test.

$$F_\phi = s_1^2/s_2^2$$

The basic assumption, or null hypothesis (H_0), is that there is no significant difference between the variance (s^2) of the two sets of data. Hence, if such a hypothesis is true, the ratio of two values for variance will be unity or almost unity. The values for s_1 and s_2 are calculated from a limited number of replicates and, as a result, are only approximate values. The values calculated for F will vary from unity even if the null hypothesis is true. Critical values for F (F_{crit}) are available for different degrees of freedom; and if the test value for F exceeds F_{crit} with the same degrees of freedom, then the null hypothesis can be rejected.

Accuracy is the closeness of the mean of a set of replicate analyses to the true value of the sample. Often, it is only possible to assess the accuracy of one method relative to another by comparing the means of replicate analyses by the two methods using the t test. The basic assumption, or null hypothesis, made is that there is no significant difference between the mean value of the two sets of data. This is assessed as the number of times the difference between the two means is greater than the standard error of the difference (t value).

$$t_\phi = (\bar{x}_i - \bar{x}_2)/[\{\Sigma\, (x_1 - \bar{x}_1)^2 + \Sigma\, (x_2 - \bar{x}_2)^2\}^{1/2}]/n(n-1)$$

The critical value of the t test can be abbreviated as $t_{\alpha(2),v}$, where $\alpha(2)$ refers to the two-tailed probability of α and $v = n - 1$ (degree of freedom). For the two-tailed t test, compare the calculated t value with the critical value from the t distribution table. In general, if $|t| \geqslant t_{\alpha(2),v}$, then reject the null hypothesis. When comparing the means of replicate determinations, it is desirable that the number of replicates be the same in each case.

The sensitivity of a method is defined as its ability to detect small amounts of the test substance. It can be assessed by quoting the smallest amount of substance that can be detected. The specificity is the ability to detect only the test substance. It is important to appreciate that specificity is often linked to sensitivity. It is possible to reduce the sensitivity of a method with the result that interference effects become less significant and the method is more specific.

2.1.2. Analysis of Variance, ANOVA

We need to become familiar with the topic of *analysis of variance*, often abbreviated ANOVA, in order to test the null hypothesis (H_0): $\mu_1 = \mu_2 = \cdots = \mu_k$, where k is the number of experimental groups, or samples. In the ANOVA, we assume that $\sigma_1^2 = \sigma_2^2 = \cdots = \sigma_k^2$, and we estimate the population variance assumed common to all k groups by a variance obtained using the pooled sum of squares (within-groups SS) and the pooled degree of freedom (within-groups DF):

$$\text{within-groups SS} = \sum_{i=1}^{k} \left[\sum_{j=1}^{n} (X_{ij} - \bar{X}_i)^2 \right]$$

and

$$\text{within-groups DF} = \sum (n_i - 1)$$

These two quantities are often referred to as the *error sum of squares* and the *error degrees of freedom*, respectively. The former divided by the latter is a statistical value that is the best estimate of the variance, σ^2, common to all k populations:

$$\sigma^2 = \sum_{i=1}^{k} \left[\sum_{j=1}^{n} (X_{ij} - \bar{X}_i)^2 \right] \bigg/ \sum_{i=1}^{k} (n_i - 1)$$

The amount of variability among the k groups is important to our hypothesis testing. This is referred to as the *group sum of squares* and can be denoted as

$$\text{among-group SS} = \sum_{i=1}^{k} n_i (\bar{X}_i - \bar{X})^2$$

and the *groups degrees of freedom* is

$$\text{among-group DF} = k - 1$$

We also consider the variability present among all N data, that is,

$$\text{total SS} = \sum_{i=1}^{k} \sum_{j=1}^{n} (X_{ij} - \bar{X})^2$$

and

$$\text{total DF} = N - 1$$

In summary, each deviation of an observed datum from the grand mean of all data is attributable to a deviation of that datum from its group mean plus the deviation of that group mean from the grand mean, that is,

$$(X_{ij} - \bar{X}) = (X_{ij} - \bar{X}_i) + (\bar{X}_i - \bar{X})$$

Furthermore, the sums of squares and the degree of freedom are additive,

$$\text{total SS} = \text{group SS} + \text{error SS} = \sum_{i=1}^{k} \sum_{j=1}^{n} X_{ij}^2 - (\Sigma \Sigma X_{ij})^2/N$$

$$\text{total DF} = \text{group DF} + \text{error DF}$$

Computationally,

$$\text{total SS} = \sum_{i=1}^{k} \sum_{j=1}^{n} X_{ij}^2 - C$$

where

$$C = (\Sigma \Sigma X_{ij})^2/N$$

and

$$\text{groups SS} = \sum_{i=1}^{k} \left[\left(\sum_{j=1}^{n} X_{ij} \right)^2 \Big/ n_i \right] - C$$

$$\text{error SS} = \sum_{i=1}^{k} \sum_{j=1}^{n} X_{ij}^2 - \sum_{i=1}^{k} \left[\left(\sum_{j=1}^{n} X_{ij} \right)^2 \Big/ n_i \right] = \text{total SS} - \text{groups SS}$$

Dividing the group SS or the error SS by the respective degrees of freedom results in a variance referred to as *mean squared deviation from the mean* (*mean square*, MS):

$$\text{groups MS} = \text{groups SS}/\text{groups DF}$$

$$\text{error MS} = \text{error SS}/\text{error DF}$$

Table 2.1 summarizes the single factor ANOVA calculations. The test for the equality of means is a one-tailed variance ratio test, where the groups MS is placed in the numerator so as to inquire whether it is significantly larger than the error MS:

$$F = \text{groups MS}/\text{error MS}$$

The critical value for this test is $F_{\alpha(1),(k-1),(N-k)}$. If the calculated F is at least as large as the critical value, then we reject H_0.

It has become uncommon for ANOVA with more than two factors to be analyzed on a computer, owing to considerations of time, ease, and accuracy. It will presume that established computer programs will be used to perform the necessary mathematical manipulation of ANOVA.

TABLE 2.1. Single Factor ANOVA Calculations

Source of Variation	Sum of Squares, SS	Degree of Freedom, DF	Mean Square, MS
Total $[X_{ij} - \bar{X}]$	$\sum\limits_{i=1}^{k} \sum\limits_{j=1}^{n} X_{ij}^2 - C$	$N - 1$	
Group (i.e., among group) $[\bar{X}_i - \bar{X}]$	$\sum\limits_{i=1}^{k} \left[\left(\sum\limits_{j=1}^{n} X_{ij} \right)^2 \Big/ n_i \right] - C$	$k - 1$	Groups SS/groups DF
Error (i.e., within groups) $[X_{ij} - \bar{X}_I]$	Total SS — groups SS	$N - k$	Error SS/error DF

Note: $C = (\Sigma \Sigma X_{ij})^2/N$; $N = \sum\limits_{i=1}^{k} n_i$; k is the number of groups; n_i is the number of data in group i.

2.1.3. Simple Linear Regression and Correlation

The relationship between two variables may be one of dependency. That is, the magnitude of one of the variable (the dependent variable) is assumed to be determined by the magnitude of the second variable (the independent variable). Sometimes, the independent variable is called the predictor or regressor variable, and the dependent variable is called the response or criterion variable. This dependent relationship is termed *regression*. However, in many types of biological data, the relationship between two variables is not one of dependency. In such cases, the magnitude of one of the variables changes with changes in the magnitude of the second variable, and the relationship is *correlation*. Both simple linear regression and simple linear correlation consider two variables. In the simple regression, the one variable is linearly dependent on a second variable, whereas neither variable is functionally dependent upon the other in the simple correlation.

It is very convenient to graph simple regression data, using the abscissa (X axis) for the independent variable and the ordinate (Y axis) for the dependent variable. The simplest functional relationship of one variable to another in a population is the simple linear regression:

$$Y_i = \alpha + \beta X_i$$

Here, α and β are population parameters (constants) that describe the functional relationship between the two variables in the population. However, in a population the data are unlikely to be exactly on a straight line, thus Y may be related to X by

$$Y_i = \alpha + \beta X_i + \varepsilon_i$$

where ε_i is referred to as an error or *residual*.

Generally, there is considerable variability of data around any straight line. Therefore, we seek to define a so-called "best-fit" line through the data. The criterion for "best-fit" normally utilizes the concept of *least squares*. The criterion of least squares considers the vertical deviation of each point from the line ($Y_i - Y_i'$) and defines the best-fit line as that which results in the smallest value for the sum of the squares of these deviations for all values of Y_i' with respect to Y_i. That is, $\Sigma_{i=1}^{n}(Y_i - Y_i')^2$ is to be minimum where n is the number of data points. The sum of squares of these deviations is called the *residual sum of squares* (or the *error sum of squares*). Because it is impossible to possess all the data for the entire population, we have to estimate parameters α and β from a sample of n data, where n is the number of pairs of X and Y values. The calculations required to arrive at such estimates and to execute the testing of a variety of important hypotheses involve the computation of sums of squared deviations from the mean. This requires calculation of a quantity referred to as the sum of the cross-products of deviations from the mean:

$$\sum xy = \sum (X_i - \bar{X})(Y_i - \bar{Y}) = \sum X_i Y_i - \{[\Sigma X_i)(\Sigma Y_i)\}/n$$

The parameter β is termed the *regression coefficient*, or the *slope* of the best-fit regression line. The best estimate of β is

$$b = \Sigma xy/\Sigma x^2 = \{\Sigma (X_i - \bar{X})(Y_i - \bar{Y})\}/\Sigma (X_i - \bar{X})^2$$

$$= [\Sigma X_i Y_i - \{(\Sigma X_i)(\Sigma Y_i)\}/n][\Sigma X_i^2 - (\Sigma X_i)^2/n]$$

Although the denominator in this calculation is always positive, the numerator may be either positive, negative, or zero. The regression coefficient expresses what change in Y is associated, on the average, with a unit change in X.

A line can be defined uniquely, by stating, in addition to β, any one point on the line, conventionally on the line where $X = 0$. The value of Y in the population at this point is the parameter α, which is called the Y *intercept*. The best estimate of α is

$$a = \bar{Y} - b\bar{X}$$

By specifying both a and b, a line is uniquely defined. Because a and b are calculated using the criterion of least squares, the residual sum of squares from this line is the smallest. Certain basic assumptions are met with respect to regression analysis (Graybill and Iyer, 1994; Sen and Srivastava, 1997).

1. For any value of X there exists in the population a normal distribution of Y values and, therefore, a normal distribution of ε's.

2. The variances of these population distributions of Y values (and of ε's) must all be equal to one another, that is, homogeneity of variances.

3. In the population, the mean of the Y's at a given X lies on a straight line with all other mean Y's at the other X's; that is, the actual relationship in the population is linear.

4. The values of Y are to have come at random from the sampled population and are to be independent of one another.

5. The measurements of X are obtained without error. In practice, it is assumed that the errors in X data are at least small compared with the measurement errors in Y.

For ANOVA, the overall variability of the dependent variable termed the *total sum of squares* is calculated by computing the sum of squares of deviations of Y_i values from $\bar{Y}$:

$$\text{total SS} = \Sigma\, y^2 = \Sigma\, (Y_i - \bar{Y})^2 = \Sigma\, Y_i^2 - (\Sigma\, Y_i)^2/n$$

Then, one determines the amount of variability among the Y_i values that results from there being a linear regression; this is termed the *linear regression sum of squares.*

$$\text{regression SS} = (\Sigma\, xy)^2/\Sigma\, x^2 = \Sigma\, (Y_i' - \bar{Y})^2$$

$$= [\Sigma\, X_i Y_i - \{(\Sigma\, X_i)(\Sigma\, Y_i)\}/n]^2/[\Sigma\, X_i^2 - (\Sigma\, X_i)^2/n]$$

The value of the regression SS will be equal to that of the total SS only if each data point falls exactly on the regression line. The scatter of data points around the regression line is defined by the *residual sum of squares*, which is calculated from the difference in the total and linear regression sums of squares:

$$\text{residual SS} = \Sigma\, (Y_i - Y_i')^2 = \text{total SS} - \text{regression SS}$$

The degrees of freedom associated with the total variability of Y_i values are $n - 1$. The degrees of freedom associated with the variability among Y_i's due to regression are always 1 in a simple linear regression. The residual degrees of freedom are calculable as

$$\text{residual DF} = \text{total DF} - \text{regression DF} = n - 2$$

Once the regression and residual mean squares are calculated ($\text{MS} = \text{SS/DF}$), the null hypothesis may be tested by determining

$$F = \text{regression MS/residual MS}$$

This calculated F value is then compared to the critical value, $F_{\alpha(1), v_1 v_2}$, where $v_1 = \text{regression DF} = 1$, and $v_2 = \text{residual DF} = n - 2$. The residual mean square is often written as $s_{Y.X}^2$, a representation denoting that it is the variance of Y after taking into account the dependence of Y on X. The square root of this quantity — that is, $S_{Y.X}$ — is called the *standard error of estimate* (occasionally termed the *standard error of the regression*). The ANOVA calculations are summarized in Table 2.2.

The proportion of the total variation in Y that is explained or accounted for by the fitted regression is termed the *coefficient of determination*, r^2, which may be thought of as a measure of the strength of the straight line relationship:

$$r^2 = \text{regression SS/total SS}$$

The quantity r is the *correlation coefficient* which is calculated as

$$r = \Sigma\, xy/(\Sigma\, x^2\, \Sigma\, y^2)^{1/2}$$

TABLE 2.2. ANOVA Calculations of Simple Linear Regression

Source of Variation	Sum of Squares, SS	DF	Mean Square, MS
Total $[Y_t - \bar{Y}]$	$\sum y^2$	$n - 1$	
Linear regression $[Y_i' - \bar{Y}]$	$(\sum xy)^2/\sum x^2$	1	Regression SS/regression DF
Residual $[Y_i - Y_i']$	Total SS-regression SS	$n - 2$	Residual SS/residual DF

The quantity is readily computed as

$$r = [\Sigma\, XY - (\Sigma\, X\, \Sigma\, Y)/n]/[\{\Sigma\, X^2 - (\Sigma\, X)^2/n\}\{\Sigma\, Y^2 - (\Sigma\, Y)^2/n\}]^{1/2}$$

A regression coefficient, b, may lie in the range of $\infty \geqslant b \geqslant -\infty$, and it expresses the magnitude of a change in Y associated with a change in X. But a correlation coefficient, r, is unitless and $1 \geqslant r \geqslant -1$. Thus, the correlation coefficient is not a measure of quantitative change of one variable with respect to the other, but it is a measure of intensity of association between the two variables.

2.1.4. Multiple Regression and Correlation

If we have simultaneous measurements for more than two variables (where number of variables $= m$), and one of the variables is assumed to be dependent upon the others, then we are dealing with a *multiple regression* analysis.

$$Y_j = \alpha + \beta_1 X_{1j} + \beta_2 X_{2j} + \beta_3 X_{3j} + \cdots + \beta_m X_{mj}$$

The underlying assumptions are as follows:

1. The observed values of the dependent variable (Y) have come at random from a normal distribution of Y values at the observed combination of independent variables.
2. All such normal distributions have the same variance.
3. The values of Y are independent of each other.
4. The error in measuring the X's is small compared to the error in measuring Y.

If none of the variables is assumed to be functionally dependent on any other, then we are dealing with *multiple correlation*, a situation requiring the assumption that each of the variables exhibits a normal distribution at each combination of all the other variables.

The computational procedures required for most multiple regression and correlation analyses are difficult, and demand computer capability to perform the necessary operation. A computer program for multiple regression and correlation analysis will typically include an analysis of variance (ANOVA) of the regression (Table 2.3).

TABLE 2.3. ANOVA Calculations of Multiple Regression

Source of Variation	Sum of Squares, SS	DF	Mean Squares, MS
Total	$\sum(Y_j - \bar{Y}_j)^2$	$n - 1$	
Regression	$\sum(Y'_j - \bar{Y}_j)^2$	m	Regression SS/regression DF
Residual	$\sum(Y_i - Y'_i)^2$	$n - m - 1$	Residual SS/residual DF

Note: n = total number of data points (i.e., total number of Y values) and m = number of independent variables in the regression.

If we assume Y to be functionally dependent on each of the X's, then we are dealing with multiple regression. If no such dependence is implied as the case of multiple correlation, then any of the $M = m + 1$ variables could be designated as Y for the purpose of utilizing the computer program. In either situation, we can test the interrelationship among the variables as

$$F = \text{regression MS/residual MS} = \{R^2/(1 - R^2)\}(\text{residual DF/regression DF})$$

The ratio

$$R^2 = \text{regression SS/total SS}$$

is the *coefficient of determination* for a multiple regression or correlation. In the regression situation, it is an expression of the proportion of the total variability in Y attributable to the dependence of Y on all the X's as defined by the regression model fit to the data. In the case of correlation, R^2 may be considered to be amount of variability in any one of the M variables that is accounted for by correlating it with the other $M - 1$ variables. The square root of the coefficient of determination is referred to as the *multiple correlation coefficient*.

$$R = (R^2)^{1/2}$$

The square root of the residual mean square is the *standard error of estimate* for the multiple regression:

$$s_{Y \cdot 1,2,3,\ldots,m} = (\text{residual MS})^{1/2}$$

The subscript $(Y \cdot 1, 2, 3, \ldots, m)$ refers to the mathematical dependence of variable Y on the independent variables 1 through m.

2.2. BIOCHEMICAL DATA ANALYSIS WITH SPREADSHEET APPLICATION

A spreadsheet is a collection of data entries: text, numbers, or a combination of both (alphanumeric) arranged in rows and columns used to display, manipulate, and analyze data (Atkinson et al., 1987; Diamond and Hanratty, 1997). Microsoft Excel (http://www.microsoft.com/) is a spreadsheet software package that allows you to do the following:

- Manipulate data through built-in or user-defined mathematical functions.
- Interpret data through graphical displays or statistical analysis.
- Create tables of numeric, text, and other formats.
- Graph tabulated input in various formats.
- Perform various curve-fitting procedures, either a built-in or a user-defined add-on Solver.
- Write user-defined macros for applications routines to automate or enhance a spreadsheet for a particular purpose.

The user is referred to the *Microsoft Excel User's Guide* or online Help for information.

Launching Microsoft Excel brings you into the workspace of Excel and a new workbook. A workbook is a collection of related spreadsheets organized in rows (number headings) and columns (letter headings). The intersection of a row and a column is a cell that is addressed by row and column headings. To select a group of cells, click on the first cell and drag down to the last cell. This collection of cells is known as a range. Inserting cells, rows, and columns is done through either the Insert or shortcut Edit menu. Excel inserts the new column to the left of the highlighted column. New rows are inserted above the highlighted row. At the bottom of the spreadsheets are sheet tabs representing the worksheets of the workbook. Click on these tabs one at a time to move from one sheet to another within the same workbook. The Excel commands are either in the form of a drop-down menu or in the form of icon buttons grouped as toolbars.

2.2.1. Mathematical Operations

There are three types of data that can be entered in the cells of the spreadsheet; number, date/time, and text. For multiple entries of serial number with constant increment, enter the initial value into the first cell, highlight the cells, and select Edit → Fill → Series. To fill multiple entries of the same value into a range, enter the value into the first cell, move the pointer to the right-hand corner of the active cell to activate fill handle (a bold crosshair), and drag it through the range. Data are edited with the usual *cut, copy,* and *paste* operations in the Edit menu and the decimal places of scientific numbers is controlled via dialog box in Format → Cells.

The fundamental operation of a spreadsheet is performing calculations on data. Excel performs mathematical operations through formula and functions. Formulas are written by using the formula bar and beginning with an equal sign (=). Functions can be either user-defined or built-in Excel functions. Because the formulas and functions are acting on cells of the spreadsheet, the variables used therein are the cell references of those cells (Figure 2.1). This is where an understanding of the difference between relative and absolute (preceded by a $ sign) references is important. Excel's built-in functions are accessed through the Function Wizard tool (f_x) found on the standard toolbar. To use a wizard, just follow the instructions

Microsoft Excel - Gly_ion.xls

File Edit View Insert Format Tools Data Window Help

Arial   10   B I U

H25   =

	A	B	C	D	E	F	G
1	pK1=	2.4					
2	pK2=	9.7					
3	.				Mole fraction of ionic forms:		
4					1+	0	1-
5	pH	[G]/[G+]	[G-]/[G]	1/[G+]	[G+]	[G]	[G-]
6	2.00	=10^(A6-B1)	=10^(A6-B1)	=1+B6*(1+C6)	=1/D6	=B6*E6	=C6*F6
7	5.00	=10^(A7-B1)	=10^(A7-B2)	=1+B7*(1+C7)	=1/D7	=B7*E7	=C7*F7
8	7.00	=10^(A8-B1)	=10^(A8-B2)	=1+B8*(1+C8)	=1/D8	=B8*E8	=C8*F8
9	9.00	=10^(A9-B1)	=10^(A9-B2)	=1+B9*(1+C9)	=1/D9	=B9*E9	=C9*F9
10	10.00	=10^(A10-B1)	=10^(A10-B2)	=1+B10*(1+C10)	=1/D10	=B10*E10	=C10*F10
11							
12	note: 1/[G+]=1+[G]/[G+](1+[G-]/[G])						

	A	B	C	D	E	F	G
1	pK1=	2.4					
2	pK2=	9.7					
3					Mole fraction of ionic forms:		
4					1+	0	1-
5	pH	[G]/[G+]	[G-]/[G]	1/[G+]	[G+]	[G]	[G-]
6	2.00	3.981E-01	1.995E-08	1.398E+00	7.153E-01	2.847E-01	5.681E-09
7	5.00	3.981E+02	1.995E-05	3.991E+02	2.506E-03	9.975E-01	1.990E-05
8	7.00	3.981E+04	1.995E-03	3.989E+04	2.507E-05	9.980E-01	1.991E-03
9	9.00	3.981E+06	1.995E-01	4.775E+06	2.094E-07	8.337E-01	1.663E-01
10	10.00	3.981E+07	1.995E+00	1.192E+08	8.386E-09	3.339E-01	6.661E-01
11							
12							

Figure 1.1. Mathematical operation with Excel. Glycine ionizes according to

$$^+H_3N\text{—CHC—OH} \rightleftharpoons {}^+H_3N\text{—CHC—O}^- \rightleftharpoons H_2N\text{—CHC—O}^-$$

$$G^+ \qquad\qquad G \qquad\qquad G^-$$

where $[G^+] = 1/\{1 + 10^{(pH-2.4)}[1 + 10^{(pH-9.7)}]\}$, $[G] = 10^{(pH-2.4)}$ $[G^+]$, and $[G^-] = 10^{(pH-9.7)}$ $[G]$. The calculation of ionic species of glycine ($pK_1 = 2.4$ and $pK_2 = 9.7$) at different pH values using Excel is illustrated with formula input and calculated values output.

in the dialog boxes, moving through them one step at a time. Excel has many mathematical, statistical, and scientific functions. These have the general syntex: = Function Name (arguments). For example, to calculate the mean and standard deviation, type Mean: in B1 and S.D: in C1, select B2 and C2, and enter the formulas = average(range) and = stdev(range). The calculated Mean and Standard Deviation are placed in the cells B2 and C2, respectively.

2.2.2. Statistical Functions

Statistical functions are selected from two menus within Excel. The first approach is through the Function Wizard. Click the Function Wizard, f_x to activate the Function Wizard dialog box; select Statistical under the Function Category list and select Statistical Functions under the Function Name (refer to Help for the definition and usage of a function). The second approach is from Tools → Data Analysis. If Data Analysis is not present when the Tools command is selected, this means that the Analysis ToolPak was not loaded during Excel installation. The Analysis ToolPak can be loaded from Tools → Add-Ins.

To perform F test or t test:

- Enter the data into a workbook.
- Select Tools in the menu bar and select Data Analysis.
- From the box, select F-Test or t-Test Two-Sample for Variances and click OK.
- Enter variable 1 and variable 2 (in absolute format, e.g., $C$4:$C$10).
- Under the output options, select New Worksheet Ply to report the results on a new worksheet.

A report is generated, which contains all the required information to interpret the data.

To perform ANOVA:

- Enter the data into a workbook.
- Select Tools in the menu bar and select Data Analysis.
- Select ANOVA: Single Factor from the list of options to bring up the dialog box.
- Enter the input range covering the entire data set (in absolute format, e.g., $B$2:$D$6)
- Check whether the data are group in columns or rows.
- Alpha can be set (default is 0.05).
- Select the output range (select New Worksheet Ply to report on a new worksheet).
- Click on the OK button.

The ANOVA report as shown in Figure 2.2 is generated. The comparison between the F-test result (F) and the critical value (F_{crit}) provide a decision concerning the similarity/difference of the groups.

2.2.3. Regression Analysis

Traditional approaches to experimental data processing are largely based on linearization and/or graphical methods. However, this can lead to problems where the model describing the data is inherently nonlinear or where the linearization process introduces data distortion. In this case, nonlinear curve-fitting techniques for experimental data should be applied.

Excel provides some built-in tools for fitting models to data sets. By far the most common routine method for experimental data analysis is linear regression, from which the best-fit model is obtained by minimizing the least-squares error between the y-test data and an array of predicted y data calculated according to a linear

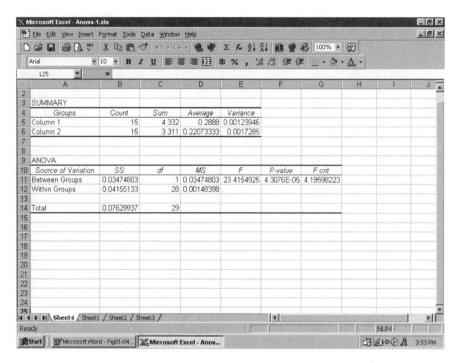

Figure 2.2. ANOVA (single factor) with Excel. Output of ANOVA data analysis of NAD(H) assays of 15 samples from each of two tissues using Excel is shown. The difference in NAD(H) content of the two tissues are indicated by a small P value and $F > F_{crit}$ (reject H_0).

equation with common x values. Linear regression can be accessed in Excel by LINEST function or *via* Data Analysis tool.

LINEST allows for detailed multiple linear regression:

- Select a blank sheet and enter the data in arrays.
- Activate a free cell.
- Click the Function Wizard icon and select Statistical and LINEST.
- Enter the array for the *known _y*'s, for example, (A1:A10), and *known _x*'s, for example, (B1:B10), (C1:C10), (D1:D10), but leave the *const* and *stats* boxes blank. Click OK.

The LINEST result is returned at the assigned (activated) free cell. To display the Slope and Intercept:

- Enter LINEST {m,b} in the new cell.
- Highlight the range of cells corresponding to $2 \times n$ matrix, where n is the number of parameters, x. Select LINEST from the Function Wizard and enter the input as before.
- Do not press Return, but instead click the mouse in the formula bar and place the cursor at the end of the entry.
- Press CTRL + SHIFT + ENTER. The values of the slope (coefficients) and the intercept have been entered into the highlighted cell range.

The detailed linear regression analysis is obtained via the Tools menu as follows:

- Select Tool→Data Analysis→Regression. This opens the Regression dialog box.
- Enter the cell ranges for the array of y and x values using absolute format — for example, $A$2:$A$17 in the appropriate boxes.
- Check Confidence Level (e.g., 95%).
- Enter the output range where you want the regression analysis report to be copied (check New Worksheet Ply for reporting on a new worksheet).
- Check the options to display the plots (e.g., Residual plots and Line Fit plots) and click OK.

The detailed statistical report (Figure 2.3) includes the slope and intercept coefficients for the "best-fit" line, the standard error in these coefficients, and a listing of the residues. The goodness of fit is evaluated from the high correlation coefficient, R^2 ($R^2 = 1.00$ for a perfect fit), and residuals evenly scattered about zero along the entire range.

Perhaps the most common situation involving graphing scientific data is to generate a linear regression plot with y error bars. In most situations, the error in the x data is regarded as being so much smaller than that of the y data that it can

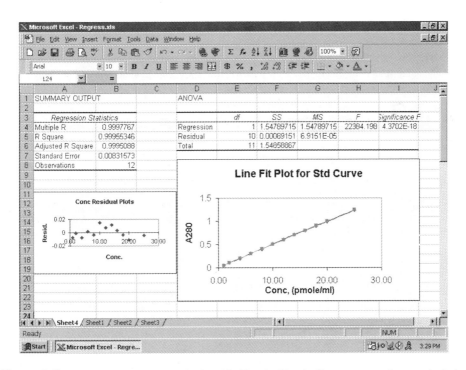

Figure 2.3. Linear regression analysis with Excel. Simple linear regression analysis is performed with Excel using Tools → Data Analysis → Regression. The output is reorganized to show regression statistics, ANOVA residual plot and line fit plot (standard error in coefficients and a listing of the residues are not shown here).

be effectively ignored. Excel allows several methods for generating error bars where *custom error* by which the experimental standard deviation of several estimates of a value can be used. This is accomplished by obtaining the mean values, =AVERAGE { }, and the standard deviations, =STADEV { }, for experimental data. Generate the regression line for the mean values and add error bar via Insert→Error Bars, given by the standard deviations.

2.2.4. Use of Statistical Packages

A number of custom-made statistical software packages are available commercially. The student is urged to be familiar with at least one of them because of their versatility and efficiency.

   SPSS. The statistical analysis software, SPSS (http://www.spsscience.com), is a common statistical package available in most of university networks, and the student could learn its use by following the tutorial session of the Spss program. Open and click Spsswin.exe to start the SPSS program (Figure 2.4). Go to Help and select the Spss Tutorial. From the Main menu page, follow the session. The biochemical data for statistical analysis can be entered directly by starting File, and then choose New and Data. However, it can be advantageous for the student to prepare data files with Excel (filename.xls) in advance. In this case, start File and then choose Open and Data. Click File type to select Excel (filename.xls). From the menu bar, select Statistics to initiate the data analysis.

   SyStat. SyStat is a stand-alone statistical package of SPSS Inc. (http://www.spsscience.com), that performs comprehensive statistical analysis. The user's

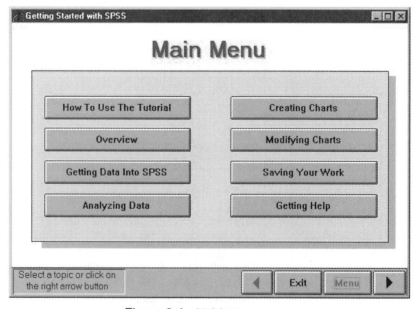

Figure 2.4. SPSS Home page.

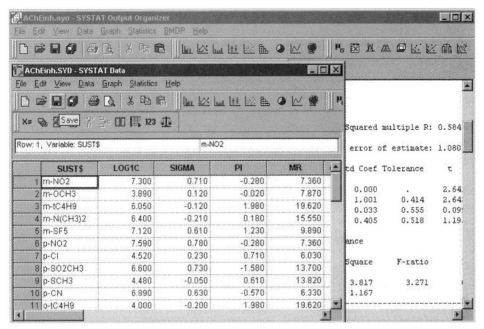

Figure 2.5. Linear regression analysis with Systat. The front window shows the input data for multiple linear regression analysis and the back window shows the statistical results.

guide, *SYSTAT Getting Started*, should be consulted. There are three types of windows (main, data and graph windows), each with its own menus. The main window displays prints and saves results from statistical analyses. The data window (opened by the Data command of the File menu from the main window) allows entering, editing, and viewing of data. The graph window (opened from Graph menu or by double-clicking a graph button from the main window) provides facilities for editing graphs. The new data can be entered directly (via File→New→Data) or imported from various spreadsheets (via File→Open→Data). Selecting Microsoft Excel and entering filename.xls imports the Excel data file. To define variable names, double click the variable heading to bring up the variable properties box (for a numeric data type, the string variable ends with $). A statistical analysis is initiated by selecting the analytical options from Statistic menu of the data window (Figure 2.5). The analytical results are displayed in the output pan by selecting it from the organization pan of the main window. The data file is saved from the data window as filename.syd, and the analysis results (outputs) are saved from the main window as filename.syo.

The online statistical calculations can be performed at http://members.aol.com/johnp71/javastat.html. To carry out linear regression analysis as an example, select "Regression, correlation, least squares curve-fitting, nonparametric correlation," and then select any one of the methods (e.g., Least squares regression line, Least squares straight line). Enter number of data points to be analyzed, then data, x_i and y_i. Click the Calculate Now button. The analytical results, a (intercept), b (slope), f (degrees of freedom), and r (correlation coefficient) are returned.

2.3. BIOCHEMICAL DATA MANAGEMENT WITH DATABASE PROGRAM

The data are the unevaluated, independent entries that can be either numeric or non-numeric—for example, alphabetic or symbolic. Information is ordered data which are retrieved according to a user's need. If raw data are to be effectively processed to yield information, they must first be organized logically into files. In computer files, a field is the smallest unit of data which contains a single fact. A set of related fields is grouped as a record, which contains all the facts about an item in the table; and a collection of records of the same type is called a file, which contains all facts about a topic in the table.

A database system (Tsai, 1988) is a computerized information system for the management of data by means of a general-purpose software package called a database management system (DBMS). A *database* is a collection of interrelated files created with a DBMS. The content of a database is obtained by combining data from all the different sources in an organization so that data are available to all users such that redundant data can be eliminated or at least minimized. Figure 2.6 illustrates the relationship among the components of a typical database. Two files are interrelated (logically linked) if they contain common data types that can be interpreted as being equivalent. Two files are also interrelated (physically linked) if the value of a data type of one file contains the physical address of a record in the other file. A relational database stores information in a collection of files, each containing data about one subject. Because the files are logically linked, you can use information from more than one file at a time.

The Microsoft Access database (http://www.microsoft.com/) is a collection of data and objects related to a particular topic (Hutchinson and Giddeon, 2000). The data represent the information stored in the database, and the objects help users define the structure of that information and automate the data manipulating tasks. Access supports SQL (Structured Query Language) to create, modify, and manipulate records in the table to facilitate the process. It is a table-oriented processing. The user is referred to the *Microsoft Access User's Guide* or online Help for information.

Launching Microsoft Access and click New database from File menu brings you the Database Window with five tabs representing the object types that Access supports:

Table is used to define the data structure for different related topics.

Query is used to select and retrieve data stored in table(s).

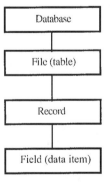

Figure 2.6. Relationship among components of databases.

Form is used to create user–interface screens for entering, displaying, and editing data.

Report is used to present information in a specified format and organization.

Macro is used to automate tasks.

Module is used as a repository of declarations and procedures to design advanced Access applications.

To build database file (table), activate *Table* tab, then:

- Click New and select Design View option.
- Define field name, data type, and optional description as a remark for the field.
- Highlight the unique field, right click, and assign the Primary key to the field.
- Save as filename.mdb.

Molecular structures can be entered as objects:

- Activate Table tab, and select Design view.
- Insert a row where the structures would appear in the field, and type the Field name.
- Select *OLE object* in the Data type column.

The database file as a table object can be opened from the activated *Table* tab for manipulation. For examples, fields (columns) can be added from the Datasheet view by clicking the right edge of the existing column and then selecting Column from Insert menu. This brings a new column (labeled Field 1) or columns (repeated selection of Column option). Double click the new column to enter the field name. Records (rows) can be added also from the Datasheet view after clicking New record button on the tool bar.

Data can be imported (or linked) from Microsoft Excel in two ways.

From Excel,

- Select MS Access Form to create form in New database.
- Select columns to be included by transferring them from Available fields to Selected fields.
- Set layout and enter filename to Finish.
- Open the Table after the import process automatically launching Access in the Access window.

From Access,

- Activate Microsoft Access window by choosing New database from File menu and entering filename.
- Select Get external data from File menu.
- Select Import or Link table.
- Select Microsoft Excel for file type and enter filename in the import (link) dialog box.
- Follow the selections to Finish.
- Repeat the Import to build the database containing more than one table for organizing data by relationship via Relationships button.

Alternatively, you can import/export data between Access and Excel by Copy and Paste operations, provided that matching records/fields (Access) and range of cells (Excel) are selected in the exchange.

To create interacting screen via activated *Form* tab:

- Click New and select Form Wizard option.
- Pick data fields by transferring available fields to selected fields.
- Choose form types: columnar, tabular, datasheet, or justified.
- Click New object button on the tool bar and select AutoForm.

The form is saved and appended to the database file dbname.mdb. Similar steps are followed to create a report after invoking Report Wizard option from the activated *Report* tab. The Wizard provides options for grouping level, sorting order, layout, and style.

To retrieve specified data via *Query* object:

- Activate Query tab and click New.
- Select Design view from New query dialog box.
- Add table(s) from where records and fields are to be retrieved after "Show Table" dialog box is removed.
- Click (pick) and drag the fields to the design query grid as shown in Figure 2.7.
- Specify Criteria for the retrieval by keying in the expression or right-click the grid and select Build to bring Expression builder.

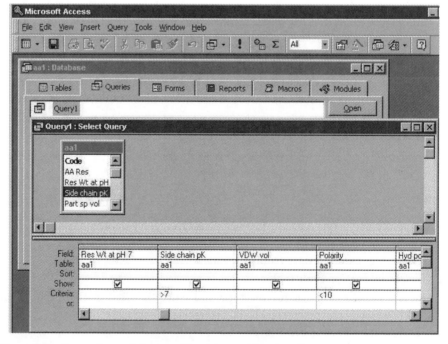

Figure 2.7. Query form of Access. Database retrieval is illustrated using Query form to retrieve amino acid database with Microsoft Access.

- Select appropriate relational operator with the expression.
- Microsoft Access uses the operators: $+$, $-$, $/$, $*$, $\&$, $=$, $>$, $<$, $<>$, And, Or, Not, Like, and ( ).
- Save the query file which is appended to database file and can be retrieved (opened) from *Query* tab.

2.4. WORKSHOPS

Use spreadsheet (Excel) or a statistical program to solve the following problems (1–8) after reviewing pertinent topics discussed in biochemistry texts (Garrett and Grisham,1999; Van Holde et al., 1998; Voet and Voet, 1995; Zubay, 1998).

1. The ionization of cysteine proceeds by the following main sequences with $pK_1 = 1.92$, $pK_2 = 8.33$, and $pK_3 = 10.78$:

The ratio of each pair of ionic forms for a given pH can be calculated by using the Henderson–Hasselbalch equation:

$$[H_2Cys^+]/[H_3Cys^{2+}] = 10^{(pH-1.92)}$$

$$[HCys]/[H_2Cys^+] = 10^{(pH-8.33)}$$

$$[Cys^-]/[HCys] = 10^{(pH-10.78)}$$

$[H_3Cys^{2+}]$ expressed as a fraction of the total concentration may be calculated from the above ratios by using the partition equation:

$$1.00 = [H_3Cys^{2+}]\{1 + ([H_2Cys^+]/[H_3Cys^{2+}])\}$$
$$+ ([HCys]/[H_2Cys^+])([H_2Cys^+]/[H_3Cys^{2+}])$$
$$+ ([Cys^-]/[HCys^+])([HCys]/[H_2Cys^+])([H_2Cys^+]/[H_3Cys^{2+}])$$

This partition equation is solved for $[H_3Cys^{2+}]$, and the concentrations of the other ionic forms can be calculated from $[H_3Cys^{2+}]$ according to

$$[H_2Cys^+] = 10^{(pH-1.92)}[H_3Cys^{2+}]$$

$$[HCys] = 10^{(pH-8.33)}[H_2Cys^+]$$

$$[Cys^-] = 10^{(pH-10.78)}[HCys]$$

Find the isoelectric point and the concentrations of all ionic species from 1.0 M cysteine solution at pH = 7.00.

2. The molecular weight is an intrinsic property of biomolecules. Sedimentation velocity is one of physicochemical methods which have been employed to determine molecular weights of many biomacromolecules. The molecular weight (M) of a biomacromolecule can be computed according to the Svelberg equation:

$$M = \{RTs\}/\{D(1 - \bar{v}\rho)\}$$

where the symbols have the following meanings: R, gas constant (8.314×10^7 erg mol^{-1} deg^{-1}); T, absolute temperature; s, sedimentation coefficient; D, diffusion coefficient; $\bar{v}$, partial specific volume of protein; and ρ, density of the solvent. The density (ρ) of water at 20°C is 0.998, and the partial specific volume is taken to be 0.735 g cm^{-3}. Construct the spreadsheet table by supplementing the following information of biomacromolecules/particles with their molecular/particle weights.

	$S_{20,w}(\times 10^{-13}$ sec$)$	$D_{20,w}(\times 10^{-7}$ cm^2/sec$)$	$\bar{v}$ (cm^3/g)
Bacteriophage T7	453	0.603	0.639
Bushy stunt virus	132	1.15	0.740
Catalase, horse liver	11.3	4.10	0.730
Chymotrypsinogen, bovine	2.54	9.50	0.721
Ferricytochrome c, bovine heart	1.91	13.20	0.707
Fibrinogen, human	7.9	2.02	0.706
Immunoglobin G, human	6.9	4.00	0.739
Lysozyme, hen's egg white	1.91	11.20	0.741
Myoglobin, horse heart	2.04	11.30	0.741
Ribosome	82.6	1.52	0.61
RNase, bovine pancreas	2.00	13.10	0.707
Serum albumin, bovine	4.31	5.94	0.734
Tobacco mosaic virus	192	0.44	0.730
Urease, jack bean	18.6	3.46	0.730

3. The free energy change (ΔG) of a chemical reaction is related to its reaction quotient (Q, ratio of products to reactants) by

$$\Delta G = \Delta G^0 + RT \ln Q$$

where ΔG^0 is the standard free energy change and R is the gas constant (8.3145 J mol^{-1} deg^{-1}). At equilibrium ($\Delta G = 0$), the standard free energy is related to the equilibrium constant (K_{eq}) by

$$\Delta G^0 = -RT \ln K_{eq}$$

Biochemists have adopted a modified standard state (1 atm for gases and pure solids or 1 M solutions at pH 7.0) using symbols $\Delta G^{0'}$ for the standard free energy change, K'_{eq} for the equilibrium constant, and so on, such that

$\Delta G^{0'} = \Delta G^0 + RT \ln[H^+]$ for a reaction in which H^+ is produced, and

$\Delta G^{0'} = \Delta G^0 - RT \ln[H^+]$ for a reaction in which H^+ is consumed

The free energy change in biochemical systems becomes

$$\Delta G' = \Delta G^{0'} + RT \ln Q'$$

and

$$\Delta G^{0'} = -RT \ln K'_{eq}$$

For coupled reaction system in which a product of one reaction is a reactant of the subsequent reaction, the overall free energy change is the sum of the free energy changes of all coupled reactions:

$$\Delta G^{0'} \text{ (overall)} = \sum \Delta G_i^{0'}$$

The equilibrium constants for the enzymatic reactions of the tricarboxylic acid cycle at pH 7.0 are given below:

Reaction	Enzyme	K_{eq}
1	Citrate synthase	3.2×10^5
2	Aconitase	0.067
3	Isocitrate dehydrogenase	29.7
4	α-Ketoglutarate dehydrogenase complex	1.8×10^5
5	Succinyl-CoA synthetase	3.8
6	Succinate dehydrogenase	0.85
7	Fumerase	4.6
8	Malate dehydrogenase	6.2×10^{-6}

Calculate the standard free energy for each step and the overall $\Delta G^{0'}$ for one pass around the cycle, that is,

$$\text{Acetyl-CoA} + 3NAD^+ + [FAD] + GDP + H_3PO_4 + 2H_2O$$

$$\rightarrow \text{CoASH} + 3NADH + [FADH_2] + GTP + 2CO_2 + 3H^+$$

4. For biochemical reactions undergoing oxidation–reduction (redox reaction), the free energy change is related to the electromotive force (emf) or redox potential ($\Delta E'$) of the reaction:

$$\Delta G' = -nF\Delta E'$$

where n denotes number of electrons involved in the reaction and F is the faraday (1 F = 96,494 C mol^{-1} = 96,494 J V^{-1} mol^{-1}). The Nernst equation,

$$\Delta E' = \Delta E'_0 - (RT/nF) \ln Q'$$

describes the redox potential of a redox reaction with its standard redox potential $(\Delta E_0')$ which can be calculated from the standard reduction potentials (E_0') of its component half-reactions

$$\Delta E_0' = E_0' \text{ (acceptor)} - \Delta E_0' \text{ (donor)}$$

and related to the standard free energy changes by

$$\Delta E_0' = -\Delta G^{0'}/nF = -(RT/nF) \ln K_{eq}$$

From the standard reduction potentials of half-reactions listed below, construct the plausible electron transfer system for the oxidation of ethanol to acetaldehyde and water.

Half-Reaction	$\varepsilon^{0'}$ (volts)
$\frac{1}{2}O_2 + 2H^+ + 2e^- = H_2O$	0.825
Cytochrome a_3 $(Fe^{3+}) + e^- = $ cytochrome a_3 (Fe^{2+})	0.385
Cytochrome a $(Fe^{3+}) + e^- = $ cytochrome a (Fe^{2+})	0.29
Cytochrome c $(Fe^{3+}) + e^- = $ cytochrome c (Fe^{2+})	0.254
Cytochrome c_1 $(Fe^{3+}) + e^- = $ cytochrome c_1 (Fe^{2+})	0.22
Cytochrome b $(Fe^{3+}) + e^- = $ cytochrome b (Fe^{2+})	0.077
Ubiquinone $+ 2H^+ + 2e^- = $ ubiquinol	0.045
$FAD + 2H^+ + 2e^- = FADH_2$ (in flavoprotein)	-0.04
Acetaldehyde $+ 2H^+ + 2e^- = $ ethanol	-0.197
$NAD^+ + H^+ + 2e^- = NADH$	-0.315

Calculate $\Delta E_0'$ and $\Delta G^{0'}$ for each redox step, and identify the sites that may couple with phosphorylation (ATP formation) assuming $\Delta G^{0'} = -30.5$ kJ mol^{-1} for the hydrolysis of ATP to ADP.

5. Amino acid analyses of acetylcholine esterase (AchE) and acetylcholine receptor (AchR) of *Electrophorus electicus* have been performed by various investig-

	Ala	Cys	Glu	Lys	Tyr
AchE	5.5	1.1	9.4	4.3	3.8
	6.2	1.6	10.4	4.6	3.6
	5.6	1.2	9.2	4.5	3.5
	5.8	1.3	9.8	4.4	3.4
	6.0	1.4	10.3	4.6	3.8
	5.8	1.5	10.0	4.5	3.9
AchR	7.5	1.8	12.8	5.7	5.0
	5.4	1.7	9.0	6.3	3.8
	5.9	1.4	10.5	5.8	4.2
	6.4	1.7	10.9	6.0	4.4
	6.9	1.9	11.4	6.2	4.7
	7.0	1.6	9.8	5.9	4.5

ators. The compositions (moles %) of selected amino acids are compared in the Table. Perform ANOVA to deduce whether AChE and AChR are identical or different protein molecule(s) based on the amino acid compositions.

6. In an early work on the compositions of DNA, Chargaff (1955) noted that the ratios of adenine to thymine and of guanine to cytosine are very close to unity in a very large number of DNA samples. This observation has been used to support the helical structure of DNA proposed by Watson and Crick (1953). From the base compositions of DNA and RNA given in the following tables, deduce the statistical significance for the statement that the base ratios (A/T(U) and G/C are unity for DNA but vary for RNA (*Note:* Calculate the ratios first and then perform statistical analysis on the ratios).

The Base Composition (%) of Some DNA

Tissue	Adenine	Guanine	Cytosine	Thymine
Calf thymus	27.3	22.7	21.6	28.4
Calf thymus	28.2	21.5	22.5	27.8
Beef spleen	27.9	22.7	20.8	27.3
Beef spleen	27.7	22.1	21.8	28.4
Beef liver	28.8	21.0	21.1	29.0
Beef pancreas	27.8	21.9	21.7	28.5
Beef kidney	28.3	22.6	20.9	28.2
Beef sperm	28.7	22.2	22.0	27.2
Sheep thymus	29.3	21.4	21.0	28.3
Sheep liver	29.3	20.7	20.8	29.2
Sheep spleen	28.0	22.3	21.1	28.6
Sheep sperm	28.8	22.0	21.0	27.2
Man thymus	30.9	19.9	19.8	29.4
Man liver	30.3	19.5	19.9	30.3
Man spleen	29.2	21.0	20.4	29.4
Man sperm	30.9	19.1	18.4	31.6
Herring sperm	27.8	22.2	22.5	27.5
Salmon sperm	29.7	20.8	20.4	29.1
Abacia lixula sperm	31.2	19.1	19.2	30.5
Sarcina lutea	13.4	37.1	37.1	12.4
Wheet germ	27.3	22.7	22.8	27.1
Yeast	31.3	18.7	17.1	32.9
Pneumococcus type III	29.8	20.5	18.0	31.6
Vaccinia virus	29.5	20.6	20.0	29.9

The Base Composition (%) of Some RNA

Tissue	Adenine	Guanine	Cytosine	Uracil
Bakers' yeast	25.1	30.1	20.1	24.7
Bakers' yeast	25.7	28.1	21.8	24.4
Escherichia coli	27.0	27.5	23.0	22.5
Rabbit liver	19.3	32.6	28.2	19.9
Rat liver	19.1	33.5	26.6	20.8
Rat liver	19.0	34.0	30.2	16.8

The Base Composition (%) of Some RNA

Tissue	Adenine	Guanine	Cytosine	Uracil
Sheep liver	20.4	39.4	25.8	14.4
Man liver	11.4	44.4	31.6	12.6
Calf liver	21.6	40.6	23.3	14.5
Calf liver	19.5	35.0	29.1	16.4
Calf pancreas	14.2	48.6	23.6	13.4
Calf spleen	17.9	35.2	31.6	15.3
Calf thymus	18.5	43.9	23.6	12.0
Tobacco mosaic virus	29.8	25.4	18.5	26.3
Cucumber virus	25.7	25.5	18.2	30.6
Tomato bushy stunt virus	27.6	27.6	20.4	24.4
Turnip yellow mosaic virus	22.6	17.2	38.0	22.2
Southern bean mosaic virus	25.9	25.9	23.0	25.2
Potato X	34.4	21.4	22.8	21.4
Sarcina lutea	16.7	28.4	32.9	22.0

7. Heating DNA solution produces an increase in UV absorption (hyperchromic effect) and a decrease in viscosity, resulting in the disruption of the hydrogen bonds between the two strands of the double helix. A plot of optical absorbance at 260 nm versus temperature is known as melting curve, and the temperature corresponding to the midpoint of the increase in absorbance is referred to as the melting temperature, T_m (Marmur and Doty, 1962). A sample of DNA purified from rabbit liver nuclei is dissolved in a solution containing 0.15 M NaCl and 0.015 M sodium citrate. The change in absorbance at 260 nm is followed while the temperature is increased from 55°C to 98°C as shown:

T,°C:	60	65	70	75	80	82.5	85	87.5	90	92.5	95	98
A_{260}:	1.00	1.00	1.01	1.02	1.06	1.09	1.12	1.15	1.18	1.21	1.23	1.24

Estimate T_m of the sample by plotting the melting curve. Perform regression analysis to deduce the correlation between T_m and the base composition of DNA by comparing the following data of DNAs from other sources:

DNA Sample	T_m (°C)	Base Composition (%)			
		A	G	C	T (HMC for T_2)
Poly(dAT)	65.6	50.0	0	0	50.0
T2 bacteriophage	78.0	18.0	32.0	32.0	18.0
Yeast	84.0	32.9	17.1	17.1	32.9
Calf thymus	86.7	28.2	21.8	21.8	28.2
Salmon sperm	87.1	29.1	20.9	20.9	29.1
Rabbit liver	?	28.0	22.0	22.0	28.0
Escherichia coli	88.3	25.0	25.0	25.0	25.0

DNA Sample	$T_m (°C)$	Base Composition (%)			
		A	G	C	T (HMC for T_2)
S. marcescens	93.5	21.0	29.0	29.0	21.0
Micrococcus lysodeikticus	95.4	16.0	34.0	34.0	16.0
Mycobacterium phlei	96.2	14.5	36.5	36.5	14.5
Poly(dGC)	105.8	0	50.0	50.0	0

8. Many biochemical and toxicological properties of compounds X_i depend on solute–solvent interaction can be rationalized in terms of the linear solvation-energy relationship (LSER) (Kamlet et al., 1981):

$$X_i = X_0 + m(V/100) + s\pi^* + b\beta + a\alpha = X_m + s\pi^* + b\beta + a\alpha$$

where V, π^*, β, and α are the solute van der Waals molar volume, the solovatochromic parameters, a measure of basicity (hydrogen acceptor), and a measure of acidity (hydrogen donor) of compounds, X_i, respectively. The toxicity of various chemical solvents (Abraham and McGowen, 1982) toward organisms expressed as log $(1/C)$, where C is the LD_{50} molar concentration, can be correlated with LSER parameters:

Solvent	Log $(1/C)$	π^*	β	α
n-Hexane	4.06	−0.08	0	0
Methanol	0.41	0.60	0.62	0.98
Ethanol	0.76	0.54	0.77	0.86
1-Propanol	1.20	0.51	0.83	0.80
2-Propanol	1.08	0.46	0.95	0.78
1-Butanol	2.40	0.46	0.88	0.79
tert-Butanol	1.71	0.41	1.01	0.62
Ethylene glycol	0.14	0.85	0.52	0.92
Chloroform	2.87	0.76	0	0.34
Diethyl ether	1.77	0.27	0.47	0
Acetone	0.78	0.72	0.48	0.07
2-Butanone	1.26	0.67	0.48	0.05
Acetophenone	2.79	0.90	0.49	0
Ethyl formate	1.30	0.61	0.46	0
Ethyl acetate	1.68	0.55	0.45	0
Butyl acetate	2.49	0.46	0.44	0
Ethyl propionate	2.10	0.50	0.42	0
Dimethylacetamide	0.47	0.88	0.76	0
Acetonitrile	0.69	0.85	0.31	0.15
Benzene	2.85	0.59	0.10	0
p-Xylene	3.64	0.43	0.12	0
Anisole	2.90	0.73	0.22	0
Pyridine	1.69	0.87	0.64	0
Quinoline	2.97	0.64	0.64	0

Perform the regression analyses to assess the contribution of LSER parameters for LD_{50} (log $1/C$) of the sample organism.

9. General constituents and approximate energy values of common foods are given below. Apply Access to design a database such that the information can be retrieved according to sources of macronutrients for food types (e.g., protein sources or carbohydrate sources for cereal, fish, fruit, dairy, meat/poultry, vegetable).

Food	Water (%)	Protein (g)	Lipid (g)	Carbohydrate, g		Mineral (g)	Energy (cal)
				Total	Fiber		
Apple	84.8	0.2	0.6	14.1	1.0	0.3	56
Asparagus	91.7	2.5	0.2	5.0	0.7	0.6	26
Avocado	71.8	1.8	14.0	7.4	1.5	1.1	149
Bacon	19.3	8.4	69.3	1.0	0	2.0	665
Banana	75.7	1.1	0.3	22.2	0.5	0.8	85
Barley	11.1	8.2	1.0	78.8	0.5	0.9	349
Bean, cooked red	69.0	7.8	0.5	21.4	1.5	1.3	118
Beef, T-bone	47.5	14.7	37.1	0	0	0.7	397
Blueberry	85.0	0.7	0.5	13.6	1.5	0.2	35
Broccoli	89.1	3.6	0.3	5.9	1.5	1.1	32
Butter	15.5	0.6	81.0	0.4	0	2.5	716
Cabbage	92.4	1.3	0.2	5.4	0.8	0.7	24
Cantaloupe	91.2	0.7	0.1	7.5	0.3	0.5	30
Carrot	88.2	1.1	0.2	9.7	1.0	0.8	42
Cauliflower	91.0	2.7	0.2	5.2	1.0	0.9	27
Celery	94.1	0.9	0.1	3.9	0.6	1.0	17
Cherry	83.7	1.2	0.3	14.3	0.2	0.5	58
Cheese, brick	40.0	23.2	30.0	1.9	0	4.9	370
Rice, white	12.0	6.7	0.4	80.4	0.3	0.5	363
Rye, whole-grain	11.0	12.1	1.7	73.4	2.0	1.8	334
Salmon, Atlantic	63.6	22.5	13.4	0	0	1.4	217
Sardine	70.7	19.3	8.6	0	0	2.4	160
Sausage, pork	38.1	9.4	50.8	Trace	0	1.7	498
Shrimp	78.2	18.1	0.8	1.5	Trace	1.4	91
Soybean, green	69.2	10.9	5.1	13.2	1.4	1.6	134
Soybean, milk	92.4	3.4	1.5	2.2	0	0.5	33
Spinach	90.7	3.2	0.3	4.3	0.6	1.5	26
Strawberry	89.9	0.7	0.5	8.4	1.3	0.5	37
Sweet potato	70.6	1.7	0.4	26.3	0.7	1.0	114
Tomato	93.5	1.1	0.2	4.7	0.5	0.5	22
Trout, rainbow	66.3	21.5	11.4	0	0	1.3	195
Tuna	71.0	25.0	3.6	0	0	1.4	139
Turkey, light meat	73.0	24.6	1.2	0	0	1.2	116
Veal, lean loin	69.0	19.2	11.0	0	0	1.0	181
Walnut	3.5	14.8	64.0	15.8	2.1	1.9	651
Watermelon	92.6	0.5	0.2	6.4	0.3	0.3	26
Wheat, all-purpose flour	12.0	10.5	1.0	76.1	0.3	0.4	364
Yam	73.5	2.1	0.3	23.2	0.9	1.0	101
Yogurt	88.0	3.0	3.4	4.9	0	0.7	62

10. Vitamins are essential nutrients that are required for animals, usually in the minute amounts, in order to ensure healthy growth and development. Because the animal cannot synthesize them in amount adequate for its daily needs, vitamins must be supplied in the diet. The composition (per 100g food) of some vitamins in our foods are given below:

Food	Vit A (I.U.)	Vit D (μg)	Vit E (mg)	Thiam (mg)	Rflv (mg)	Niac (mg)	F.a. (μg)	P.a. (μg)	Pydx (mg)	Vit C (mg)
Apple	65		0.74	0.03	0.02	0.1	0.5		20	4.5
Banana	190		0.40	0.05	0.06	0.7	9.6		320	10
Bean, dry	20		3.60	0.55	0.15	2.3	117	850		
Beef	60	0.33	0.63	0.07	0.15	3.8	12	1,100	275	
Broccoli	2,500			0.10	0.23	0.9	33.9	1,400		113
Cabbage	130	0.005		0.05	0.05	0.3	24		205	51
Carrot	11,000	0.075	0.45	0.06	0.05	0.6	8		170	8
Cauliflower	60			0.11	0.10	0.7	29.1	920		78
Cheese	1,150	0.83		0.02	0.70		28.5	680	98	
Chicken	60		0.25	0.05	0.09	10.7	3.0	715	7.5	
Egg	1,180	6.6*	2.0	0.11	0.30	0.10	5.1	2,700	35	
Grapefruit	45		0.26	0.04	0.02	0.2	2.7		21	36
Halibut	440	3,500*		0.07	0.07	8.3			110	
Herring	110	8.2		0.02	0.15	3.6				3
Lamb			0.77	0.16	0.22	5.1	1.9	600	300	
Lettuce	1,200		0.50	0.05	0.07	0.04	29		71	7
Milk, whole	150	0.11		0.03	0.17	0.1	11	290	82	1
Orange	200		0.24	0.10	0.04	0.4	5.1	340	37	55
Peach	1,330			0.02	0.05	1.09	2.3		16	7
Pea	575		2.1	0.32	0.13	2.9	20	820	46	17
Peanut			22*	1.14	0.13	17.2	56.6	2,500	300	
Pork			0.71	0.77	0.19	4.1	5.1	920	510	
Potato			0.06	0.10	0.04	1.5	66	525	405	20
Rice				0.07	0.03	1.6			395	
Salmon	230	7.4		0.12	0.16	6.7			590	
Sardine, can	220	34.5		0.04	0.20	3.4			280	
Soybean	385		140*	0.77	0.24	0.18	3.2	185	960	14
Spinach	8,100	0.005		0.10	0.06	0.6	132		60	51
Strawberry	40			0.03	0.07	0.6	5.3		44	59
Sweet potato	14,000		4.0	0.10	0.06	0.6	12	940		22
Tomato	900		0.36	0.06	0.04	0.7	9			23
Tuna, can	80	5.8		0.05	0.12	11.9			440	
Wheat flour				0.55	0.12	4.3	38	400	490	

Note 1: Group B vitamins are thiamine (Thiam), riboflavin (Rflv), niacin (Niac), folic acid (F.a.), pantothenic acid (P.a.), and pyridoxins (Pydx).

Food	Vit A (I.U.)	Vit D (μg)	Vit E (mg)	Thiam (mg)	Rflv (mg)	Niac (mg)	F.a. (μg)	P.a. (μg)	Pydx (mg)	Vit C (mg)

Note 2: Values with * relating to vitamin D are for egg yolk and halibut liver oil, and those * relating to vitamin E are for peanut oil and soybeam oil.

Apply Access to build a database and create queries to retrieve records of (a) fishes that would supply recommended minimum daily supply of vitamin D (5 mg), (b) fruits that would supply recommended daily allowance of vitamin C (50 mg), and (c) foods of plant origin that would supply recommended daily allowance of vitamin A (800 I.U.).

REFERENCES

Abraham, M. H., and McGowen, J. C. (1982) *Chromatographia* **23**:243–246.

Anderson, D. R., Sweeney, D. J., and Williams, T. A. (1994) *Introduction to Statistics: Concepts and Applications*, 3rd edition. West Publisher, Minneapolis/St. Paul, MN.

Atkinson, D. E., Clarke, S. G., and Rees, D. C. (1987). *Dynamic Models in Biochemistry*. Benjamin/Cummings, Menlo Park, CA.

Chargaff, E. (1955) in *The Nucleic Acids*, Vol. 1 (Chargaff, E., and Davidson, J. N., Eds.), Academic Press, New York, pp. 307–373.

Diamond, D., and Hanratty, V. C. A. (1997) *Spreadsheet Applications in Chemistry Using Microsoft Excel*. John Wiley & Sons, New York.

Fry, J. C., Ed. (1993) *Biological Data Analysis: A Practical Approach*. IRL Press, Oxford.

Garrett, R. H., and Grisham, C. M. (1999) *Biochemistry*, 2nd edition. Saunders College Publishing, San Diego.

Graybill, F. A., and Iyer, H. K. (1994) *Regression Analysis: Concepts and Applications*. Duxbury Press, Belmont, CA.

Hutchinson, S. E., and Giddeon, K. (2000) *Microsoft Access 2000*. Irwin/McGraw-Hill, Boston.

Kamlet, M. J., Abboud, L. J. M., and Taft, R. W. (1981) *Prog. Phys. Org. Chem.* **13**:485–630.

Marmur, J., and Doty, P. (1962) *J. Mol. Biol.* **5**:109–118.

Milton, J. S., McTeer, P. M., and Corbet, J. J. (1997) *Introduction to Statistics*, McGraw-Hill, New York.

Sen, A. K., and Srivastava, M. (1997) *Regression Analysis: Theory, Methods, and Applications*. Springer, New York.

Tsai, A. Y. H. (1988) *Database Systems: Management and Use*. Prentice-Hall Canada, Scarborough, Ontario.

Van Holde, K. E., Johnson, W. C., and Ho, P. S. (1998) *Principles of Physical Biochemistry*. Prentice-Hall, Englewood Cliffs, NJ.

Voet, D., and Voet, J. D. (1995) *Biochemistry*, 2nd edition. John Wiley & Sons, New York.

Watson, J. D., and Crick, F. H. C. (1953) *Nature* **171**:737–738.

Williams, B. (1993) *Biostatistics*. Chapman and Hall, London.

Zar, J. H. (1999) *Biostatistical Analysis*. 4th edition, Prentice-Hall, Upper Saddle River, NJ.

Zubay, G. L. (1998) *Biochemistry*, 4th edition. W.C. Brown, Chicago.

3

BIOCHEMICAL EXPLORATION: INTERNET RESOURCES

The Internet resources have provided great impetus to an advancement of bioinformatics and computational biochemistry. They contribute greatly in the collection and dissemination of biochemical information. The Internet resources of biochemical interest and their uses to acquire biochemical information are described.

3.1. INTRODUCTION TO INTERNET

The Internet is a network of networks, meaning that many different networks operated by a multitude of organizations are connected together collectively to the Internet (Chellen, 2000; Crumlish, 1999; Marshall, 1996; Swindell et al., 1996; Wyatt, 1995). It is based on a so-called "packet-switching network" whereby data are broken up into *packets*. These packets are forwarded individually by adjacent computers on the network, acting as *routers*, and are reassembled in their original form at their destination. Packet switching allows multiple users to send information across the network both efficiently and simultaneously. Because packets can take alternate routes through the network, data transmission is easily maintained if parts of the network are damaged or not functioning efficiently. The widespread use of *Transmission Control Protocol* (TCP) together with *Internet Protocol* (IP) allowed many networks to become interconnected through devices call *gateways*. The term *Internet* derives from connecting networks, technically known as *internetworking*.

The IP address is generally associated with fully qualified domain name which users recognize and use. The name of a computer can then be <computer>.domain

TABLE 3.1. Top-Level Domain Names

Inside the United States

.com	Commercial site
.edu	Educational site
.gov	Government site
.mil	Military site
.net	Gateway or network host
.org	Organization site

Outside the United States for examples

.ca	Canada
.ch	Switzerland
.de	Germany
.fr	France
.jp	Japan
.uk	Britain

with six categories of top-level domain names in the United States. Outside the United States, the top-level domain names are replaced with a two-letter code specifying the country (Table 3.1).

If your computer at work is on a local area network (LAN), you may already have access to an external connection. Almost all universities have permanent connections between their internal systems and the Internet. If you don't have access to an institutional connection, you have to obtain an access from an Internet service provider (ISP). For a full Internet connection, you must have software to make your computer use the Internet protocol, TCP/IP, and if your connection is through a modem this software will probably use PPP (point to point protocol).

List Servers and Newsgroups. A list server/mail server contains discussion group created to share ideas and knowledge on a subject; LISTSERV is the most common list server program. A message sent to a list is copied and then forwarded by e-mail to every person who subscribes to the list, thereby providing an excellent resource for distributing information to a group with a shared interest. Discussion groups are usually created and sometimes monitored by someone with an interest in that subject. You join the list by sending an appropriately worded e-mail request to the list. The program automatically reads your e-mail message, extracts your address and adds you to the circulation list.

Unlike list servers, which disseminate information on a specific topic from one person to many, newsgroup servers (e.g., USENET) provide access to thousands of topic-based discussion group services that are open to everyone. Access is provided through a local host or news server machine.

Telnet. Telnet is a program using the communication protocol of the Internet (TCP/IP) to provide a connection onto remote computers. You can use Telnet to contact a host machine simply by typing in the host name or IP number if you have Internet access from your computer. You will then be asked for a login identity and your password. Often buried within Telnet is a version of FTP, so you can transfer files from the TCP/IP host to your own computer.

World Wide Web. The World Wide Web (WWW) is the worldwide connection of computer servers and a way of using the vast interconnected network to find and view information from around the world. Internet uses a language, TCP/IP, for talking back and forth. The TCP part determines how to take apart a message into small packets that travel on the Internet and then reassemble them at the other end. The IP part determines how to get to other places on the Internet. The WWW uses an additional language called the *HyperText Transfer Protocol* (HTTP). The main use of the Web is for information retrieval, whereby multimedia documents are copied for local viewing. Web documents are written in *HyperText Markup Language* (HTML) which describes, most importantly, where hypertext links are located within the document. These *hyperlinks* provide connections between documents, so that a simple click on a hypertext word or picture on a Web page allows your computer to extend across the Internet and bring the document to your computer. The central repository having information the user wants is called a *server*, and your (user's) computer is a *client* of the server. Because HTML is predominantly a generic descriptive language based around text, it does not define any graphical descriptors. The common solution was to use bit-mapped graphical images in so-called GIF (graphical interchange format) or JPEG (Joint photographic experts group) format.

Every Web document has a descriptor, *Uniform Resource Locator* (URL), which describes the address and file name of the page (Figure 3.1). The key to the Web is the browser program, which is used to retrieve and display Web documents. The browser is an Internet compatible program and does three things for Web documents:

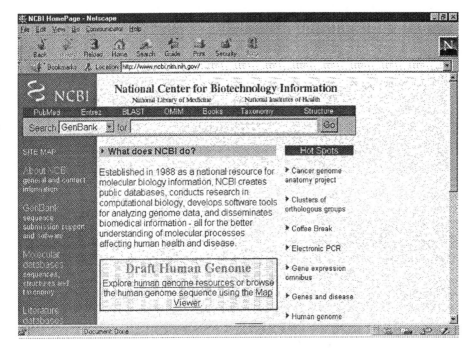

Figure 3.1. World Wide Web page of NCBI.

- It uses Internet to retrieve documents from servers.
- It displays these documents on your screen, using formatting specified in the document.
- It makes the displayed documents active.

The common browsers are Netscape Navigator from Netscape Communications (http://home.netscape.com/) and Internet Explore from Microsoft (http://www.microsoft.com/).

To request a Web page on the Internet, you either click your mouse on a hyperlink or type in the URL. The HTML file for the page is sent to your computer together with each graphic image, sound sequence, or other special effect file that is mentioned in the HTML file. Since some of these files may require special programming that has to be added to your browser, you may have to download the program the first time you receive one of these special files. These programs are called *helper applications*, *add-ons*, or *plug-ins*.

A mechanism called MIME (Multipurpose Internet Mail Extensions), which allows a variety of standard file formats to be exchanged over the Internet using electronic mail, has been adopted for use with WWW. When a user makes a selection through a hyperlink within an HTML document, the client browser posts the request to the designated web server. Assuming that the server accepts the request, it locates the appropriate file(s) and sends it with a short header at the top of each datafile/document to the client, with the relevant MIME header attached. When the browser receives the information, it reads the MIME types such as text/html or image/gif the browser have been built in such a way that they can simply display the information in the browser window. For given MIME type, a local preference file is inspected to determine what (if any) local program (known as a helper application or plug-in) can display the information, this program is then launched with the data file, and the results are displayed in a newly opened application window. The important aspect of this mechanism is that it achieves the delivery of semantic content to the user, who can specify the style in which it will be displayed via the choice of an appropriate application program.

Once connected to the Web, a variety of WWW directories and search engines are available for those using the Web in a directed fashion. First, there are Web catalogues; the best known of these is Yahoo, which organizes Web sites by subject classification. You can either scroll through these subject categories or use the Yahoo search engine. It also simultaneously forwards your search request to other leading search engines such as AltaVista, Excite, and others. Alternatively, there are the Web databases, where the contents of Web pages are indexed and searchable such as Lycos and InfoSeek. Google is extremely comprehensive. Pages are ranked based on how many times they are linked from other pages, thus a Google search would bring you to the most well-traveled pages that match your search topic. HotBot is relatively comprehensive and regularly updated. It offers form-based query tools that eliminate the need for you to formulate query statements. Major search engines on the Web are listed in Table 3.2. Many other specialized search engines can be found in EasySearcher 2 (http://www.easysearcher.com/ez2.html).

File Fetching: File Transfer Protocol. The File Transfer Protocol (FTP) allows you to download (get) resources from a remote computer onto your own, or

TABLE 3.2. Some Search Engines on the World
Wide Web

Search Engine	URL
AltaVista	http://www.altavista.digital.com
Excite	http://www.excite.com
Google	http://www.google.com
HotBot	http://www.hotbot.com
InfoSeek	http://www.infoseek.com
Lycos	http://www.lycos.com
WebCrawler	http://www.webcrawler.com
Yahoo	http://www.yahoo.com

upload (put) these from your computer onto a remote computer. Sometimes FTP
access to files may be restricted. To retrieve files from these computers, you must
know the address and have a user ID and a password. However, many computers
are set up as anonymous FTP servers, where user usually logs in as anonymous and
gives his/her e-mail address as a password. Internet browsers such as Netscape
Navigator and Internet Explorer support anonymous FTP. Simply change the URL
from http:// to ftp:// and follow it with the name of the FTP site you wish to go to.
However, you must fill in your identity under the Options menu so that the browser
can log you in as an anonymous user. The program and files on FTP sites are usually
organized hierarchically in a series of directories. Those on anonymous FTP sites
are often in a directory called pub (i.e., public). It is worth remembering that many
FTP sites are running on computers with a UNIX operating system that is case
sensitive.

Many FTP servers supply text information when you login, in addition to the
readme file. You can get help at the FTP prompt by typing 'help' or '?'. Before you
download/upload the image file, select the option of transferring from *asc* (ascii) for
text to *bin* (binary) for the image (graphic) files. Most files are stored in a compressed
or zipped format. Some programs for compressing and uncompressing come as one
integrated package, while others are two separate programs. The most commonly
used compression programs are as follows:

Suffix	Compression Program
.sit	Mac StuffIt
.sea	Mac self-extracting archive
.zip	Win PkZip, WinZip
.z	UNIX compress
.tar	UNIX tape

Internet versus Intranet. The Web of the Internet is a way to communicate
with people at a distance around the globe, but the same infrastructure can be used
to connect people within an organization known as the Intranet. Such Intranets

provide an easily accessible repository of relevant information (but isolated from the world through firewalls), capitalizing on the simplicity of the Web interface. They also provide an effective channel for internal announcements or confidential communications within the organization.

3.2. INTERNET RESOURCES OF BIOCHEMICAL INTEREST

PubMed is one of the most valuable web resources for biochemical literatures accessible from Entrez (http://www.ncbi.nlm.nih.gov/entrez/). Many well-regarded journals in biochemistry, molecular biology and related fields as well as clinical publications of interest to medical professionals are indexed in PubMed. The resource can be searched by a keyword with its boolean operators (AND, OR and NOT). Users can specify the database field to search. The Preview/Index menu allows the user to build a detailed query interactively.

The first issue for each volume of *Nucleic Acids Research* since 1996 (Volume 24) has been reserved for presenting molecular biology databases. Relevant database resources for biochemistry/molecular biology have been recently compiled (Baxevanis, 2001; Baxevanis, 2002). The resources available on the Internet are always changing. Some servers may move their URL addresses or become nonoperational while new ones may be created online. Therefore it is not the goal of this section to provide a nearly complete guide to the WWW servers, but instead to provide samples of common biochemical resources for database information or databases (Table 3.3). For a comprehensive catalog of biochemical databases, Dbcat (Discala et al., 2000) at http://www.infobiogen.fr/services/dbcat/ should be consulted.

The National Center for Biotechnology Information (NCBI) (Wheeler et al., 2001) was established in 1988 as a division of the National Library of Medicine (NLM) of the National Institutes of Health (NIH) in Bethesda, Maryland (Figure 3.1). Its mandate is to develop new information technology to aid our understanding of the molecular and genetic processes that underlie health and disease. The specific aims of NCBI include (a) the creation of automated systems for storing and analyzing biological information, (b) the development of advanced methods of computer-based information processing, (c) the facilitation of user access to databases and software, and (d) the coordination of efforts to gather biotechnology information worldwide. Entrez is the comprehensive, integrated retrieval server of biochemical information operated by NCBI.

The European Molecular Biology network (EMBnet) was established in 1988 to link European laboratories that used biocomputing and bioinformatics in molecular biology research by providing information, services, and training to users in European laboratories, through designated sites operating in their local languages. EMBnet consists of national sites (e.g., INFOBIOGEN, GenBee, SEQNET), specialist sites (e.g., EBI, MIP, UCL), and associate sites. National sites which are appointed by the governments of their respective nations have mandate to provide databases, software, and online services. Special sites are academic, industrial, or research centers that are considered to have particular knowledge of specific areas of bioinformatics responsible for the maintenance of biological database and software. Associate sites are biocomputing centers from non-European countries. The Sequence Retrieval System (SRS) (Etzold et al., 1996) developed at EBI is a network browser for databases in molecular biology allowing users to retrieve, link, and access entries from all the interconnected resources. The system links nucleic

TABLE 3.3. Some Web Sites of Biochemical Interest

WWW Site	URL
National Center for Biotechnology Information (NCBI)	http:www.ncbi.nlm.nih.gov/
Entrez browser of NCBI	http://www.ncbi.nlm.nih.gov/Entrez/
GenBank	http://www.ncbi.nlm.nih.gov/Web/Genbank/
European Molecular Biology Laboratory (EMBL)	http://www.embl-heidelberg.de/
European Bioinformatics Institute (EBI)	http://www.ebi.ac.uk/
National Biotechnology Information Facility	http://www.nbif.org/data/data.html
DNA Database of Japan (DDBJ)	http://www.ddbj.nig.ac.jp/
DBGet database link of GenomeNet, Japan	http://www.genome.ad.jp/dbget/
Expert Protein Analysis System (ExPASy)	http://expasy.hcuge.ch/
Protein Information Resource (PIR)	http://nbrfa.georgetown.edu/pir/
SWISS-PROT, Protein sequence database	http://expasy/hcuge.ch/sprot/
Munich Inform. Center for Protein Sequences (MIPS)	http://www.mips.biochem.mpg.de/
Protein Data Bank at RCSB	http://www.rcsb.org/pbd/
BSM at University College of London (UCL)	http://www.biochem.ucl.ac.ul/bsm/dbbrowser/
TIGR Database (TDB)	http://www.tigr.org/tdb/
Computation Genome Group, Sanger Centre	http://genomic sanger.ac.uk/
Bioinformatic WWW sites	http://biochem.kaist.ac.kr/bioinformatics.html
INFOBIOGEN Catalog of DB	http://www.infobiogen.fr/services/dbcat/
Links to other bio-Web servers	http://www.gdb.org/biolinks.html
Pedro's list for molecular biologists	http://www.public.iastate.edu/~pedro/
Survey of Molecular Biology DB and servers	http://www.ai.sri.com/people/mimbd/
Harvard genome research DB and servers	http://golgi.harvard.edu
Johns Hopkins University OWL Web server	http://www.gdb.org.Dan/proteins/owl.html
Human genome project information	http://www.ornl.gov/TechResources/Human Genome
IUBio archive	http://iubio.bio.indiana.edu/soft/molbio/Listing.html

acid, EST, protein sequence, protein pattern, protein structure, specialist, and/or bibliographic databases. The SRS therefore permits users to formulate queries across a range of different database types via a single interface.

GenomeNet is a Japanese network of database and computational service for genome research and related areas in molecular and cellular biology. It is operated jointly by the Institute for Chemical Research, Kyoto University and the Human Genome Center of the University of Tokyo.

3.3. DATABASE RETRIEVAL

Databases are electronic filing cabinets that serve as a convenient and efficient means of storing vast amounts of information. An important distinction exists between primary (archival) and secondary (curated) databases. The primary databases represent experimental results with some interpretation. Their record is the sequence as it was experimentally derived. The DNA, RNA, or protein sequences are the items to be computed on and worked with as the valuable components of the primary databases. The secondary databases contain the fruits of analyses of the sequences in the primary sources such as patterns, motifs, functional sites, and so on. Most biochemical and/or molecular biology databases in the public domains are flat-file databases. Each entry of a database is given a unique identifier (i.e., an entry name and/or accession number) so that it can be retrieved uniformly by the combination of the database name and the identifier.

Three integrated retrieval systems—Entrez, EBI, and DBGet—will be described briefly. The molecular biology database and retrieval system, Entrez (Schuler et al., 1996), was developed at and maintained by NCBI of NIH to allow retrieval of biochemical data and bibliographic citation from its integrated databases. Entrez typically provides access to (a) DNA sequences (from GenBank, EMBL, and DDBJ), (b) protein sequences (from PIR, SWISS-PROT, PDB), (c) genome and chromosome mapping data, (d) 3D structures (from PDB) and (e) the PubMed bibliographic database (MEDLINE). Entrez searches can be performed using one of two Internet-based interfaces. The first is a client-server implementation known as NetEntrez. This makes a direct connection to an NCBI computer. Because the client software resides on the user's machine, it is up to the user to obtain, install, and maintain the software, downloading periodic updates as new features are introduced. The second implementation is over the World Wide Web and is known as WWW Entrez or WebEntrez (simply referred to as Entrez). This option makes use of available Web browsers (e.g., Netscape or Explorer) to deliver search results to the desktop. The Web allows the user to navigate by clicking on selected words in an entry. Furthermore, the Web implementation allows for the ability to link to external data sources. While the Web version is formatted as sequential pages, the Network version uses a series of windows with faster speed. The NCBI databases are, by far, the most often accessed by biochemists, and some of their searchable fields include plain text, author name, journal title, accession number, identity name (e.g., gene name, protein name, chemical substance name), EC number, sequence database keyword, and medical subject heading.

The Entrez home page (Figure 3.2) is opened by keying in URL, http://www.ncbi.nlm.nih.gov/Entrez/. An Entrez session is initiated by choosing one of the available databases (e.g., click Nucleotides for DNA sequence, Proteins for protein sequence or 3D structures for the 3D coordinates) and then composing a Boolean query designed to select a small set of documents after choosing Primary accession for the Search Field. The term comprising the query may be drawn from either plain text or any of several more specialized searchable fields. Once a satisfactory query is selected, a list of summaries (brief descriptions of the document contents) may be viewed. Successive rounds of neighboring (to related documents of the same database types) or linking (to associated records in other databases) may be performed using this list. After choosing an appropriate report format (e.g., GenBank, fasta), the record may be printed and saved.

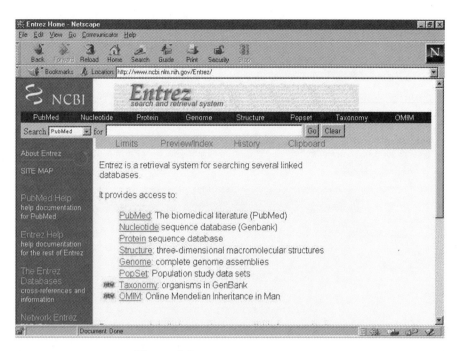

Figure 3.2. Home page of Entrez.

One of the central activities of the European Bioinformatics Institute (EBI) (Emmert et al., 1994) is development and distribution of the EMBL nucleotide sequence database (Stoesser et al., 2001). This is a collaborative project with GenBank (NCBI, USA) and DDBJ (DNA database of Japan) to ensure that all the new and updated database entries are shared between the groups on a daily basis. The search of sequence databases and an access to various application tools can be approached from the home page of EBI at http://www.ebi.ac.uk/ (Figure 3.3).

The Sequence Retrieval System (Etzold et al., 1996) is a network browser for databases at EBI. The system allows users to retrieve, link, and access entries from all the interconnected resources such as nucleic acid, EST, protein sequence, protein pattern, protein structure, specialist/boutique, and/or bibliographic databases. The SRS is also a database browser of DDBJ, ExPASy, and a number of servers as the query system. The SRS can be accessed from EBI Tools server at http://www2.ebi.ac.uk/Tools/index.html or directly at http://srs6.ebi.ac.uk/. The SRS permits users to formulate queries across a range of different database types via a single interface in three different methods (Figure 3.4):

Quick search: This is the fastest way to generate a query. Click the checkbox to select databank(s)—for example, EMBL for nucleotide sequences or SWISSPROT for amino acid sequences. Enter the query word(s), accession ID, or regular expression and then click the Go button to start the search.

Standard query: Select the databanks and click Standard button under Query forms. This opens the standard query form where the user is given choices of data fields to search, operator to use, wild card to append, entry type, and result views.

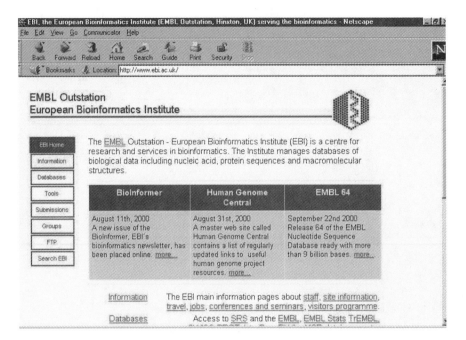

Figure 3.3. Home page of EBI.

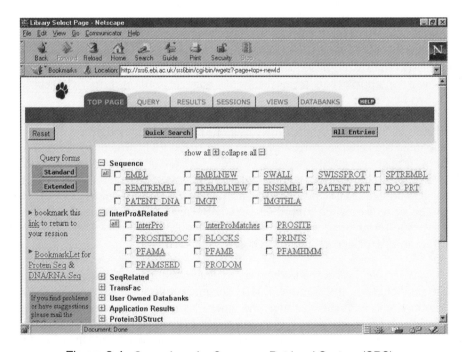

Figure 3.4. Query form for Sequence Retrieval System (SRS).

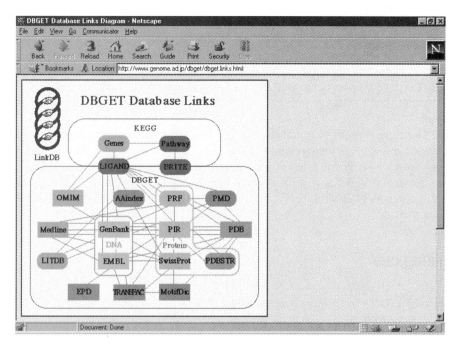

Figure 3.5. DBGet link of DDBJ.

There are four textboxes with corresponding data field selectors. After entering querystrings to textboxes and choosing data fields, select sequence formats (embl, fasta or genbank) and then click the Submit Query button to begin the search.

Extended query: Select the databanks and click Extended button to open the extended query form. Enter query string to the Description field, name of the organism to the Organism field, and data string (yyyymmdd) in the Date field. Click Submit Query button to initiate the search.

The GenomeNet facility of DDBJ (Tateno et al., 2000) consists of four components: (1) DBGet/LinkDB with integrated database and retrieval system, (2) KEGG, which represents *Kyoto Encyclopedia of Genes and Genomes* (Kanehisa and Goto, 2000), (3) sequence interpretation tools, and (4) collection of genome-related databases in Japan. A DBGet query can be made from the DBGet links diagram (Figure 3.5) by selecting the database or an organism of interest as shown after keying in http://www.genome.ad.jp/dbget/.

DBGet is a comprehensive database retrieval system supporting a wide range of databases (locally or hyperlinked) including GenBank, EMBL, DDBJ, SWISS_PROT, PIR, PDB, PDBSTR, EPD, TRANSFAC, PROSITE, LIGAND, PATHWAY, AAindex, LITDB, and MEDLINES. The user is given a query form with a choice of either *bfind* mode or *bget* mode to initiate search/retrieval by specifying keywords. When an entry is retrieved, the highlighted links given by the original database providers can be followed to extract related entries.

3.4. WORKSHOPS

1. What information is available from MEDLINE of Entrez? Retrieve a document reporting recent work on a genomic analysis of a human disease.

2. Retrieve one sequence (either DNA or protein) each from three major database retrieval systems — Entrez, SRS of EBI, and DBGet — and compare their outputs.

3. Retrieve one primary database of protein sequence and its secondary databases and discuss their relationships.

4. Search WWW for a list of available gene identification servers.

5. Search WWW for a list of available secondary sequence databases of proteins.

6. Search WWW for carbohydrate database servers. What information do these servers provide?

REFERENCES

Baxevanis, A. D. (2001) *Nucleic Acids Res.* **29**:1–10.

Baxevanis, A. D. (2002) *Nucleic Acids Res.* **30**:1-12.

Chellen, S. S. (2000) *The Essential Guide to the Internet.* Routledge, New York.

Crumlish, C. (1999) *The Internet.* Sybex, San Francisco.

Discala, C., Benigni, X., Barillot, E., and Vaysseix, G. (2000) *Nucleic Acids Res.* **28**:8–9.

Emmert, D. B., Stoehr, P. J., Stoesser, G., and Cameron, G. N. (1994) *Nucleic Acids Rese.* **22**(17):3445–3449.

Etzold, T., Ulyanov, A., and Argos, P. (1996) *Meth. Enzymol.* **266**:114–128.

Kanehisa, M., and Goto, S. (2000) *Nucleic Acids Res.* **28**:27–30.

Marshall, E. L. (1996) *A Student's Guide to the Internet.* Millbrook Press, Brookfield, CT.

Schuler, G. D., Epstein, J. A., Okawa, H., and Kans, J. A. (1996) *Meth. Enzymol.* **266**:141–162.

Stoesser, G., Baker, W., van den Broek, A., Camon, E., Garcia-Pastor, M., Kanz, C., Kulilova, T., Lombard, V., Lopez, R., Parkinson, H., Redaschi, N., Sterk, P., Stoehr, P., and Tuli, M. A. (2001) *Nucleic Acids Res.* **29**:17–21.

Swindell, S. R., Miller, R. R., and Myer, G., Eds. (1996) *Internet for the Molecular Biologist.* Horizon Scientific Press, Washington, D.C.

Tateno, Y., Miyazaki, S., Ota, M., Sugawara, H., and Gojobori, T. (2000) *Nucleic Acids Res.* **28**:24–26.

Wheeler, D. L., Church, D. M., C., Lash, A. E., Leipe, D. D., Madden, T. L., Pontius, J. U., Schuler, G. D., Schriml, L. M., Tatusova, T. A., Wagner, L., and Rapp, B. A. (2001) *Nucleic Acids Res.* **29**:11–16.

Wyatt, A. L. (1995) *Success with Internet.* Boyd & Fraser, Danvers, MA.

4

MOLECULAR GRAPHICS:
VISUALIZATION OF BIOMOLECULES

Computer graphics has changed the way in which chemical structures are presented and perceived. The facile conversion of macromolecular sequences into three-dimensional structures that can be displayed and manipulated on the computer screen have greatly improved the biochemist's understanding of biomolecular structures. Tools for graphical visualization and manipulation of biomolecular structures are described.

4.1. INTRODUCTION TO COMPUTER GRAPHICS

Molecular graphics (Henkel and Clarke, 1985) refers to a technique for the visualization and manipulation of molecules on a graphical display device. The technique provides an exciting opportunity to augment the traditional description of chemical structures by allowing the manipulation and observation in real time and in three dimensions, of both molecular structures and many of their calculated properties. Recent advances in this area allow visualization of even intimate mechanisms of chemical reactions by graphical representation of the distribution and redistribution of electron density in atoms and molecules along the reaction pathway.

All graphics programs must be able to import commands defining representations and translate these into a picture according to the representations specified. The graphics programs offer various choices of renderings of the model, with color coding of atoms or groups and with selective labeling. These include the following.

Line Drawings: Skeletal and Ball-and-Stick Models. Traditionally, drawings of small molecules have represented either (a) each atom by a sphere or (b) each bond by a line segment. Bond representations give a clearer picture of the topology or connectivity of a structure. In a simple picture of line drawings, there is a direct correspondence; that is, one line segment equals one bond. There are basically two ways to extract bonds from a given set of atomic coordinates:

1. *Screen the atoms by distance.* This is the most general approach. For every pair of atoms in the structure, the distance between them is calculated. If the distance is less than the sum of the van der Waals radii of the two atoms, a bond between them is assumed. For proteins or nucleic acids, this approach can be specialized by checking only atoms in the same residue/nucleotide, plus the atoms in the peptide/nucleotide bonds between successive residues/ nucleotides.

2. *Create an explicit list of bonded pairs.* For protein containing only standard amino acids and common ligands, this can be done once and for all. For each residue type, one can make a list of pairs of atom names, each pair corresponding to a bond. Then in drawing a picture of a protein for each residue, one can search the coordinates of each pair of atoms in the list of bonds and add the appropriate line segment to the drawing. Similar considerations apply to nucleic acids. In pdb files, explicit connectivity lists are provided.

A ball-and-stick drawing is a simple skeletal model, in which the representation of the bond is generalized to a cylinder, and a disc is added at the position of each atom. Additional information may thereby be displayed in that different atom types may be distinguished by size and shading, and bonds of different appearance may be drawn. To create the line segments corresponding to a pure skeletal model, one needs only copy the coordinates of each pair of bonded atoms as the line segment end-points. To create ball-and-stick pictures, one must:

1. Draw a circle at the position of each atom with the facility to vary the atomic color and radius.

2. Determine the line segments that represent the bond and attach these segments at the edge of the circle.

Wire models and ball-and-stick models are extremely useful because of the great detail they contain. They are particularly useful in connection with blow-ups of selected protein/nucleic acid molecules.

Shaded-Sphere Pictures. Each atom is shown to represent complete chemical detail of molecules. This category includes three basic types of picture:

1. *Line drawings:* Each atom is represented as a disc. The picture is a limit of the ball-and-stick drawings as the radius of each atomic ball is made in proportional to the van der Waals radius.

2. *Color raster devices:* A raster device can map an array stored in memory on to the screen so that the value of each element of the array controls the appearance of the corresponding point on the screen. It is possible to draw

each atom as a shaded sphere, or even to simulate the appearance of the Corey–Pauling–Koltun (CPK) physical models to maintain most of the familiar color scheme (C = black, N = blue, O = red, P = Green, and S = yellow, etc.). In such representation, atoms are usually opaque, so that only the front layer of atoms is visible. However, clipping with an inner plane or rotation can show the packing in the molecular interior.

3. *Real-time rotation and clipping:* Facilities available on vector graphics devices are very useful in connection with another technique for representing atomic and molecular surfaces. The spatter-painting of the surface of a sphere by a distribution of several hundred dots produces a translucent representation of the surface (dot-surface). It is possible to combine dot-surface representations with skeletal models to show both the topology of the molecule and its space-filling properties (Connolly, 1983). Dot-surface pictures used on an interactive graphics device with a color screen have been helpful in solving problems of docking ligands to proteins and exploring the goodness of fit in interfaces.

Basically, any color can be matched by a suitable combination of three primary colors, RGB (red, green, and blue). These three primary colors on a binary status (on/off) provide eight colors (Figure 4.1). Video monitors generate a large number of colors by combining various amount of the three primary components. The observed color of an object depends on the spectrum of the light it emits, transmits, or reflects. Observable colors may be distinguished on the basis of three characteristics:

1. *Hue* (*color*): Color in the most common colloquial sense (i.e., red, green and blue) describes different hues. For monochromic light, different wavelengths

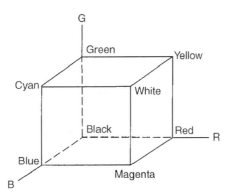

Figure 4.1. Cube model for RGB system. The RGB cube model illustrates the definition of colors by the three primary components along the three axes R, G, and B. Each color point is represented by a triple (r, g, b). The three primary colors are red (1, 0, 0), green (0, 1, 0), and blue (0, 0, 1). Other binary-status (0/1 for r, g, b) colors are cyan (0, 1, 1), magenta (1, 0, 1), yellow (1, 1, 0), white (1, 1, 1), and black at origin (0, 0, 0). Different colors are expressed by a combination of r, g, and b values varied between 0 and 1. For example, gray colors correspond to the main diagonal between black and white.

correspond to different hues, but different spectra can give the same perceived hue. A flat spectrum appears achromatic: white, gray, or black.

2. *Tone* (*value*): Roughly speaking, the total amount of light per unit area (i.e., multiplying a spectrum by a constant) changes the tone. Members of the series white → gray → black differ in tone. Empirical scales of tone are not linear in integrated intensity; moreover, changes in tone can alter perceived hue.

3. *Saturation* (*intensity or chroma*): The difference between a color and the gray with the same tone exemplifies saturation. A pure or saturated color can be diminished in saturation by adding white light and normalizing the result to the same perceptual tone.

As a result, colors may be thought of as point in a three-dimensional space, the axes of which might be the primary colors (primaries): red, green, and blue (Figure 4.1). Each color is a vector, the components of which are the intensities of the primaries required to match it. For displays generated by three-gun (red, green, and blue) monitors, these are the numbers specified. The literature of computer graphics reveals considerable efforts to achieve truly convincing representations of real objects (Roger, 1985; Wyszecki and Stiles, 1982).

4.2. REPRESENTATION OF MOLECULAR STRUCTURES

There are three levels of structural graphics: 1D formula or string/character format (e.g., SMILES string), 2D chemical structures (e.g., ISIS draw), and 3D molecular structures (e.g., PDB atomic coordinate display). For proteins and nucleic acids, the primary structures represented in coding sequences are 1D formats. The sequences represented in the chemical linkages of amino acid or nucleotide structures are 2D drawings, and the full atomic representations of the 3D structures constitute 3D graphics. The common 1D presentations for nucleotide sequences are in GenBank (Figure 4.2), EMBL (Figure 4.3), and fasta (Figure 4.4) formats. The common files for amino acid sequences are PIR (Figure 4.5), Swiss-Prot (Figure 4.6), GenPept (Figure 4.7) and fasta (Figure 4.8) formats.

```
BASE COUNT       132 a      169 c      172 g      111 t
ORIGIN
        1 tcccgctgtg tgtacgacac tggcaacatg aggtctttgc taatcttggt gctttgcttc
       61 ctgcccctgg ctgctctggg gaaagtcttt ggacgatgtg agctggcagc ggctatgaag
      121 cgtcacggac ttgataacta tcggggatac agcctgggaa actgggtgtg tgttgcaaaa
      181 ttcgagagta acttcaacac ccaggctaca aaccgtaaca ccgatgggag taccgactac
      241 ggaatcctac agatcaacag ccgctggtgg tgcaacgatg gcaggacccc aggctccagg
      301 aacctgtgca acatcccgtg ctcagccctg ctgagctcag acataacagc gagcgtgaac
      361 tgcgcgaaga agatcgtcag cgatggaaac ggcatgagcg cgtgggtcgc ctggcgcaac
      421 cgctgcaagg gtaccgacgt ccaggcgtgg atcagaggct gccggctgtg aggagctgcc
      481 gcacccggcc cgcccgctgc acagccggcc gctttgcgag cgcgacgcta cccgcttggc
      541 agttttaaac gcatccctca ttaaaacgac tatacgcaaa cgcc
//
```

Figure 4.2. GenBank format for nucleotide sequence of chicken egg-white lysozyme.

```
SQ      Sequence 584 BP;  132 A;  169 C;  172 G;  111 T;  0 other;
        tcccgctgtg tgtacgacac tggcaacatg aggtctttgc taatcttggt gctttgcttc   60
        ctgcccctgg ctgctctggg gaaagtcttt ggacgatgtg agctggcagc ggctatgaag  120
        cgtcacggac ttgataacta tcggggatac agcctgggaa actgggtgtg tgttgcaaaa  180
        ttcgagagta acttcaacac ccaggctaca aaccgtaaca ccgatgggag taccgactac  240
        ggaatcctac agatcaacag ccgctggtgg tgcaacgatg gcaggacccc aggctccagg  300
        aacctgtgca acatcccgtg ctcagccctg ctgagctcag acataacagc gagcgtgaac  360
        tgcgcgaaga agatcgtcag cgatggaaac ggcatgagcg cgtgggtcgc ctggcgcaac  420
        cgctgcaagg gtaccgacgt ccaggcgtgg atcagaggct gccggctgtg aggagctgcc  480
        gcacccggcc cgcccgctgc acagccggcc gctttgcgag cgcgacgcta cccgcttggc  540
        agttttaaac gcatccctca ttaaaacgac tatacgcaaa cgcc                    584
//
```

Figure 4.3. EMBL format for nucleotide sequence of chicken egg-white lysozyme.

```
>gi|63580|emb|V00428.1|GGLYS1 Gallus gallus mRNA coding for lysozyme
TCCCGCTGTGTGTACGACACTGGCAACATGAGGTCTTTGCTAATCTTGGTGCTTTGCTTCCTGCCCCTGG
CTGCTCTGGGGAAAGTCTTTGGACGATGTGAGCTGGCAGCGGCTATGAAGCGTCACGGACTTGATAACTA
TCGGGGATACAGCCTGGGAAACTGGGTGTGTGTTGCAAAATTCGAGAGTAACTTCAACACCCAGGCTACA
AACCGTAACACCGATGGGAGTACCGACTACGGAATCCTACAGATCAACAGCCGCTGGTGGTGCAACGATG
GCAGGACCCCAGGCTCCAGGAACCTGTGCAACATCCCGTGCTCAGCCCTGCTGAGCTCAGACATAACAGC
GAGCGTGAACTGCGCGAAGAAGATCGTCAGCGATGGAAACGGCATGAGCGCGTGGGTCGCCTGGCGCAAC
CGCTGCAAGGGTACCGACGTCCAGGCGTGGATCAGAGGCTGCCGGCTGTGAGGAGCTGCCGCACCCGGCC
CGCCCGCTGCACAGCCGGCCGCTTTGCGAGCGCGACGCTACCCGCTTGGCAGTTTTAAACGCATCCCTCA
TTAAAACGACTATACGCAAACGCC
```

Figure 4.4. Fasta format for nucleotide sequence of chicken egg-white lysozyme.

```
> SUMMARY            #length 147 #molecular_weight 16238
SEQUENCE
               5        10        15        20        25        30
    1 M R S L L I L V L C F L P L A A L G K V F G R C E L A A A M
   31 K R H G L D N Y R G Y S L G N W V C A A K F E S N F N T Q A
   61 T N R N T D G S T D Y G I L Q I N S R W W C N D G R T P G S
   91 R N L C N I P C S A L L S S D I T A S V N C A K K I V S D G
  121 N G M N A W V A W R N R C K G T D V Q A W I R G C R L
```

Figure 4.5. PIR format for amino acid sequence of chicken egg-white lysozyme.

```
Length: 147 AA     Molecular weight: 16238 Da
         10        20        30        40        50        60
          |         |         |         |         |         |
   MRSLLILVLC FLPLAALGKV FGRCELAAAM KRHGLDNYRG YSLGNWVCAA KFESNFNTQA

         70        80        90       100       110       120
          |         |         |         |         |         |
   TNRNTDGSTD YGILQINSRW WCNDGRTPGS RNLCNIPCSA LLSSDITASV NCAKKIVSDG

        130       140       147
          |         |         |
   NGMNAWVAWR NRCKGTDVQA WIRGCRL
```

Figure 4.6. Swiss-Prot format for amino acid sequence of chicken egg-white lysozyme.

```
ORIGIN
        1 mrsllilvlc flplaalgkv fgrcelaaam krhgldnyrg yslgnwvcva kfesnfntqa
       61 tnrntdgstd ygilqinsrw wcndgrtpgs rnlcnipcsa llssditasv ncakkivsdg
      121 ngmsawvawr nrckgtdvqa wirgcrl
//
```

Figure 4.7. GenPept format for amino acid sequence of chicken egg-white lysozyme.

```
>gi|63581|emb|CAA23711.1| coding sequence lysozyme [Gallus gallus]
MRSLLILVLCFLPLAALGKVFGRCELAAAMKRHGLDNYRGYSLGNWVCVAKFESNFNTQATNRNTDGSTD
YGILQINSRWWCNDGRTPGSRNLCNIPCSALLSSDITASVNCAKKIVSDGNGMSAWVAWRNRCKGTDVQA
WIRGCRL
```

Figure 4.8. Fasta format for amino acid sequence of chicken egg-white lysozyme.

Protein topology cartoons (TOPS) are two-dimensional schematic representations of protein structures as a sequence of secondary structure elements in space and direction (Flores et al., 1994; Sternberg and Thornton, 1977). The TOPS of trypsin domains as exemplified in Figure 4.9 have the following symbolisms:

1. Circular symbols represent helices (α and 3_{10}).
2. Triangular symbols represent β strands.
3. The peptide chain is divided into a number of fragments, and each fragment lies in only one domain.
4. Each fragment is labeled with an integer (i), beginning at N_i and ending at C_{i+1} with the first fragment being $N_1 \rightarrow C_2$.
5. If the chain crosses between domains, it leaves the first at C_{i+1} to join the next N_{i+2}.
6. Each secondary structure element has a direction (N to C) that is either *up* (out of the plane of the diagram) or *down* (into the plane of the diagram).
7. The direction is up if the N-terminal connection is drawn to the edge of the symbol and the C-terminal connection is drawn to the center of the symbol. Otherwise, the direction is down if the N-terminal connection is drawn to the center of the symbol and the C-terminal connection is drawn to the edge.

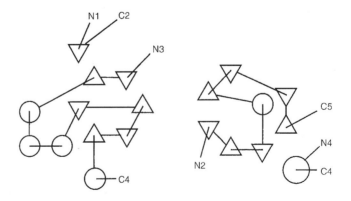

Figure 4.9. TOPS diagrams for trypsin domains.

8. For β strands, up strands are indicated by upward-pointing triangles whereas down strands are indicated by downward-pointing triangles.

The topology cartoons can be browsed and searched at TOPS server (http:// tops.ebi.ac.uk/tops/).

The most obvious data in a typical 3D structure record are the atomic coordinate data, the locations in space of the atoms of a molecule represented by (x, y, z) triples, and distances along each axis to some arbitrary origin in space. The coordinate data for each atom are attached to a list of labeling information in the structure record such as that derived from the protein or nucleic acid sequence.

Three-dimensional molecular structure database records employ two different "minimalist" approaches regarding the storage of bond data. The chemistry rule applies observable physical principles of chemistry to record molecular structures without bond information. There is no residue dictionary required to interpret data encoded by this approach, just a table of bond lengths and bond types for every conceivable pair of bonded atoms. This approach is the basis for 3D biomolecular structure file format of Protine Data Bank (Bernstein et al., 1997). The other approach is used in the database records of the Molecular Modeling Database (MMDB) at NCBI, which uses a standard residue dictionary of all atoms and bonds in the biomacromolecules of amino acid and nucleotide residues plus end-terminal variants (Hogue et al., 1996). The software that reads in MMDB data which are derived from the data of PDB can use the bonding information supplied in the dictionary to connect atoms together, without trying to enforce the rules of chemistry.

Almost all known protein structures have been determined by X-ray crystallography. A few contain details derived from neutron diffraction, and a few have been determined from nuclear magnetic resonance (NMR). Recently, the theoretical models derived from molecular modeling are added. The resolution of a structure is a measure of how much data were collected. The more data collected, the more detailed the features in the electron density map to be fitted, and, of course, the greater the ratio of observations to parameters to be determined (i.e., the atomic coordinates and R-factors). Resolution is expressed in angstroms (Å), which is a measure of distance. The lower the number, the higher the resolution. Protein structures are generally determined to a resolution between 1.7 and 3.5 Å; those determined at 2.0 Å or better are considered high-resolution. The R-factor of a structure determination is a measure of how well the model reproduces the experimental intensity data. Other things being equal, the lower the R-factor, the better the structure. The R-factor is a fraction expressed as a percentage; $R = 0\%$ would be an impossible ideal case (no disorder, no experimental error), and $R = 58\%$ for a collection of atoms placed randomly in unit cell of the crystal.

The Protein Data Bank (PDB) is the collection of publicly available structures of proteins, nucleic acids, and other biological macromolecules initiated by Brookhaven National Laboratory and now maintained by the Research Collaboratory for Structural Bioinformatics (RCSB) at http://www.rcsb.org/pdb/ (Berman et al., 2000). The PDB coordinates of biomacromolecules can be classified into the following:

1. Protein structures determined by X-ray or neutron diffraction or NMR which may include co-factors, substrates, inhibitors, or other ligands

2. Oligonucleotide or nucleic acid structures determined by X-ray crystal-lography

3. Carbohydrate structures determined by X-ray diffraction

4. Hypothetical models of protein structures

5. Bibliographic entries

Each set of coordinates deposited with the PDB becomes a separate entry. Each entry is associated with an accession PDB code with a unique set of four alpha-numeric characters. PDB and its mirror sites offer a text search engine that uses an index of all the textual information in each PDB record (e.g., PDB ID); an example of such an index is 1LYZ for hen's egg-white lysozyme. The first character is a version number. An identifier beginning with the number 0 signifies that the entry is purely bibliographic. The pdb file is a text file with an explanatory header followed by a set of atomic coordinates. The atomic coordinates are subjected to a set of standard stereochemical checks and are translated into a standard entry format; for example, Figure 4.10 shows partial coordinate file for 1LYZ.pdb or pdb1LYZ.ent.

The PDB format includes information about the structure determination, bibliographic references describing the structure, types/locations of secondary struc-tures, and the atomic coordinates. Most software programs created for molecular graphics or other computational analysis of protein structures can read files in PDB format, with file extension of either .pdb or .ent. The 3D graphical representations of these biomacromolecules can be displayed with RasMol (Sayle and Milner-White, 1995), Cn3D (Wang et al., 2000) or KineMage (Richardson and Richardson, 1992; Richardson and Richardson, 1994) as shown in Figure 4.11. For online visualizatiion of 3D structures, the MIME types for PDB format is chemical/x-pdb, which enables the display of 3D structure on the Web with RasMol. Once the atomic coordinates are known, the reader can manipulate the image in the browser to rotate the molecule, view it from a different perspective, or change the manner in which the structure is presented. A comprehensive list of chemical MIME media type (Rzepa et al., 1998) is available from http://www.ch.ic.ac.uk/chemime/.

4.3. DRAWING AND DISPLAY OF MOLECULAR STRUCTURES

4.3.1. String and Sequences

One of the common character formats or chemical nomenclature of a valence model which is recognizable by a number of 2D-structure drawing programs is SMILES (Weininger, 1988). A full SMILES language tutorial can be accessed at http://www.daylight.com/dayhtml/smiles/. The general rules for biochemical compounds are as follows:

1. *Hydrogen atom:* Hydrogen atoms are not normally specified. An explicit hydrogen specification is required for charged hydrogen, isotopic hydrogen, or chiral assignment. It is written within brackets [ ] if an isotope, chiral, or charge are inferred — that is, $[H^+]$ for a proton.

2. *Atom specification:* Elemental identity is represented by a standard atomic symbol without explicit hydrogen if the number of attached hydrogens

```
HEADER     HYDROLASE (O-GLYCOSYL)                      01-FEB-75  1LYZ        1LYZ
COMPND     LYSOZYME (E.C.3.2.1.17)                                           1LYZ
SOURCE     HEN (GALLUS $GALLUS) EGG WHITE                                    1LYZ
... ...
... ...
ATOM    258   N   PHE    34     2.966  25.379  11.370  7.00  1.50            1LYZ
ATOM    259   CA  PHE    34     3.955  25.906  11.749  6.00  1.50            1LYZ
ATOM    260   C   PHE    34     4.614  25.247  12.887  6.00  1.50            1LYZ
ATOM    261   O   PHE    34     5.801  25.379  12.887  8.00  1.50            1LYZ
ATOM    262   CB  PHE    34     3.494  27.356  12.002  6.00  1.50            1LYZ
ATOM    263   CG  PHE    34     3.362  27.949  10.612  6.00  1.50            1LYZ
ATOM    264   CD1 PHE  . 34     2.241  29.004  10.297  6.00  1.50            1LYZ
ATOM    265   CD2 PHE    34     4.614  27.949   9.728  6.00  1.50            1LYZ
ATOM    266   CE1 PHE    34     1.978  29.663   9.160  6.00  1.50            1LYZ
ATOM    267   CE2 PHE    34     4.417  28.674   8.402  6.00  1.50            1LYZ
ATOM    268   CZ  PHE    34     3.164  29.400   8.212  6.00  1.50            1LYZ
ATOM    269   N   GLU    35     3.889  24.653  13.771  7.00  1.50            1LYZ
ATOM    270   CA  GLU    35     4.680  23.796  14.971  6.00  1.50            1LYZ
ATOM    271   C   GLU    35     5.273  22.412  14.529  6.00  1.50            1LYZ
ATOM    272   O   GLU    35     6.526  22.149  14.845  8.00  1.50            1LYZ
ATOM    273   CB  GLU    35     4.087  23.599  15.982  6.00  1.50            1LYZ
ATOM    274   CG  GLU    35     3.757  23.665  17.308  6.00  1.50            1LYZ
ATOM    275   CD  GLU    35     4.680  24.522  17.940  6.00  1.50            1LYZ
ATOM    276   OE1 GLU    35     5.603  24.588  17.687  8.00  1.50            1LYZ
ATOM    277   OE2 GLU    35     4.153  25.642  18.509  8.00  1.50            1LYZ
ATOM    278   N   SER    36     4.680  21.687  13.834  7.00  1.50            1LYZ
ATOM    279   CA  SER    36     4.878  20.435  13.455  6.00  1.50            1LYZ
ATOM    280   C   SER    36     5.142  20.039  11.876  6.00  1.50            1LYZ
ATOM    281   O   SER    36     5.669  18.919  11.370  8.00  1.50            1LYZ
ATOM    282   CB  SER    36     4.285  19.446  14.024  6.00  1.50            1LYZ
ATOM    283   OG  SER    36     2.966  19.116  13.266  8.00  1.50            1LYZ
... ...
TER    1002       LEU   129                                                  1LYZ
HETATM 1003   O   HOH     1     1.437  16.676  19.902  7.36  1.59  1         1LYZ
HETATM 1004   O   HOH     2    -.616  11.133  19.523  8.12  1.80  2         1LYZ
... ...
CONECT    48  47  981                                                        1LYZ
... ...
END                                                                         1LYZ
```

Figure 4.10. PDB file (partial) for 3D structure of hen's egg-white lysozyme (1LYZ.pdb). The abbreviated file shows partial atomic coordinates for residues 34–36. Informational lines such as AUTHOR (contributing authors of the 3D structure), REVDAT, JRNL (primary bibliographic citation), REMARK (other references, corrections, refinements, resolution and missing residues in the structure), SEQRES (amino acid sequence), FTNOTE (list of possible hydrogen bonds), HELIX (initial and final residues of α-helices), SHEET (initial and final residues of β-sheets), TURN (initial and final residues of turns, types of turns), and SSBOND (disulfide linkages) are deleted here for brevity. Atomic coordinates for amino acid residues are listed sequentially on ATOM lines. The following HETATM lines list atomic coordinates of water and/or ligand molecules.

conforms to the lowest normal valence consistent with explicit bonds — that is, C(4), N(3,5), O(2), P(3,5), and S(2,4,6).

3. *Bond specification:* Single, double, triple, and aromatic bonds are represented by the symbols —, =, #, and :, respectively, for example, CC=O for acetaldehyde. Generally, single and aromatic bond symbols are omitted.

4. *Branch specification:* Branches are enclosed in nested or stacked parentheses — for example, C(C)CC(N)C(=O)O for valine.

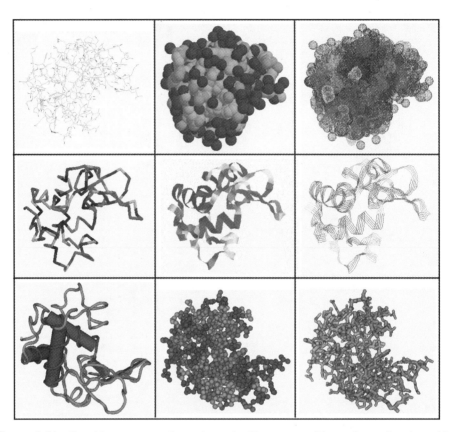

Figure 4.11. Graphic representations of protein 3D structure. Three-dimensional graphics of hen's egg-white lysozyme as visualized with RasMol (first and second rows, 1LYZ.pdb) and Cn3D (third row, 1LYZ.val) are shown from left to right (color type) in wireframe (atom), spacefill (atom), dots (residue), backbone (residue), ribbons (secondary structure), strands (secondary structure), secondary structure (secondary structure), ball-and-stick (residue), and tubular (domain) representations.

5. *Ring specification:* Ring closure bonds are specified by appending matching digits to the specifications of the joined atoms, with the bond symbol preceding the digit—for example, N1CCCC1C(=O)O for proline.

6. *Aromaticity* (A ring having sp^2 hybridized carbons with $4N + 2$ p-electrons): Aromatic atoms are specified with lowercase atomic symbols with appending matching digits following the joined atomic symbols—for example, Oc1ccccc1CC(N)C(=O)O or OC1=CC=CC=C1CC(N)C(=O)O for tyrosine.

7. *Disconnection:* The period or dot is used to represent disconnections—for example, C=COP(=O)(=O)OC(=O)[O⁻].[Na⁺] for sodium phosphoenolpyruvate.

8. *Isotope:* Isotopic specification is indicated by prefixing the atomic symbol with a number equal to the integral isotopic mass—for example, [²H] for deuterium and [¹³C] for carbon-13.

9. *Isomerism* (*geometric*): Configuration around double bonds is specified by the characters / and \ indicating relative directionality between the connected (by double bond) atoms — for example,

$$CCCCCCCC/C=C\backslash CCCCCCCC(=O)O$$

for oleic acid with *cis* double bond.

10. *Isomerism* (*chiral*): The most common type of chirality in biochemistry is tetrahedral. The tetrahedral chiral specification (@ or @@) is written as an atomic property following the atomic symbol of the chiral atom. Looking at the chiral center from the direction of the "from" atom (preceding the chiral atom), @ (or @1) means "the other" three atoms (following the chiral atom) are listed *anti-clockwise*; @@ (or @2) means *clockwise*. If the chiral atom has a nonexplicit hydrogen, it will be listed inside the chiral atom's brackets as [C@H] — for example,

$$[C@2H]O1([C@2H](O)[C@2H](O)[C@2H](O)[C@2H]1CO$$

for *β*-D-glucopyranose that has all hydroxyl groups in equatorial configuration.

The default sequence format (nucleotide/amino acid) retrieved from three integrated database retrieval systems are GenBank/GenPept from Entrez as well as DBGet, and EMBL/Swiss-Prot from EBI. Though different formats can be specified at the time of retrieval, these formats can be interconverted by the use of ReadSeq facility at http://dot.imgen.bcm.tmc.edu:9331/seq-util/Options/readseq.html. The formats supported by ReadSeq are IG/Stanford, GenBank/GB, NBRF, EMBL, GCG, DNASrider, Fitch, Pearson.Fasta, Philip3.2, Philip. PIR/CODATA, MSF, ASN.1, and PAUP/NEXUS.

4.3.2. Drawing of Molecular Structures

The 1D nucleotide/amino acid sequences in character format (without index, e.g., fasta format) can be converted into the 2D chemical structures with ISIS Draw, which can be downloaded from MDL Information System at http://www.mdli.com/download/isisdraw.html for academic use. Install the package by issuing Run command, C:\Isis\Draw23.exe. Launch IsisDraw to open the Draw window.

Retrieve nucleotide/amino acid sequence file in fasta format (remove the heading, >line) or prepare text file of sequence in one-letter characters. Rename the file as seqname.seq. Prepare to import the sequence by checking (√) Show sequence bond, Show leaving groups, Amino acid-/DNA-/RNA-1 letter from Sequence options of Chemistry menu. Invoke File→Import→Sequences. Select Amino acid-/DNA-/RNA-1 letter. The sequence (with bonds and leaving groups attached) should appear within the draw window. Mark the whole sequence with Select All from the Edit menu or by using Lasso tool. From the Chemistry menu, select Residue→Expand, the 1D (text string) sequence is transformed into the 2D molecular structure. Save it as struname.skc (e.g., heptapeptide, STANLEY as stanley.skc) as shown in Figure 4.12.

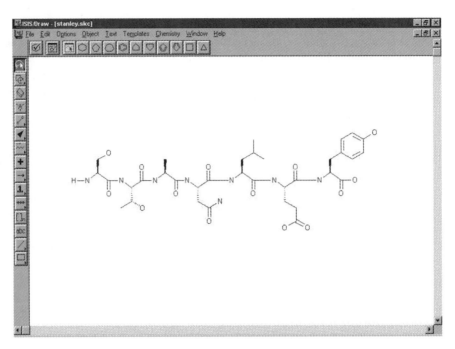

Figure 4.12. Two-dimensional structure sketch with ISIS Draw. The two-dimensional structure of a heptapeptide, SerThrAlaAsnLeuGluTyr (without hydrogens), is sketched on the ISIS display window after importing the sequence file in text format (STANLEY).

To draw 2D molecular structures, the users should refer to ISIS Draw Quick Start (ISIS Draw help) for operating instructions. Draw the basic framework from template tools (horizontal template-tool icons and template pages from Template menu) and drawing tools (vertical drawing-tool icons). The small triangular sign on the drawing tool icons indicates additional tools available for selection. For example, pressing on the Single bond tool provides selection for drawing a double bond or a triple bond. Verify the chemistry of sketch by clicking run Chemisrtry inspector icon (or select Chemistry inspector from the Chemistry menu). To ensure uniform bond lengths and angles for the sketched molecule, select the molecule and then choose Object→Clean Molecule. Save the sketch as struname.skc.

To place template, click one of the template-tool icons or an atom/bond in the structural fragment/molecule on the template page and place the template anywhere inside the window by clicking an empty area. The template can be fused/attached to an existing bond/atom by simply clicking the bond/atom. To draw bonds, click a bond tool (Single/double/triple bond or Up wedge/down wedge/either/up bond/down bond), and then click the drawing area or drag the mouse from an existing atom to add a bond. To sprout a bond from an atom, click a bond tool and then click the atom. To draw chain in one direction/ring of specific shape, click chain/multibond tool. To draw atoms, click Atom tool and enter atom symbol or choose one from the drop-down list. Arrow tool provides options for drawing a variety of arrows (e.g., unidirection, equilibrium, double-head, and electron-shift arrows, etc., after pressing arrow tool then choosing one of the arrows) for chemical reactions. Use Lasso select tool to select any structure/structural component for

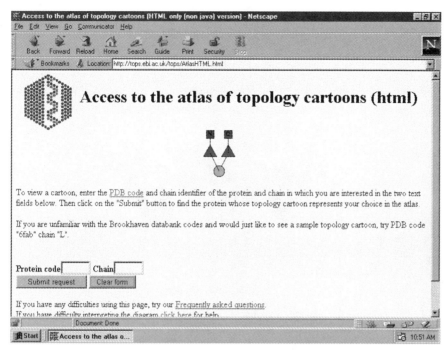

Figure 4.13. Home page for an access to TOPS cartoons.

editing or relocating. To delete atom/bond/object one at a time, click Eraser tool and then the atom/bond/object. Text tool appends text description to the structures/reactions.

To search for TOPS cartoons at http://tops.ebi.ac.uk/tops/, select Browse the Atlas of topology cartoons and Browse HTML page version to open the query form (Figure 4.13). Enter PDB ID on the Protein code query box (Chain query box can be left blank). The search may request a choice of the chain (if more than one chains are available) and returns TOPS atlas information listing the protein of your choice and representative protein in atlas. Click to view the TOPS cartoon(s) of the representative protein. Right click on the diagram to save the TOPS cartoon as cartoon.gif.

4.3.3. Display of 3D Structures with Molecular Graphics Programs

For the 3D view of ISIS/Draw, ACD/3D Viewer Add-in can be installed. Retrieve ACD/3D Viewer for ISIS/Draw from http://www.acdlabs.com/downloar/download.cgi and installed it as an Add-in according to instructions. To view 3D structure which is opened/sketched on the ISIS/Draw window, select ACD/3D Viewer tool from Object menu to open ACD/3D Viewer window and subsequent display of the 3D structure. The 3D structure can be optimized and can be saved only in .s3d format, which is not recognizable by other modeling packages.

The 2D structures of ISIS draw can be transformed into the 3D structures by WebLab Viewer Lite, which can be downloaded free for academic use from Accelrys Inc. at http://www.accelrys.com/viewer/viewlite/index.html. Select Download View-

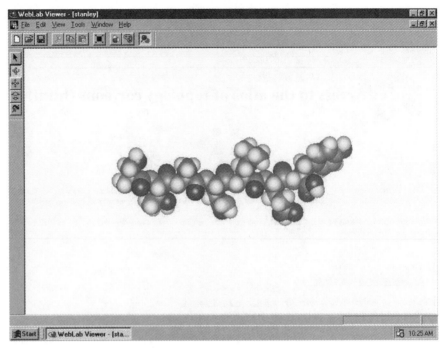

Figure 4.14. Conversion of 2D structure into 3D structure. The 2D structure file from ISIS draw (stanley.skc) is converted into the 3D structure with WebLab Viewer Lite. It should be noted that the atomic coordinate file does not contain ATOM columns with residue ID.

lite to register and download. Open the file by selecting MDL (*.skc); the 2D structure (e.g., hexapeptide, stanley.skc) is converted into the 3D structure (Figure 4.14) whose coordinate file can be saved as struname.pdb (e.g., stanley.pdb).

Alternately, the commercial molecular modeling software programs such as ChemOffice (http://www.camsoft.com) can be used. The ISIS draw in sketch format (struname.skc) is first converted to ChemDraw format (struname.cdx), which is then transformed into 3D structure (struname.c3d) with Chem 3D (Chapter 14) and saved as PDB format (struname.pdb).

The common atomic coordinate files for 3D structure in biochemistry is PDB format. The pdb files of polysaccharides, proteins, and nucleic acids can be retrieved from the Protein Data Bank at RCSB (http://www.rcsb.org/pdb/). On the home page (Figure 4.15), enter PDB ID (check the box "query by PDB id only") or keywords (check the box "match exact word") and click Find a structure button. Alternatively, initiate search/retrieval by selecting SearchLite. On the query page, enter the keyword (e.g., the name of ligand or biomacromolecule) and click Search button. Select the desired entry from the list of hits to access Summary information of the selected molecule. From the Summary information, select Download/Display file and then PDB Text and PDB noncompression format to retrieve the pdb file. In order to display 3D structure online, choose View structure followed by selecting one of 3D display options. The display can be saved in .jpg or .gif image format.

Most of molecular modeling software programs accept the pdb files (struname.pdb). RasMol, which is one of the most widely used molecular graphics freeware, can be downloaded from http://www.umass.edu/microbio/rasmol/

index2.htm. In addition to PDB (struname.pdb or struname.ent) file, RasMol also read Alchemy, Sybtk MOL2, MDL mol, CHARMm, and MOPAC files. Launch RasMol (double click rswin.exe or rw32b2a.exe) to open the display window. Open the pdb file from File menu. The 3D structure can be displayed as wireframe, backbone, sticks, spacefill, ball and stick, ribbons, strands and cartoons (Figure 4.16). The display can be exported as bmp, gif, epsf, ppm, and rast graphics.

KineMage (kinetic image) is an interactive 3D structure illustration software that can be downloaded from http://orca.st.usm.edu/~rbateman/kinemage/. It is adapted for the structure representation of biological molecules by many biochemical textbooks. The program consists of two components: PREKIN and MAGE. The PREKIN program interprets struname.pdb file to kinemage struname.kin file that is then displayed and manipulated with the MAGE program. To start the PREKIN program, click Proceed to enter an output file name. This opens a dialog box, "Starting ranges." Accepting the default (Backbone browsing script) saves the script producing Cα, disulfides for all subunits in the file to struname.kin. To start the MERGE program, click Proceed and select Open new file from File menu to open struname.kin with three windows (caption, display, and text). The 3D structure with connected series of alpha carbons is shown in the display window. To highlight the secondary structures, choose Selection of build-in scripts from the dialog box, "Starting ranges," to open Build-in scripts box. Select ribbon: HELIX, SHEET from pdb to save as ribbon.kin. The MERGE program opens ribbon.kin as shown in Figure 4.17.

Cn3D is a molecular graphics program that interprets structure files in MMDB (ASN.1) format (struname.val or struname.cgi) of Entrez/MMDB (Wang et al.,

Figure 4.15. Home page of PDB at Research Collaboratory for Structural Bioinformatics.

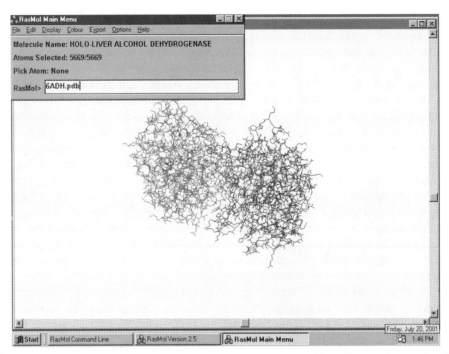

Figure 4.16. Graphic display of 3D structure with RasMol. The display shows the 3D structure of liver alcohol dehydrogenase complex (6ADH.pdb) with two subunits and bound NAD^+. The protein molecule is visualized with RasMenu.

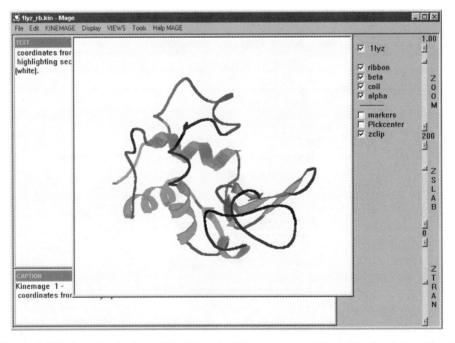

Figure 4.17. Graphic display of KineMage in ribbon representation. The $C\alpha$ chain of hen's egg-white lysozyme (1LYZ.kin derived from 1LYZ.pdb) is displayed in ribbons showing secondary structure features.

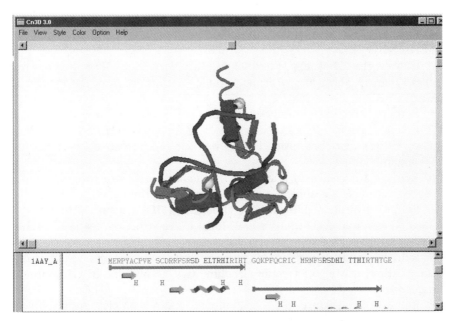

Figure 4.18. Graphic display of macromolecular interaction with Cn3D. The display window of Cn3D illustrates the 3D structure of Zn finger peptide fragments (secondary structure features) bound to the duplex oligonucleotides (brown backbone). Zinc atoms are depicted as spheres. The alignment window shows the amino acid sequence depicting the secondary structures (blue helices and arrows for α-helical and β-strand structures, respectively) and interacting (thin brown arrows) residues. The structure file, 1A1K.val, is derived from 1AAY.pdb.

2002). Cn3D can be accessed online from Entrez at http://www.ncbi.nlm.nih.gov/ Entrez or downloaded from http://www.ncbi.nlm.nih.gov/Structure/CN3D/cn3d.html to be installed and used locally. This program accepts the coordinate file in MMDB ASN.1 format (*.val or *.cgi) but can be saved as PDB format (*.pdb) or KineMage format (*.kin). It is a structure-sequence interactive program. In addition to structure view in the Graphic window, Cn3D provides the sequence view in the Sequence window which can be activated via View→Sequence Window (Figure 4.18). This enables the user to view the structure and sequence interactively. Select the region of protein molecule in the structure view by double clicking of the mouse, and both the region in the Graphic window and the amino acid residue(s) in the Sequence window are highlighted (yellow) and vice versa. Using this tool, it is possible to map the interaction sites between structure and sequence. The view menu of the Graphic window also provides an option for Animation, and the Sequence window offers options for alignment (Align menu). The style menu enables the user to display the structure in secondary structure, wireframe, neighbor, tabular, spacefill or ball-and-stick modes (Figure 4.11).

4.4. WORKSHOPS

1. Write SMILES strings for fumaric acid, D-gluconic acid, cholesterol, histidine, and AMP.

2. Use ISIS/Draw to sketch the above biochemical compounds.

3. Use ISIS/Draw to sketch maltotriose (malt3) and an octapeptide, GRAPHICS.

4. Convert the 2D sketch of matl3 and GRAPHICS into 3D and save them as pdb files. Compare these files with the pdb file you retrieve from the Protein Data Bank at RCSB.

5. Retrieve protein topology cartoons (TOPS) for alcohol dehydrogenase (3BTO), concanavalin (2CNA), lysozyme (1LZ1), papain (1PPN), phosphofructokinase (1PFK), and rhodopsin (1BRD) or their representative TOPS. Compare their domain structures.

6. Retrieve a nucleic acid pdb file and visualize its 3D structure with RasMol. Save the structure as a graphic file in GIF format for printing.

7. Retrieve a protein pdb file and visualize its 3D structure with RasMol in different representations (Display menu). Identify the structural features or characteristics for which each display is best illustrative.

8. Retrieve a pdb file of an enzyme–substrate complex and visualize its 3D structure with KineMage. Save the structure as a graphic file in GIF format for printing.

9. Retrieve a structure file of a protein–DNA complex from Molecular Modeling Database at NCBI and visualize its 3D structure with Cn3D. Identify the interaction between the protein and DNA molecules.

10. Retrieve atomic coordinate files of two metalloenzymes, alcohol dehydrogenase (ADH) and Fe-superoxide dismutase (SOD), from PDB or MMDB. The subunit structure of ADH displays two metal ions, one catalytic and the other structural. The catalytic metal atom is chelated to Cys46, His67, Cys174, and a water molecule. Identify the catalytic metal atom and measure the approximate geometry of its chelation. The dimeric Fe-SOD similarly contains two Fe atoms per monomer. Search the literature to supplement the 3D structure view and present your findings regarding the function and geometry of the Fe atoms of Fe-SOD.

REFERENCES

Berman, H., Westbrook, J., Feng, Z., Gilliland, G., Bhat, T. N., Weissig, H., Shindyalov, I. N., and Bourne, P. E. (2000) *Nucl. Acids Res.* **28**:235–242.

Bernstein, F. C., Koetzle, T. F., Williams, G. J. B., Meyer, E. F. Jr., Brice, M. D., Rogers, J. R., Kennard, O., Shimanouchi, T., and Tasumi, M. (1997) *J. Mol. Biol.* **112**:535–542.

Connolly, M. L. (1983) *Science* **221**:709–713.

Flores, T. P., Moss, D. M., and Thornton, J. M. (1994) *Prot. Eng.* **7**:31–37.

Henkel, J. G., and Clarke, F. H. (1985) *Molecular Graphics on the IBM PC Microcomputer.* Academic Press, Orlando, FL.

Hogue, C. W. V., Ohkawa, H., and Bryant, S. H. (1996) *Trends Biochem. Sci.* **21**:226–229.

Richardson, D. C., and Richardson, J. S. (1992) *Protein Sci.* **1**:3–9.

Richardson, D. C., and Richardson, J. S. (1994) *Trends Biochem. Sci.* **19**:135–138.

Roger, D. F. (1985) *Procedure Elements for Computer Graphics*, McGraw-Hill, New York.

Rzepa, H. S., Murray-Rust, P., and Whitaker, B. J. (1998) *J. Chem. Inf. Comput. Sci.* **38**:976–982.

Sayle, R. A., and Milner-White, E. J. (1995) *Trends Biochem. Sci.* **20**:374–376.

Sternberg, M. J. E., and Thornton, J. M. (1977) *J. Mol. Biol.* **110**:269–283.

Wang, Y., Anderson, J. B., Chen, J., Geer, L. Y., He, S., Hurwitz, D. I., Liebert, C. A., Madej, T., Marchler, G. H., Marchler-Bauer, A., Panchenko, A. R., Shoemaker, B. A., Song, J. S., Thiessen, P. A., Yamashita, R. A., and Bryant, S. H. (2002) *Nucleic Acids Res.* **30**:249–252.

Wang, Y., Geer, L. Y., Chappey, C., Kans, J. A., and Bryant, S. H. (2000) *Trends Biochem. Res.* **25**:300–302.

Weininger, D. (1988) *J. Chem. Inf. Comput. Sci.* **28**:31–36.

Wyszecki, G., and Stiles, W. S. (1982) *Color Science: Concepts and Methods, Quantitative Data and Formula*, 2nd edition. John Wiley & Sons, New York.

5

BIOCHEMICAL COMPOUNDS: STRUCTURE AND ANALYSIS

All living organisms are composed of the same types of substances, namely water, inorganic ions, and organic compounds. The organic compounds including carbohydrates, lipids, proteins, and nucleic acids are often called biomolecules. Biochemical studies of biomolecules start with isolation and purification, the chromatographic techniques that are the most commonly employed. The most informative method to investigate structures of biomolecules is spectroscopic techniques. The computational adjuncts to these techniques are presented. The Internet search for biomolecular structures and information is described.

5.1. SURVEY OF BIOMOLECULES

The bulk of all cells is water, in which relatively small amounts of inorganic ions and several organic compounds are dissolved. The number of distinct organic chemical species in a cell is large, but most may be classified as carbohydrates, lipids, proteins, nucleic acids, or derivatives thereof. Those few compounds not so classified account for only a small fraction of the mass of the cell and are metabolically derived from substances in one of the four major classes. The four major classes of organic compounds have markedly different structures, properties, and biological functions, but each class contains two types of compounds: monomeric species with molecular weights of about 10^2 to 10^3 and polymeric biomolecules (biomacromolecules), with molecular weights of about 10^3 to 10^{10}. The Web site of International Union of

Biochemistry and Molecular Biology (http://www.chem.qmw.ac.uk/iubmb/) is an excellent starting point for the classification and nomenclature of biochemical compounds.

5.1.1. Carbohydrates

The simple sugars (monosaccharides, glycoses), their oligomers (oligosaccharides), and their polymers (polysaccharides, glycans) are collectively known as carbohydrates. They constitute a major class of cell components (Boons, 1998; Collins, 1987; Rademacher et al., 1988). Monosaccharides, commonly named sugars, are aldoses (polyhydroxyaldehydes) or ketoses (polyhydroxyketones) with a general formula, $[C(H_2O)]_n$. Sugars contain several chiral centers at secondary hydroxy carbons resulting in numerous diastereomers. These various diastereomers are given different names; for example, ribose (Rib) and xylose (Xyl) are two of the four diastereomeric aldopentoses; galactose (Gal), glucose (Glc), and mannose (Man) are three of the eight diastereomeric aldohexoses; fructose (Fru) is the most common one of the four diastereomeric ketohexoses. Each of the sugars exists as enantiomeric pairs, D and L forms, according to the configuration at the chiral center farthest from the carbonyl group. The carbonyl functional group forms a cyclic hemiacetal with one of the hydroxy groups to yield a five-member *furanose* ring (for pentoses) or a six-member *pyranose* ring (for hexoses). The cyclization of sugars yields a new chiral center at the carbonyl carbon atom (anomeric carbon atom), giving rise to α- and β-anomers that mutarotate in solution via an open-chain structure. The common pyranoses occur in the chair conformation.

Many sugar derivatives occur in nature. Among these are aldonic acids such as 6-phosphogluconic acid, uronic acids such as glucuronic acid, and galacturonic acid. The hydroxyl group at the 2 position of glucose and galactose may be replaced by an amino group to form glucosamine (GlcN) and galactosamine (GalN), respectively. The amino group of these amino sugars is often acetylated to yield *N*-acetylglucosamine and *N*-acetylgalactosamine.

Oligosaccharides (sugar units $\leqslant 10$) and polysaccharides are formed from joining the monosaccharide units by glycosidic linkages. Several disaccharides are important to the metabolism of plants and animals. Examples are maltose [αD-Glc-(1→4)-αD-Glc], lactose [βD-Gal-(1→4)-αD-Glc], sucrose [αD-Glc-(1→2)-βD-Fru], and trehalose [αD-Glc-(1→1)-D-αGlc]. Polysaccharides are present in all cells and serve in various functions. Despite the variety of different monomer units and types of linkage present in carbohydrate chains, the conformational possibilities of oligo- and polysaccharides are limited. The sugar ring is a rigid unit and the connections of two torsion angles, ϕ and ψ (Figure 5.1a), which are conveniently taken as 0° when the two midplanes as defined by the linked glycosidic atoms of the sugar rings are coplanar. A systematic examination of the possible value for ϕ and ψ for β-1,4-linked glucose units in cellulose (β-glucan) shows that these angles are constrained to an extremely narrow range around 180°, placing the monomer units in an almost completely extended conformation. Each glucose unit is flipped over 180° from the previous one so that the plane of the rings extends in a zigzag manner. In glycogen and starch (α-glucan), glucose units are joined in α-1,4-linkages, and the chain tends to undergo helical coiling. The helix contains 6 residues per turn with a diameter of nearly 14 nm.

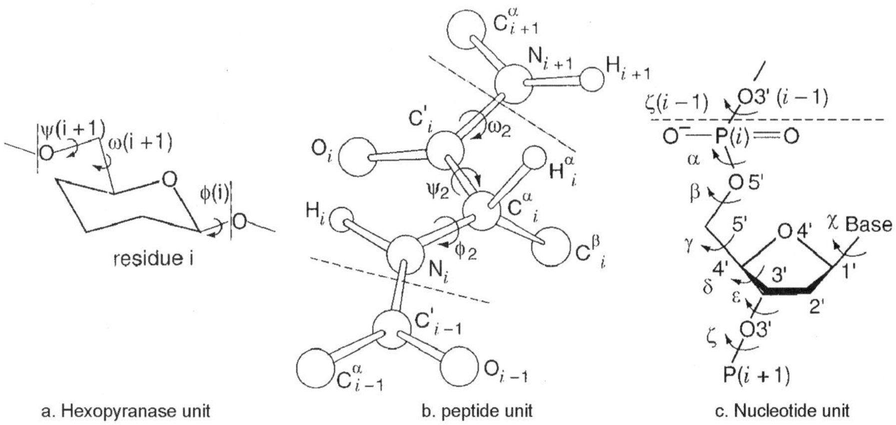

Figure 5.1. Notation for torsion angles of biopolymer chains. Torsion angles (ϕ and ψ) that affect the main chain conformations of biopolymers are shown for polysaccharide (a), polypeptide (b), and polynucleotide (c) chains according to the IUBMB notation. The two torsion angles, ϕ and ψ, specified around the phosphodiesteric bonds of nucleic acids correspond to α and ξ, respectively. Reproduced from IUBMB at http://www.chem.gmw. ac.uk/iubmb.

Many oligosaccharides and polysaccharides are often attached to cell surface or secreted in the forms of glycoproteins. This may be done via O-glycosidic linkages to hydroxy groups of serine, threonine, and hydroxylysine or in N-glycosidic linkages to the amide nitrogen of asparagine (Sharon, 1984). Because sugar chains can occur between hydroxyls of different positions or configurations, the oligo- and polysaccharide molecules are branched with varied linkage (unlike linear nucleic acids with 3',5'-diphosphoesteric linkages or linear proteins with amide linkages). O-Linked glycans contain an N-acetylgalactosamine residue at their reducing termini conferring particular physicochemical properties to the proteins. N-Linked glycans act as recognition signals of cell surface proteins. They contain the triasaccharide core, Manβ1→4GlcNACβ1→4GlcNAc, from which various mono- bi- tri-, tetra-, or penta-antennary sugar chains are attached (Kobata, 1993; Rademacher et al., 1988).

5.1.2. Lipids

Lipids refer to a large, heterogeneous group of substances classified together by their hydrophobic properties (Gurr, 1991; Vance and Vance, 1991). Most lipids are derived from linking together small molecules such as fatty acids, or they are derived from acetate units by complex condensation reactions. Fatty acids such as palmitic (C_{16}), palmitoleic (9E-C_{16}), stearic (C_{18}), oleic (9E-C_{18}), linoleic (9E,12E-C_{18}), linolenic (9E,12E, 15E-C_{18}), and arachidonic (5E,8E,11E,14E-C_{20}) acids are commonly found in lipids. They have even carbon number and may contain double bonds in *cis* (E) configuration. The hydrocarbon chains of the fatty acids tend to assume the extended conformation, but double bonds induce kinks and bends. Triglycerides, which are commonly found in adipose fats and vegetable oils, are formed from esterification of glycerol by three fatty acids. The phosphatides are fatty acid derivatives of *sn*-glycerol-3-phosphate which form phosphoesteric linkages with choline, serine, or inositol to yield phospholipids. A second group of phospholipids, the sphingomyelins, contain sphingosine as the central unit. Phospholipids are amphipathic (molecules containing both hydrophobic and hydrophilic ends) and are important constituents of biomembranes.

Lipids are the only class of four major biomolecules which do not undergo polymerization, but they may associate together to form large aggregate such as liposomes or membranes. Biomembranes (Jain, 1988) are made up of protein and lipid with the ratio (by weight) varying from 0.25 to 3.0 and having a typical value of 1:1. According to lipid bilayer model for membrane structure, hydrophobic bonding holds the fully extended hydrocarbon chains together while the polar groups of the phospholipid molecules interact with proteins that line both sides of the lipid bilayer. Phospholipids, which make up from 40% to over 90% of the membrane lipid, are predominated by five types: phosphatidylcholine, phosphatidylethanolamine, phosphatidylserine, diphosphatidylglycerol, and sphingomyelin.

Polyprenyl compounds are formed by condensation of isoprene units that are derived from acetate by reductive trimerization and decarboxylation. These compounds include terpenes, carotenoids, sterols, and steroids (Coscia, 1984; Hobkirk, 1979).

5.1.3. Proteins

The complete hydrolysis of proteins produces 20 α-amino acids that also occur as free metabolic intermediates. As free acids, they exist mostly as dipolar ions (zwitterions). Except for glycine, they contain chiral α-carbons and therefore exist in the D- and L-enantiomeric pair, of which the L-isomers are the monomeric units of proteins. They are differentiated structurally by their side-chain groups with varying chemical reactivities that determine many of the chemical and physical properties of proteins. These side-chain groups include:

- Hydrogen — that is, glycine (G).
- Alkyl groups — for example, alanine (A), valine (V), leucine (L), and isoleucine (I), which participate in hydrophobic interactions and tend to cluster inside protein molecules.
- Aromatic rings — for example, phenylalanine (F), tyrosine (Y), and tryptophane (W), which give proteins characteristic UV absorption at 280 nm and form effective hydrophobic bonds especially in bonding to flat molecules.
- Hydroxyl functionality — for example, serine (S) and threonine (T), which provide glycosylation and phosphorylation sites and are found at the active center of some enzymes.
- Carboxylic group — for example, aspartic acid (D) and glutamic acid (E), which provide anionic charges on the surface of proteins and constitute catalytic residues of glycosidases.
- Amide derivative — for example, asparagine (N) and glutamine (Q), which may participate in hydrogen bonding and provide glysosylation site.
- Sulfhydryl functionality — for example, cysteine (C) and methionine (M), of which cysteine is noted for its easy formation of disulfide linkages and its

involvement in the activities of oxidoreductases.

- Amino group — for example, lysine (K) provides cationic charge on the surface of proteins and is the site of glycation by blood glucose.
- Imidazole ring — for example, histidine (H), which may participate in general acid and general base catalyses. It is found at the catalytic site of many enzymes.
- Guanidino group — for example, arginine (R), which may involve in the binding of phosphate groups of nucleotide coenzymes.
- Imino structure — for example, proline (P), whose amino group forms imino ring with the rigid conformation therefore its presence disrupts the formation of regular secondary structures.

Proteins are the cornerstones of cell structure and the agents of biological function (Brandén and Tooze, 1991; Creighton, 1993; Fersht, 1999; Murphy, 2001; Schulz and Schirmer, 1979). They function as enzymes that catalyze virtually all cellular reactions, serve as regulators of these reactions, act as transporters, function as transducers of energy/electrical pulses/mechanical motion/light, form an essential biological defense system, and provide structure and storage materials of the cells.

Proteins are formed by joining α-amino acids together *via* peptide linkages (polypeptide chains) with monomer units in the chain known as amino acid residues. A polypeptide chain usually has one free terminal amino group (N-terminal) and a terminal carboxyl group (C-terminal) at the other end, though sometimes they are derivatized. Polypeptide chains formed by polymerization of α-amino acids in specific sequences provide the *primary structure* of proteins. Knowledge of amino acid sequences is of major importance in understanding the behavior of specific proteins.

A polypeptide (protein) can be thought of as a chain of flat peptide units for which each peptide unit is connected by the α-carbon of an amino acid. This carbon provides two single bonds to the chain, and rotation can occur about both of them (except proline). To specify the conformation of an amino acid unit in a protein chain, it is necessary to specify torsion angles about both of these single bonds (Figure 5.1b). These torsion angles are indicated by the symbols ϕ (around C_α–N bond) and ψ (around C_α–C_o bond), and they are assigned the value $0°$ for the fully extended chain. Both ϕ and ψ can vary by rotation, resulting in a large number of conformations. Because no two atoms may approach one another more closely than is allowed by their van der Waals radii, the restricted rotation around the two single bonds at C_α atom imposes steric constraints on the torsion angles, ϕ and ψ, of a polypeptide backbone that limits the number of permissible conformations. The whole range of possible combination of ϕ and ψ are plotted (ϕ versus ψ) in the conformational map (Ramachandran plot) indicating the allowable combinations of the two angles within the blocked areas (Ramachandran and Sasisekharan, 1968).

A polypeptide chain twisted by the same amount about each of its C_α atoms assumes a helical conformation, the most important of which is α helical structure. The α helix (3.6_{13} helix) is found when a stretch of consecutive amino acid residues all have $\phi = -57°$ and $\psi = -47°$ twisting right-handed 3.6 residues per turn with hydrogen bonds between C=O of residue n and NH of residue $n + 4$. The β strands are aligned close to each other such that hydrogen bonds can form between C=O of one β strand and NH on an adjacent β strand. The β sheets that are formed from

TABLE 5.1. Regular Secondary Structures of Proteins

Secondary Structure	Optimal Dihedral Angles	
	ϕ	ψ
Right-handed α helix (3.6_{13})	-57	-47
Right-handed 3_{10} helix	-49	-26
Right-handed π helix (4.4_{16})	-57	-70
2.2_7 Ribbon	-78	$+59$
Left-handed α helix	$+57$	$+47$
Collagen triple helix	-51	$+153$
Parallel β sheet	-119	$+113$
Antiparallel β sheet	-139	$+135$

several β strands are pleated with side chains pointing alternatively above and below the β sheet. The β strands can interact in a parallel or an antiparallel manner. The parallel β sheet has evenly spaced hydrogen bonds that angle across between the β strands while the antiparallel β sheet has narrowly spaced hydrogen bond pairs that alternate with widely spaced pairs. Almost all β sheets in proteins have their strands twisted in a right-hand direction. These repeated local structures are known as *secondary structures*, some of which (with the optimal ϕ and ψ values) are listed in Table 5.1.

The regular secondary structures, α helices and β sheets, are connected by coil or loop regions of various lengths and irregular shapes. A variant of the loop is the β turn or reverse turn, where the polypeptide chain makes a sharp, hairpin bend, producing an antiparallel β turn in the process.

The secondary structures are combined with specific geometric arrangement to form compact globular structure known as *tertiary structure*. The fundamental unit of tertiary structure is the *domain*, which is defined as a polypeptide chain or a part of a polypeptide chain that can independently fold into a stable tertiary structure (Murphy, 2001). Domains are also units of function, and often the different domains of a protein are associated with different functions. Polypeptide chains, especially of regulatory proteins, often aggregate by specific interactions to form oligomeric structures. These oligomeric proteins are said to exhibit *quarternary structure*. The association of proteins with other biomacromolecules to form complexes of cellular components is referred to as *quinternary structure*.

Both internal structure and overall size and shape of proteins vary enormously. Globular proteins vary considerably in the tightness of packing and the amount of internal water of hydration. However, a density of ~ 1.4 g cm^{-3} is typical.

5.1.4. Nucleic Acids

The nucleic acids (Blackburn and Gait, 1995; Bloomfield et al., 1999; Saenger, 1983), deoxyribonucleic acids (DNA), and ribonucleic acids (RNA) are polymers of nucleotides which are made up of three parts: a purine or pyrimidine base, D-2-deoxyribose for DNA or D-ribose for RNA, and phosphoric acid. The nucleotides are joined through phosphodiesteric linkages between the 5′-hydroxyl of the sugar in one nucleotide and the 3′-hydroxyl of another. Two purine bases, adenine (A) and guanine (G), occur in both DNA and RNA. However, the presence of pyrimidine bases differ such that cytosine (C) and thymine (T) are found in DNA while cytosine and uracil (U) are found in RNA. In addition, some minor bases are present in transfer RNA (tRNA).

Deoxyribonucleic acid is the genetic material such that the information to make all the functional macromolecules of the cell is preserved in DNA (Sinden, 1994). Ribonucleic acids occur in three functionally different classes: messenger RNA (mRNA), ribosomal RNA (rRNA), and transfer RNA (tRNA) (Simons and Grunberg-Manago, 1997). Messenger RNA serves to carry the information encoded from DNA to the sites of protein synthesis in the cell where this information is translated into a polypeptide sequence. Ribosomal RNA is the component of ribosome which serves as the site of protein synthesis. Transfer RNA (tRNA) serves as a carrier of amino acid residues for protein synthesis. Amino acids are attached as aminoacyl esters to the 3′-termini of the tRNA to form aminoacyl-tRNA, which is the substrate for protein biosynthesis.

In the nucleic acids, the furanose ring of ribose or deoxyribose can exist in several envelope and skew conformations. It is ordinarily C2′ or C3′ that is out of the plane of the other four ring atoms. If this carbon atom lies above the ring toward the base, the ring conformation is known as *endo*; if the atom lies below the ring away from the base, the conformation is known as *exo*. The C2′-*endo* and C3′-*endo* conformations are most common in individual nucleotides. The orientation of the base with respect to the sugar is specified by the torsion angle χ of the bond connecting N1 (pyrimidines) or N9 (purines) of base to C1′ of sugar (Figure 5.1c). The zero χ angle is often assigned to the conformation in which the C2–N1 bond of pyrimidine or the C4–N9 bond of purine is cis to the C1′–O1′ bond of the sugar. Measured values of χ vary among different nucleotides, a typical value being $\sim 127°$. In this *anti* conformation the CO and NH groups in the 2 and 3 positions of the pyrimidine ring or the 1, 2, and 6 positions of the purine ring are away from the sugar ring, while in the *syn* conformation they lie over the ring. The *anti* conformation is the one commonly present in most free nucleotides and nucleic acids. The two torsion angles $\phi(\alpha)$ and $\psi(\xi)$ are specified around the phosphodiesteric bonds, O(5′)–P and P–O(3′), respectively, of the elongated polynucleotide chains. The B form of DNA has $\phi = -96°$ and $\psi = -46°$.

One of the most exciting biological discoveries is the recognition of DNA as a double helix (Watson and Crick, 1953) of two antiparallel polynucleotide chains with the base pairings between A and T, and between G and C (Watson and Crick's DNA structure). Thus, the nucleotide sequence in one chain is complementary to, but not identical to, that in the other chain. The diameter of the double helix measured between phosphorus atoms is 2.0 nm. The pitch is 3.4 nm. There are 10 base pairs per turn. Thus the rise per base pair is 0.34 nm, and bases are stacked in the center of the helix. This form (B form), whose base pairs lie almost normal to the helix axis, is stable under high humidity and is thought to approximate the conformation of most DNA in cells. However, the base pairs in another form (A form) of DNA, which likely occurs in complex with histone, are inclined to the helix axis by about 20° with 11 base pairs per turn. While DNA molecules may exist as straight rods, the two ends bacterial DNA are often covalently joined to form circular DNA molecules, which are frequently supercoiled.

While RNA molecules usually exist as single chains, they often form hairpin loops consisting of double helices in the A conformation (Moore, 1999). The best-known forms of RNA are the low-molecular-weight tRNA molecules. In all of them the bases can be paired to form a cloverleaf structure with three hairpin loops and sometimes a fourth. The cloverleaf structure of tRNA is further folded into an L-shape conformation with the anticodon triplet and the aminoacyl attachment CCA forming the two ends.

5.1.5. Minor Biomolecules

Vitamins and hormones are minor organic biomolecules, but both of them are required by animals for the maintenance of normal growth and health. They differ in that vitamins are not synthesized by animals and must be supplied in diets while hormones are secreted by specialized tissues and carried by the circulatory system to the target cells somewhere in the body to initiate/stimulate specific biochemical or physiological activities. Vitamins (Dyke, 1965) can be classified as water-soluble (B vitamins and vitamin C) or fat-soluble (vitamins A, D, E, and K) and act as cofactors for numerous enzyme catalyzed reactions or cellular processes. Hormones (Nornam and Litwack, 1997) can be classified structurally as follows:

- amino acid derivatives—for example, epinephrine, norepinephrine, and thyroxine;
- polypeptides—for example, adrenocorticotropic hormone and its releasing factor, atrial natriuretic factor, bradykinin, calcitonin, cholecystokinin, chorionic gonadotropin and its releasing factor, enkephalins, folicle-stimulating hormone, gastric inhibitory peptide, gastrin, glucagon, growth hormone and its releasing factor, insulin, luteinizing hormone, oxytocin, parathyroid hormone, prolactin, secretin, somatomedins, thyrotropin and its releasing factor, and vasopressin; or
- steroids—for example, androgens, estrogens, glucocorticoids, mineralocorticoids, and progesterones.

They function as regulators or messengers of specific cellular processes.

5.2. CHARACTERIZATION OF BIOMOLECULAR STRUCTURES

One of the approaches to understanding biological phenomena has been to purify an individual chemical component from a living organism and to characterize its chemical structure or biological activity. The most commonly used techniques for purifying biomolecules are chromatographic and electrophoretic methods (Deutscher, 1990; Heftmann, 1983), and the most facile approaches to characterization of molecular structures are spectroscopic methods (Brown, 1998; McHale, 1999).

5.2.1. Chromatographic Purification

Chromatography (Edward, 1970; Millner, 1999; Poole and Poole, 1991)—in particular, high-performance liquid chromatography (HPLC)—is an ideal technique for purification and analysis of biomolecules. The position or retention time of a chromatographic peak is governed mainly by the fundamental thermodynamics of solute partitioning between mobile and stationary phases. The parameters of interest are the capacity factor (k) and the resolution (R). The capacity factor is estimated by

$$k = t_r/t_0 - 1 \quad \text{or} \quad k = v_r/v_0 - 1$$

where t_r is retention time and v_r is elution volume of the peak measured at the peak maximum, and t_0 is dead time and v_0 is void volume of the column. The resolution is calculated according to

$$R = (t_{r1} - t_{r2})/\{2(W_1 - W_2) \quad \text{or} \quad R = (v_{r1} - v_{r2})/\{2(W_1 - W_2)$$

where t_{r1} and t_{r2} are retention times, and v_{r1} and v_{r2} are elution volumes. W_1 and W_2 are full widths at the peak base of the first and second peaks, respectively.

The modes of liquid chromatographic separation can be classified as follows:

1. *Gel-filtration/permeation chromatography:* The technique is normally used for the separation of biological macromolecules and polymers. It separates compounds on the basis of size. Solutes are eluted in the order of decreasing molecular size.

2. *Adsorption chromatography:* The process can be considered as a competition between the solute and solvent molecules for adsorption sites on the solid surface of adsorbent to effect separation. In normal phase or liquid–solid chromatography, relatively nonpolar organic eluents are used with the polar adsorbent to separate solutes in order of increasing polarity. In reverse-phase chromatography, solute retention is mainly due to hydrophobic interactions between the solutes and the hydrophobic surface of adsorbent. Polar mobile phase is used to elute solutes in order of decreasing polarity.

3. *Partition chromatography:* A partition packing consists of a liquid phase coated on an inert solid. The separation is effected by the interaction of the solute between the mobile phase and the liquid stationary phase.

4. *Ion-exchange chromatography:* Ion-exchange chromatography separates compounds on the basis of their molecular charges. Compounds capable of ionization, particularly zwitterionic compounds, separate well on ion-exchange column. The separation proceeds because ions of opposite charge are retained to different extents. The resolution is influenced by (a) the pH of the eluent which affects the selectivity and (b) the ionic strength of the buffer which mainly affects the retention.

5. *Affinity chromatography:* Affinity chromatography is a type of absorption chromatography in which the molecule to be purified is specifically and reversibly adsorbed by a complementary binding ligand immobilized on the matrix. It is used to purify biomolecules on the basis of their biological function or specific chemical structure. Purification is often of the order of several thousandfold, and recoveries of the purified biomolecule are generally very high.

5.2.2. Migratory Separation in Electric Field

Charged biomolecules — especially amino acids, peptides, proteins (at pH other than their isoelectric points), nucleotides, and nucleic acids (at pH above 2.4) — migrate in an electric field with the rates of migration dependent upon their charge densities. Zone electrophoresis is the separation of charged molecules in a supporting medium resulting in the migration of charged species in discrete zones. Various gels (such as agar, agarose, and polyacrylamide) used as the supporting media may also exert a molecular sieving effect. This allows gel electrophoresis to separate charged biomolecules according to their mobilities (size, shape, and charge), the applied current, and the resistance of the medium. Polyacrylamide gels can be prepared with pores of the same order of size as protein/oligonucleotide molecules and are hence very effective in the fractionation of proteins and oligonucleotides (Hames and Rickwood, 1981). On the other hand, agarose gels are used to separate larger molecules or complexes such as certain nucleic acids and nucleoproteins (Rickwood and Hames, 1982).

The gel concentration required for polyacrylamide gel electrophoresis (PAGE) to achieve optimal resolution of two proteins (or nucleic acids) can be determined by measuring the relative mobility of each protein in a series of gels of different acrylamide concentrations to construct a Ferguson plot ($\log_{10}R_f$ versus %T) according to

$$\log_{10}R_f = Y_0 + K_r(\%T)$$

where %T refers to percent total monomer (acrylamide + bisacrylamide used) and R_f is the distance migrated by protein/distance migrated by marker protein or tracking dye. Each Ferguson plot is characterized by its slope K_r, a retardation coefficient which is related to the molecular size of the protein, and its ordinate intercept Y_0, which is a measure of the mobility of the protein in solution and is related to its charge.

Under appropriate conditions, all reduced polypeptides bind the same amount of sodium dodecylsulfate (SDS), that is, 1.4 g SDS/g polypeptide. Furthermore, the reduced polypeptide–SDS complexes form rod-like particles with lengths proportional to the molecular weight of the polypeptides. This forms the basis for the empirical estimation of the molecular weight (M_r) of proteins using SDS-PAGE (Weber et al., 1972) according to

$$R_f = \alpha + \beta \log M_r$$

where R_f are relative mobilities of the sample and reference proteins in the same run.

Electrofocusing (EF) is a charge fractionation technique that separates molecules predominantly by the difference in their net charge, not by size. Thus EF can be considered as an electrophoretic technique by which amphoteric compounds are fractionated according to their isoelectric points (pIs) along a continuous pH gradient maintained by ampholyte buffers. This is contrary to zone electrophoresis, where the constant pH of the separation medium establishes a constant charge density at the molecule and causes it to migrate with constant mobility. The charge of an amphoteric compound in EF decreases according to its titration curve, as it moves along the pH gradient approaching its equilibrium position at pI where the molecule comes to a stop. Proteins with difference in pIs of 0.02 pH units can be resolved by EF.

5.2.3. Spectral Characterization

Spectroscopic methods are used at some point in the structural characterization of biomolecules (Bell, 1981; Campbell and Dwek, 1984; Gendreau, 1986). These methods are usually rapid and noninvasive, generally require small amount of samples, and can be adapted for analytical purposes. Spectroscopy is defined as the study of the interaction of electromagnetic radiation with matter, excluding chemical effects. The electromagnetic spectrum covers a very wide range of wavelengths (Figure 5.2).

Interaction of radiation with matter may occur in a number of ways. X-ray diffraction depends on elastic scattering of the radiation. The molecule may either emit radiant energy at the expense of its internal energy or it may absorb radiant energy, being promoted to an excited state. According to the quantum theory, the energy content of a molecule is confined to certain discrete values (*energy level*). Furthermore, a molecule will absorb radiation only when the frequency (v) of the radiation is related to the energy difference (ΔE) between two energy levels by the equation $\Delta E = hv$, where h is Planck's constant (6.67×10^{-27} erg sec). Because the relative positions of the energy levels depend characteristically on the molecular structure, absorption spectra provide subtle tools for structural investigation. A molecule can become excited in a variety of ways, corresponding to absorption in different regions of the spectrum. The various processes give rise to different spectroscopic methods as summarized in Table 5.2.

Ultraviolet (UV) and visible spectra, also known as electronic spectra, involve transitions between different electronic states. The accessible regions are 200–400 nm for UV and 400–750 nm for visible spectra. The groups giving rise to the electronic transitions in the accessible regions is termed *chromophores*, which include aromatic amino acid residues in proteins, nucleic acid bases, NAD(P)H, flavins, hemes, and some transition metal ions. Two parameters characterize an absorption band, namely the position of peak absorption (λ_{max}) and the extinction coefficient (ε), which is related to concentrations of the sample by the Beer–Lambert law:

$$A = \log(I_0/I_s) = \varepsilon cl$$

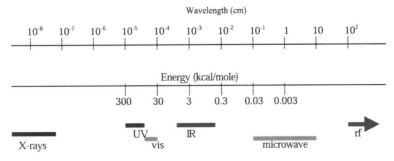

Figure 5.2. Electromagnetic radiation and its corresponding spectra.

TABLE 5.2. Spectroscopic Methods

Spectroscopic Method	Principle	Approximate Sample Size	Information Obtained and Applications
Ultraviolet and visible spectroscopy	Absorption of UV and visible radiation leading to electronic excitation	0.1–10 mg	Presence and nature of unsaturation, especially conjugated double bonds and aromatic systems. Aqueous solutions can be used. Quantitative analysis of proteins and nucleic acids DNA conformation.
Infrared spectroscopy	Absorption of IR radiation leading to vibrational excitation	1–10 mg	Presence and environment of functional groups, especially X—H or multiple bonds, for example, C=O. Diagnosis of finer structural detail such as conformation, intramolecular hydrogen bonding. Identification of unknown by fingerprinting. Studies of macromolecular dynamics by H–D exchange.
Raman spectroscopy	Scattering of visible or UV light with abstraction of some of the energy leading to vibrational excitation	0.05–10 g	As for IR but with sampling restriction. However, aqueous solution can be used. Functional groups that have weak IR absorption often give strong Raman spectra.
Fluorescence spectroscopy	Emission of radiation when a molecule in an excited electronic state returns to the ground state	1 μM solution	Environment, relative abundance, and interactions of fluorophore. Quantitation of proteins and fluorophoric compounds. Ligand binding studies.
Nuclear magnetic resonance	Absorption of radiation giving rise to transitions between different spin orientations of nuclei in a magnetic field	0.1–1 M solution	Environment of nuclei, hence unambiguous detection of certain functional groups and information about the environment. Determination of proton sequences and hence of relative points of attachment of functional groups. Inference to molecular configuration and conformation. Studies of macromolecular structures and ligand interactions.

Method	Basis	Sample	Information obtained
Electron spin resonance	Absorption of radiation giving rise to transitions between opposite spin orientations of unpaired electrons in a magnetic field	10^{-10} mole of radical	Detection and estimation of free radicals and diagnosis of their structure and electron distribution.
Optical rotatory dispersion	Rotation of the plane polarized light by asymmetric molecules in solution without and with variation in wavelength	1–200 mg/0.5–20 mg	Determination of relative and absolute configurations of asymmetric centers. Location of functional groups in certain types of compounds. Information about conformation.
Circular dichroism	Difference in intensity of absorption of right- and left-circularly polarized light by functional groups in asymmetric environment	1–15 mg	Applications similar to but more powerful than ORD especially for functional groups such as C=O. Conformational analysis of biomacromolecules in solutions.
X-ray diffraction	Interference between scattered X rays caused by atomic electrons	1–10 mg	Determination of complete molecular structure and stereochemistry. Such structural analysis gives bond lengths and angles as well as distances between nonbonded atoms in crystals.
Electron diffraction	Interference between scattered electron beam caused by electrostatic field	1–10 mg	Determination of complete molecular structure of fairly simple molecules.
Neutron diffraction	Interference between scattered neutron beam caused by atomic nuclei	0.1–1 g	Particularly useful for the location of hydrogen atoms in a molecule.
Mass spectrometry	Determination of the mass/charge ratio and relative abundance of the ions formed upon electron bombardment	0.01–1 mg	Accurate molecular weight determination. Elucidation of structure, especially the nature of the skeleton and the length of side chains. Sequence determination.

where I_0 and I_s are the incident and transmitted radiation, respectively, and l is the length of the cell through which radiation travels. The biochemical applications of UV and visible spectroscopy are determination of concentrations and interactions of ligands with biomacromolecules. The sensitivity of UV and visible spectra to the solvent environment of the chromophore leads to shifts in λ_{max} and ε, and it is the basis of solvent perturbation spectra in the structural studies of biomacromolecules (Beechem and Brand, 1985).

Infrared (IR) spectra provide information on molecular vibration. The common region of IR spectrum is 1400–4000 cm^{-1}. The main experimental parameter is the frequencies (e.g., v for stretching, δ for in-plane bending, and γ for out-plane bending) of the absorption bands characterizing functional groups — in particular, $v_{C=O}$ and v_{N-H} for biomolecules. Water interferes with IR spectrum. The main biochemical applications involve studying ligand binding to macromolecules, probing hydrogen bonds, and molecular conformation in nonaqueous or oriented samples (Singh, 2000).

Fluorescence is the emission of radiation that occurs when an excited molecule returns to the ground state. The molecular group giving rise to fluorescence is termed *fluorophore*, which includes tryptophan residue in proteins, NAD(P)H, and chlorophyll. The measurable parameters are the quantum yield, the intensity, and the position of peak emission (λ_{max}), which are sensitive to environment. The quantum yield (ϕ_F) is the fraction of molecules that becomes de-excited by fluorescence and is defined as:

$$\phi_F = \tau/\tau_F$$

where τ is the observed lifetime of the excited state and τ_F is the theoretical lifetime. Applications include ligand binding, probing of environment, and measurement of distance between fluorophores.

Nuclear magnetic resonance (NMR) is a technique that detects nuclear-spin reorientation in an applied magnet field. The system requires that the nuclei display unpaired spin state, S such as ^{1}H, ^{13}C, and ^{31}P of biochemical interest. The spinning nucleus generates a magnetic field and thus has an associated magnetic moment which interacts with the applied field. The parameters of NMR are chemical shift, spin–spin coupling, area or intensity of the signal, and two relaxation times (T_1 and T_2) with T_2 related to the line width for freely tumbling molecules in solution. Chemical shifts (δ) that measure the resonance frequencies of nuclei are affected by the magnetic shielding of their circulating electrons and are dependent on the molecular environment. The chemical shifts are normally recorded in the NMR spectra in field-independent units with a reference to the frequency (v_{ref}) of the common reference compound:

$$\delta\,(\text{ppm}) = (v - v_{ref})\text{Hz}/\text{operating frequency}, \qquad \text{MHz} \times 10^6$$

They are useful for structural inference. The magnetic interaction between neighboring nuclei is called spin–spin coupling. Its magnitude, known as the spin–spin coupling constant (J), is dependent on the degree of delocalization of electrons between the interacting nuclei, and thus it is informative of the neighboring groups. The intensity gives the relative quantities of the resonance nuclei. The dynamic state of the resonance nucleus can be inferred from its T_2. 2D-NMR has yielded valuable structural information on biomacromolecules in solutions (Bax, 1989).

Various applications of NMR in biochemistry include structural identification of biomolecules, chemistry of individual groups in macromolecules, structural and dynamic information of biomacromolecules, metabolic studies, and kinetic and association constants of ligand bindings to macromolecules (Wüthrich, 1986).

The optical activity arises from the chiral centers of chemical compounds or interactions of asymmetrically placed neighboring groups in macromolecules. An optically active molecule interacts differentially with left- and right-circularly polarized light. This interaction can be detected either by optical rotatory dispersion (ORD) or by circular dichroism (CD). ORD spectrum records a differential change in velocity of the two beams of the polarized light and is characterized by $[\alpha]_\lambda$, which is the specific rotation at a given wavelength or the molar rotation $[\phi]_\lambda$. CD spectrum records a differential absorption of each beam of the polarized light and is characterized by ΔA (the differential absorption of the two beams) or the molar ellipticity θ_m. Main applications of ORD or CD spectroscopy are the determination of the secondary structure of biomacromolecules and detection of their conformational changes.

X-ray and neutron diffraction patterns can be detected when a wave is scattered by a periodic structure of atoms in an ordered array such as a crystal or a fiber. The diffraction patterns can be interpreted directly to give information about the size of the unit cell, information about the symmetry of the molecule, and, in the case of fibers, information about periodicity. The determination of the complete structure of a molecule requires the phase information as well as the intensity and frequency information. The phase can be determined using the method of multiple isomorphous replacement where heavy metals or groups containing heavy element are incorporated into the diffracting crystals. The final coordinates of biomacromolecules are then deduced using knowledge about the primary structure and are refined by processes that include comparisons of calculated and observed diffraction patterns. Three-dimensional structures of proteins and their complexes (Blundell and Johnson, 1976), nucleic acids, and viruses have been determined by X-ray and neutron diffractions.

The mass spectrometry does not involve an interaction between electromagnetic radiation and sample molecule. The functions of a mass spectrometer are to produce positive ions from the sample under investigation, to resolve these ions into a series of ion beams that are homogeneous with respect to their mass/charge ratio (m/e), and to measure the relative abundance of the ions in these beams. The main applications include the molecular weight and structural determinations (Bieman, 1992). Mass spectrometry has emerged as the method for rapid analyses of protein sequences and annotation of their databases (Mann and Pandey, 2001).

5.3. FITTING AND SEARCH OF BIOMOLECULAR DATA AND INFORMATION

5.3.1. Chromatographic and Spectroscopic Peak Fitting

The success of chromatographic and spectroscopic techniques depends largely on the resolution and analysis of chromatographic/spectroscopic peaks. Automatic peak fitting software, PeakFit (http://www.spsscience.com), uses the following routines for finding hidden peaks, thus enhancing the resolution and facilitating the analysis of experimental data:

- AutoFit Peaks I, *Residual method*: A residual is the difference in y value between a data point and the sum of component peaks evaluated at the data point's x value. Hidden peaks are revealed by positive residuals.
- AutoFit Peaks II, *Second derivative method*: A smooth second derivative of the data will contain local minima at peak positions. The second derivative method requires a constant x-spacing operated in the time domain.
- AutoFit Peaks III, *Deconvolution method*: Deconvolution is a mathematical procedure that is used to remove the smearing or broadening of peaks arising because of the imperfection in an instrument's measuring system. Hidden peaks that display no maxima may do so once the data have been deconvoluted and filtered. This method requires a uniform x-spacing operated in the frequency domain.

Some of the common PeakFit functions that can be selected from the program are as follows:

Chromatography:	Haarhoff–Van der Linde (HVL)
	Gidding
	Nonlinear chromatography (NLC)
	Exponentially modified Gaussian (EMG)
Spectroscopy:	Gaussian, amplitude and area
	Lorentzian, amplitude and area
	Voigt, amplitude and area

The hierarchy of processing in PeakFit generally follows the order listed below:

1. *Baseline fitting*
2. *Width and shape*
3. *Smooth option* — Savitzky–Golay as the default is adequate. For start, choose *AI Expert* to let the program seeks an optimum smoothing level.
4. *Peak function family* — Specify either chromatography or spectroscopy.
5. *Peak function type* — Select one of the peak functions appropriate for chromatography or spectroscopy.
6. *Amp %*
7. *AutoFit* — Start PeakFit with full graphical update or fast numerical update.

Prepare chromatographic/spectroscopic data files in fileneme.dat or filename.txt (ascii), preferably filename.xls (Excel) format and save to your disk. Launch PeakFit. From the File menu, invoke Import, then enter (or browse) A:\filename.xls. Select columns for x–y data by highlighting column data for x and y, respectively. Enter titles and then click OK. Click toolbar "AutoFit Peaks I: Residual" initially then follow the hierarchy of fitting process (Figure 5.3). Click toolbar AutoFit Peaks I, II, or III of your choice. Select the default "Savitzky–Golay smoothing algorithm" and click AI expert. Invoke Review Fit to view and save the fitted data. For the peak function family, choose either chromatography or spectroscopy and an appropriate fitting function.

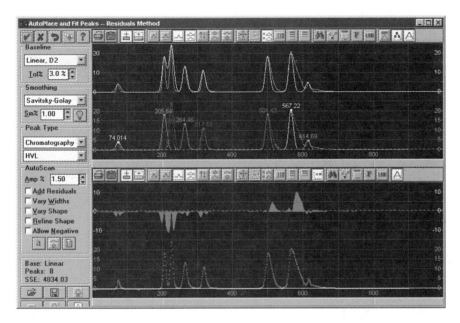

Figure 5.3. Peak analysis of liquid chromatogram. The chromatographic separation of a protein sample is analyzed with PeakFit using Haarhoff–Van der Linde (HVL) function by AutoFit Peaks I menu.

5.3.2. Search Databases for Biochemical Compounds

Most biochemical databases request the user to enter keywords to search/retrieve information concerning biomolecules (unless the identifiers of the compounds are known). The keyword is normally the IUBMB (International Union of Biochemistry and Molecular Biology) name or common name of the compound. In particular, the linkage and conformational designations of oligomeric compounds needs to be specified. The IUBMB nomenclature for biochemical compounds can be accessed from the IUBMB site at http://www.chem.qmw.ac.uk/iubmb/. Some useful databases for biochemical compounds are listed in Table 5.3.

The collection and categorization of biomolecules can be found at Klotho server (http://ibc.wustl.edu/klotho/). On the Klotho home page, click Compound Listing to open the alphabetical compound list (Figure 5.4). Enter the compound name and click Search Klotho or scan the list for the desired compound that is displayed (in the .gif format). Activate (requiring PDB plug-in on the user's computer) the molecular view by clicking on Interactive Viewer. Left press the mouse to move/ rotate the molecular view. Right click the view window to bring up the pop-up option menu. The molecule can be viewed in wireframe (default), sticks, ball and stick, or space-fill model by choosing the Display option. Select File→Save Molecule As to save the molecule in PDB (.pdb) format (which is the format for most molecular modeling programs) or MDL (.mol) format. This is a handy way to obtain 3D structures of biochemical molecules for modeling. To save 2D structure, select Edit→Transfer to ISIS Draw to convert 3D view into 2D view as molecule.skc (ISIS/Draw needs to be installed on the user's computer). The Options toggle-check for displays of hydrogen bond, disulfide bond, and dot surface according to van der

TABLE 5.3. Some Databases for Biochemical Compounds

Resource	URL
ChemFinder: Chemical structures, properties	http://www.chemfinder.com
Klotho: General, metabolites	http://ibc.wustl.edu/klotho/
Monosaccharide database	http://www.cermav.cnrs.fr/databank/mono/
GlycoSuiteDB: Glycan structures	http://www.glycosuite.com/
Lipid Bank: Comprehensive lipid information	http://lipid.bio.m.u-tokyo.ac.ip/
LIPID: Membrane lipid structures	http://www.biochem.missouri.edu/LIPIDS/membrane_lipid.html
Mptopo: Membrane protein topology	http://blanco.biomol.uci.edu/mptopo/
AAindex: Parameters for amino acids	http://www.genome ad.jp/dbget-bin/
Entrez: Sequences of proteins/nucleic acids	http://www.ncbi.nlm.nih.gov/Entrez
EBI: Sequences of proteins/nucleic acids	http://www.ebi.ac.uk/
PDB: 3D structures of biomacromolecules	http://www.rcsb.org/pdb/
Histone database	http://genome.nhgri.nih.gov/histones/
RNA structure database	http://rnabase.org
RNA modification database	http://medlib.med.utah.edu/RNAmods
European large subunit rRNA database	http://rrna.uia.ac.be/lsu/index.html
European small subunit rRNA database	http://rrna.uia.ac.be/ssu/index.html
tRNA sequence database	http://www.uni-bayreuth.de/departments/biochemie/trna/
Merck manual: Vitamin/hormone function	http://www.merck.com/pubs/mmanual/

Waals radii or Connolly/Richards solvent (1.2 Å). Select Rotation and then Start to initiate an automatic rotation until Stop command is issued.

To obtain useful information on chemical (including biochemical) compounds from ChemFinder at http://www.chemfinder.com, enter the compound name and click Search. The server returns with synonymous names, chemical formula and structure, and onsite information, as well as links for further biochemical information. The structure can be saved in .cdx format (ChemDraw) and viewed with Chem3D.

The monosaccharide database at http://www.cermav.cnrs.fr/databank/mono/ is the resource site for monosaccharides. Click Databank to open the query/result windows. Select the desired compound from Choose sugar type (e.g., Ribo-, Gluco-) and the subsequent pop-up list (of the sugar type). The search result (Figure 5.5) is returned with the 2D molecular structure and the choices for 3D structural view (double click to enlarge the view window), along with atomic coordinate files in PDB or MDL format. GlycoSuiteDB (Cooper et al., 2001) at http://www.glycosuite.com/ provides annotated glycan (polysaccharide) structures that can be queried by:

• mass or mass range,
• attached protein by protein name, keyword or Swiss-Prot accession number,
• taxonomy via selection from the scrolling list box,

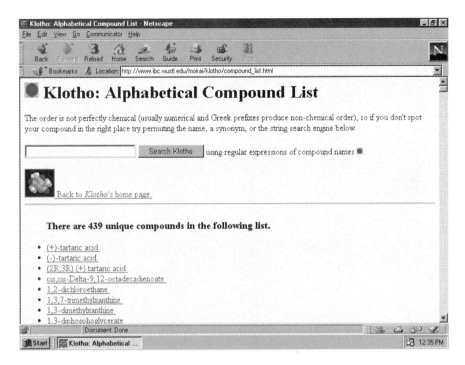

Figure 5.4. Alphabetical list of biochemical compounds at Klotho. The 1D (Smile strings), 2D, and 3D structures of metabolites can be viewed/retrieved from Klotho server.

- composition by entering the number of units for each monosaccharide field (i.e., Hex, HexNAc, dHex, Pent, NeuAc, and NeuGc for hexose, N-acetyl-hexosamine, deoxyhexose, N-acetylneuramicnic acid, and N-glycolylneuramic-nic acid, respectively),
- tissue/cell type via selection from the scrolling list box,
- linkage (i.e., N-linked, O-linked or C-linked),
- GlycoSuiteDB accession number, or
- advance query (all categories selected by the user).

To query by "composition," enter the desired number for monosaccharide units and click the "perform search" button. From the returned list (arranged by taxanomy and tissue/cell types), choose the entry (or entries) with the matched composition(s) from the desired biological source(s) and then click either the "refine selection" button to trim the hit list or the "show glycan entries" button to display the search results. Clicking the "show glycan entries" button returns the list of glycan structures with pertaining information (glycan structure as exemplified in Figure 5.6: taxonomy, source, attached protein, linkage, glycosylation site, mass, composition, method of identification, and references).

The Lipid Bank at http://lipid.bio.m.u-tokyo.ac.jp/ provides a wide range of information for lipids and their derivatives. Click Next Page at the home page to open Lipid Menu (Table 5.4).

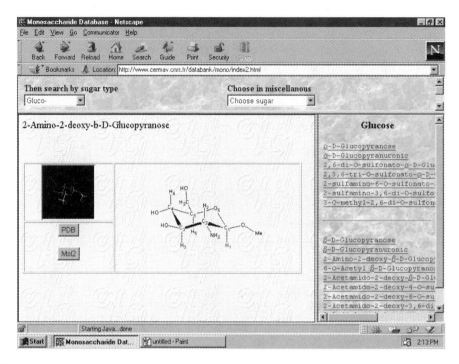

Figure 5.5. Search result of monosaccharide database. The web site provides 2D and 3D structures of monosaccharides. The chair conformer of methyl 2-amino-2-deoxy-β-D-glucopyranoside in which all hydroxy (1-methoxy and 2-amino) groups are equatorial is displayed.

TABLE 5.4. Lipid Menu from Lipid Bank

Click at the square box to open the query form for the lipid class of compounds

☐ ACYLGLYCEROLS	☐ HOPANOIDS
☐ BILE ACIDS (CHOLANOIDS)	☐ ISOPRENOIDS
DERIVED LIPIDS	☐ LIPOAMINO ACIDS
☐ LONG-CHAIN ALCOHOL	☐ LIPOPOLYSACCHARIDES
☐ LONG-CHAIN ALDEHYDE	☐ LIPOPROTEINS
☐ LONG-CHAIN BASE and CERAMIDE	
☐ ETHER TYPE LIPIDS	☐ MYCOLIC ACIDS
FAT SOLUBLE VITAMINS	PHOSPHOLIPIDS
☐ CAROTENOID	☐ GLYCEROPHOSPHOLIPID
☐ COENZYME Q	☐ PAF
☐ VITAMIN A	☐ SPHINGOPHOSPHOLIPID
☐ VITAMIN D	
☐ VITAMIN E	☐ PROSTANOIDS
☐ VITAMIN K	☐ STEROIDS
GLYCOLIPIDS	☐ WAXES
☐ glycoSPHINGOLIPID	
☐ glycoGLYCEROLIPID and OTHERS	

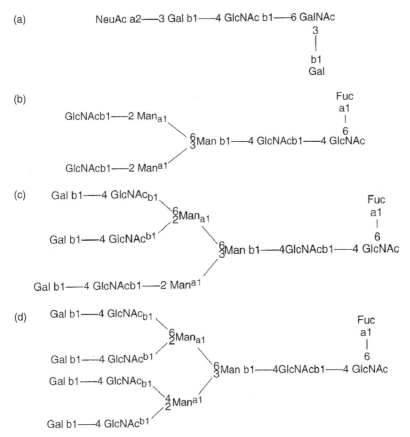

Figure 5.6. Examples of glycans retrieved from GlycoSuiteDB. Examples of various antennary structures are illustrated for human glycans retrieved from GlycoSuiteDB. (a) Monoantenary $Hex_2HexNAc_2NeuAc_1$, O-linked to mucin of intestine, colon, and mucosa. (b) Bianternary $Hex_3HexNAc_4dHex_1$, N-linked to saposin A of spleen lymphatic system. (c) Triternary $Hex_6HexNAc_5dHex_1$, N-linked to plasma coagulation factor X. (d) Tetraantennary $Hex_7Hex_6dHex_1$, N-linked to saposin B of liver. Abbreviations used are: Ac, acetyl; Fuc, L-fucose (6-deoxy-L-galactopyranose); Gal, D-galactopyranose; GalNAc, 2-acetamido-2-deoxy-D-galactopyranose; GlcNAc, 2-acetamido-2-deoxy-D-glucopyranose; Man, D-Mannopyranose; and NeuAc, *N*-acetylneuraminic acid.

Select lipid category from the Lipid Menu to open the query form. The database search can be conducted via Keyword (lipid name, source, biological activity, metabolism, or genetic information), Classification (lipid class from classification list), Numeric Attributes (number of carbons, double bonds, etc.), and Linkage Position (carbon position attached to specific residue). Choosing the desired lipid class from the Classification list (the easiest approach) returns a hit list of compounds. Select the desired compound by clicking the View button on the left-hand side of the compound. Tabulated information including names, formula, molecular structure (which can be downloaded in ChemDraw format), biological activity, physical and chemical properties, spectra data, source, chemical synthesis, metabolism, genetics, and cited references is returned.

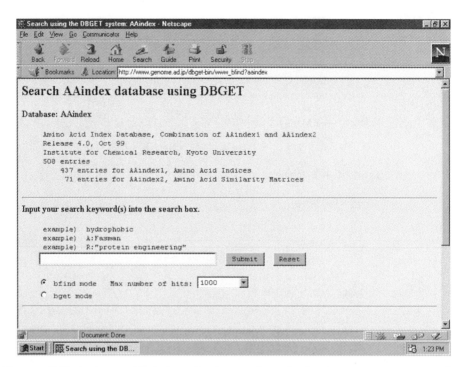

Figure 5.7. Amino acid index databases for physicochemical properties. The physicochemical properties, conformational parameters, and mutational indexes can be retrieved from AAindex database using keyword search.

The AAindex database (Tomii and Kaneshisa, 1996) of DBget, which can be accessed from http://www.genome.ad.jp/dbget/, provides physicochemical properties, conformational propensities, and mutation matrices of amino acids. To search AAindex database (Figure 5.7), enter keywords (e.g., chemical shifts, volume), choose the *bfind* mode and click Submit. The search returns a list of hits from which you select the desired entry. The amino acid index data for each entry starts with an index I in the following alphabetical order: Ala, Arg, Asn, Asp, Cys, Gln, Glu, Gly, His, Ile (in the first line), and Leu, Lys, Met, Phe, pro, Ser, Thr, Trp, Tyr, Val (in the second line). Alternatively, entering accession number in the *bget* mode will return the index data. The integrated sequence and structural information of proteins can be viewed at http://www.protein.bio.msu.su/issd.

RNA modification database at http://medlib.med.utah.edu/RNAmods furnishes information concerning naturally modified nucleosides in RNA (Limbach et al., 1994). After clicking Search, select individual files (pick corresponding radios under categories: base type, RNA source, and phylogenetic occurrence) to be accessed, and output options (common name, structures, and/or mass values). Enter partial name (i.e., substituent name such as methyl, thio, etc.) and choose the parent base(s) from which modified nucleosides are derived (Figure 5.8). Click Search button. The tabulated search results based on selected files are returned according to the requested output options.

The amino acid sequences of proteins and nucleotide sequences of DNA can be retrieved from the integrated database retrieval systems; Entrez (http://

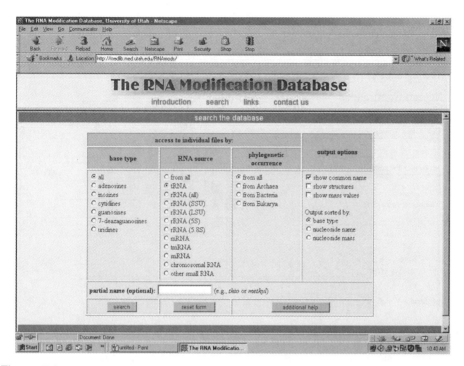

Figure 5.8. Search form for RNA modification database. To search the database, only one radio from each category can be selected, however, multiple selections from the output (show) boxes are possible. Partial name is optional, and more than one partial name can be entered if they are separated by a space.

www.ncbi.nlm.nih.gov/Entrez/) of NCBI, EBI (http://www.ebi.ac.uk/), and DDBJ (http://www.genome.ad.jp/dbget/). The three-dimensional structures of proteins, nucleic acids, and polysaccharides can be retrieved from PDB (http://www.rcsb.org/pdb/). These are described in Chapters 3 and 4. The structural information on protein topology of biomembrane can be obtained from the Mptopo site (http://blanco.biomol.uci.edu/mptopo/).

The biomedical information for vitamins and hormones can be obtained from the Merck manual online at http://www.merck.com/pubs/mmanual/. For example, entering vitamin returns a list of vitamins from which an entry describing deficiency, dependence, and/or toxicity of the vitamin can be selected, viewed, and saved.

5.3.3. Spectral Information

National Institute of Materials and Chemical Research, Japan maintains Spectral Database Systems (SDBS) of organic compounds including a number of biochemical compounds (Yamamoto et al., 1988). The database can be accessed at http://www.aist.go.jp/RIODB/SDBS/menu-e.html. To begin searching spectra, click Search compounds/Search NMR & MS/Display spectra on the SDBS home page (Figure 5.9). Enter compound name and/or molecular formula. The wild card, % can be used to search spectra data by molecular formula. Clicking the SDBS number that

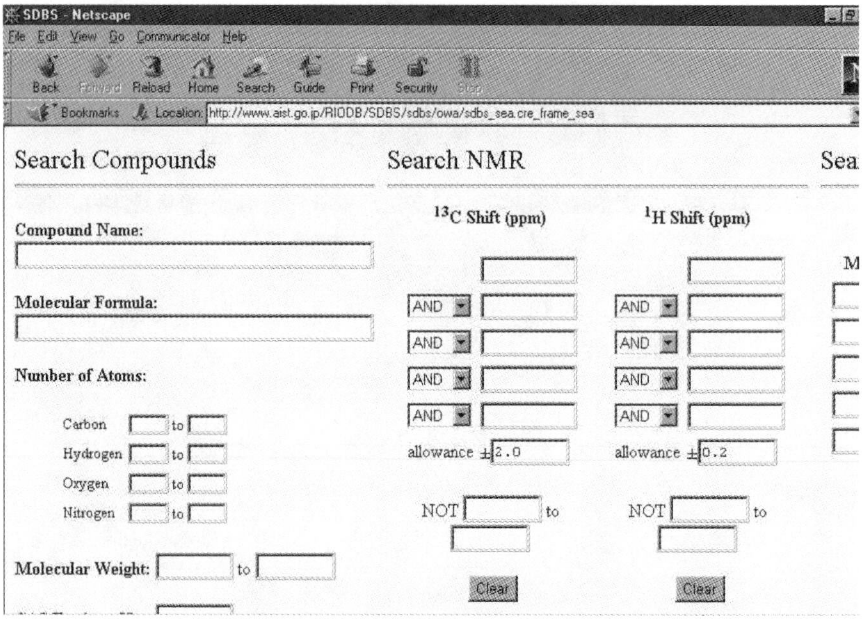

Figure 5.9. Spectral search at Spectral Database Systems (SDBS). The infrared (IR), nuclear magnetic resonance (^{1}H-NMR and ^{13}C-NMR), electron spin resonance (ESR), and mass (MS) spectra of organic compounds and common biochemical compounds can be viewed/retrieved from SDBS.

corresponds to the desired compound on the response page returns the information page from which the desired spectrum is selected (Figure 5.10).

The infrared (IR) spectra (in liquid film, KBr disk, nujor mull, or CCl_{14}/CS_2 solution) are recorded with an FT spectrometer in the region of 4000–400 cm^{-1} with an optical resolution of 0.5 cm^{-1}. The proton magnetic resonance (^{1}H-NMR) spectra (in $CDCl_3$ or D_2O/DMSO at the sample concentration of less than 100 mg/ml) are obtained at the resonance frequency of 90 MHz with the digital resolution of 0.0625 Hz or 400 MHz at the digital resolution of 0.125 Hz. The carbon-13 nuclear magnetic resonance (^{13}C-NMR) spectra (in $CDCl_3$ or D_2O/DMSO at the sample concentration of about 25 mol%) are obtained under ^{1}H decoupled condition. The flip angle is 22.5–45.0 degrees, and the pulse repetition time is 4–7 sec with the digital resolution of 0.025–0.045 ppm. The mass (MS) spectra are obtained by the electron impact method with the electron accelerating voltage of 75 V and the ion accelerating voltage of 8–10 V. The NMR and IR spectra are appended with information on peak and frequency analyses. Clicking *peak data* on the mass spectral page returns the table of mass numbers and relative intensities.

Alternatively, a search for the structure of an unknown compound with spectral information can be conducted by entering the peak positions in ppm of C^{13} and/or ^{1}H NMR spectra or sets of mass number and relative intensity of MS spectrum. NMR data for some proteins can be obtained from BioMagResBank (BMRB) at http://www.bmrb.wisc.edu/pages/homeinfo.html.

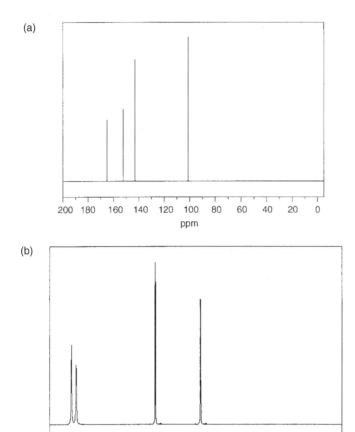

Figure 5.10. Sample spectra retrieval from SDBS. (a) ^{13}C-NMR spectrum in DMSO-d$_6$. (b) ^{1}H-NMR (400 MHz) spectrum in DMSO-d$_6$. (c) Mass spectrum. (d) Infrared spectrum in KBr. Sample spectra (including spectral analysis) of uracil are retrieved from Spectral Database Systems. The structure of uracil (molecular weight = 112) is represented with the number corresponding to the position of carbons and the alphabet denoting the position of protons to facilitate NMR assignments:

C Position	Carbon-13 NMR Chemical shift (ppm)	H Position	Proton NMR Chemical shift (ppm)
1	165.09	a	11.02
2	152.27	b	10.62
3	142.89	c	7.41
4	101.01	d	5.47

The major mass spectral peaks and their relative intensities based on 100 for the molecular ion (M+) are m/e = 18.0 (12), 28.0 (34), 42.0 (47), 69 (64), and 112.0 (100). The retrieved uracil spectra (C-13 NMR, proton NMR, mass, and infrared) are depicted.

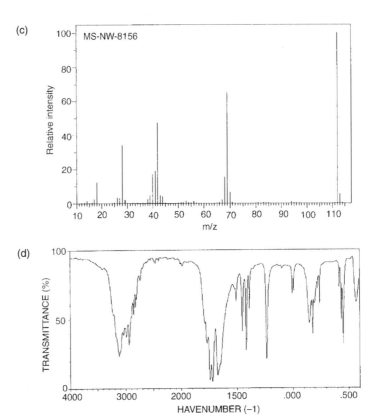

Figure 5.10. Continued.

5.4. WORKSHOPS

1. Perform peak fits for the liquid chromatographic separation of proteins as shown below:

Fractions	A280	Fractions	A280	Fractions	A280	Fractions	A280
20	0	55	1.851	87	0.117	119	0.768
22	0	56	1.068	88	0.235	120	0.832
24	0	57	0.695	89	0.478	121	1.016
26	0	58	0.268	90	0.760	122	1.272
27	0.04	59	0.102	91	0.996	123	1.668
28	0.08	60	0.066	92	1.122	124	1.962
29	0.22	61	0.034	93	1.016	125	1.855
30	0.54	62	0.026	94	0.836	126	1.792
31	0.28	63	0.014	95	0.545	127	1.662
32	0.11	64	0.032	96	0.482	128	1.536
33	0.06	65	0.064	97	0.482	129	1.472
34	0.04	66	0.141	98	0.482	130	1.407
35	0	67	0.256	99	0.482	131	1.343
36	0	68	0.442	100	0.456	132	1.279

Fractions	A280	Fractions	A280	Fractions	A280	Fractions	A280
37	0.006	69	0.665	101	0.249	133	1.151
38	0.088	70	0.908	102	0.128	134	0.962
39	0.326	71	1.139	103	0.064	135	0.896
40	0.698	72	1.304	104	0.026	136	0.768
41	1.193	73	1.096	105	0.023	137	0.602
42	1.624	74	0.832	106	0.071	138	0.489
43	1.828	75	0.603	107	0.128	139	0.382
44	1.647	76	0.424	108	0.332	140	0.297
45	1.195	77	0.298	109	0.639	141	0.212
46	0.726	78	0.192	110	0.984	142	0.135
47	0.382	79	0.137	111	1.322	143	0.086
48	0.218	80	0.098	112	1.842	144	0.047
49	0.197	81	0.064	113	1.458	145	0.023
50	0.355	82	0.045	114	1.088	146	0.011
51	0.729	83	0.030	115	0.924	147	0.006
52	1.426	84	0.024	116	0.832	148	0.002
53	1.943	85	0.319	117	0.768	149	0
54	2.034	86	0.053	118	0.703	150	0

2. Perform peak fit for the spectrum of FAD chromophore of a flavoprotein:

Wave Number	A	Wave Number	A	Wave Number	A
14025.0	0.000	20825.0	0.448	27625.0	0.463
14425.0	0.000	21225.0	0.513	28025.0	0.454
14825.0	0.000	21625.0	0.565	28425.0	0.422
15225.0	0.000	22025.0	0.584	28825.0	0.394
15625.0	0.000	22425.0	0.546	29225.0	0.362
16025.0	0.000	22825.0	0.498	29625.0	0.325
16425.0	0.000	23225.0	0.430	30025.0	0.298
16825.0	0.004	23625.0	0.357	30425.0	0.271
17225.0	0.008	24025.0	0.281	30825.0	0.214
17625.0	0.026	24425.0	0.253	31225.0	0.178
18025.0	0.054	24825.0	0.224	31625.0	0.153
18425.0	0.112	25225.0	0.201	32025.0	0.130
18825.0	0.187	25625.0	0.198	32425.0	0.117
19225.0	0.195	26025.0	0.235	32825.0	0.094
19625.0	0.198	26425.0	0.292	33225.0	0.072
20025.0	0.252	26825.0	0.377	33625.0	0.052
20425.0	0.375	27225.0	0.458	34025.0	0.034

3. A dehydrogenase is purified to homogeneity from fish liver. The atomic absorption spectrometry shows that the enzyme contains 0.316% (by weight) of zinc. This enables calculation of the minimal molecular weight (M_{min}) of the metalloenzyme according to

$$M_{min} = \{(\text{atomic weight of Zn}/(\% \text{ Zn})\} \times 100$$

Two empirical approaches, SDS polyacrylamide electrophoresis (SDS-PAGE) and gel (Sephadex G200) filtration chromatography, are used to determine the subunit molecular weight (M_{su}) and Stokes' radius (r) of the dehydrogenase. In the presence of sodium dodecylsulfate (SDS), proteins undergo dissociation and unfolding, therefore their electrophoretic mobilities (x) are related to the dissociated subunit weight (M_{su}) (Weber et al., 1972) by

$$\log M = \alpha - \beta x$$

where α and β are constants for a given gel at a given electric field. In gel filtration chromatography the elution constant, K_e, of protein is related to its radius of an equivalent hydrodynamic sphere known as the Stokes' radius (r) (Siegel and Monty, 1966) by

$$(-\log K_e)^{1/2} = A + Br$$

where A and B are constants and K_e is determined experimentally from the elution volume (V_e) in relation to the total bed volume (V_t) and the void volume (V_0) of the column according to $K_e = (V_e - V_0)/(V_e - V_t)$. The electrophoretic mobilities and elution constants of the dehydrogenase and marker proteins of known subunit weights and Stokes' radii are given, respectively, below:

SDS-PAGE in 10% Separating Polyacrylamide Gel

Protein	Electrophoretic Mobility	Subunit Weight (kdal)
Lysozyme	0.81	14.4
Hemoglobin	0.77	16.0
Chymotrypsinogen	0.56	25.7
Triosephosphate isomerase	0.54	26.5
Yeast alcohol dehydrogenase	0.43	35.0
Aldolase	0.37	40.0
Horse liver alcohol dehydrogenase	0.34	42.0
Hexokinase	0.26	51.0
Catalase	0.22	57.5
Purified dehydrogenase	0.36	?

Sephadex G200 Filtration Chromatography

Protein	Elution Constant K_e	Stokes' Radius (nm)
Cytochrome c	0.59	1.64
Ribonuclease	0.56	1.92
Trypsin	0.51	1.94
Chymotrypsin	0.45	2.09
Monomeric serum albumin	0.19	3.50
Yeast alcohol dehydrogenase	0.062	4.53
Catalase	0.038	5.20
Purified dehydrogenase	0.22	?

Perform regression analyses of the SDS-PAGE and the gel filtration chromatographic results to evaluate the subunit weight (M_{su}) and the Stocks' radius (r) of the purified dehydrogenase.

The native molecular weight (M) of 83.0 kdal is determined independently with electron spray ionization mass spectrometry. If the protein is spherical and unhydrated, its radius (r_0) can be estimated by

$$r_0 = [(3\bar{v}M)/(4\pi N)]^{1/3}$$

where $\pi = 3.1416$ and N is the Avogadro's number (6.022×10^{23} mol^{-1}). From the molecular weight and the Stokes' radius a useful shape parameter, the frictional ratio (f/f_0), which expresses the deviation of the protein molecule from the unhydrated spherical shape, can be calculated according to

$$f/f_0 = r/[(3\bar{v}M)/(4\pi N)]^{1/3}$$

Evaluate the frictional ratio by taking the partial specific volume to be 0.735 g cm^{-3}. Deduce the structural features of the dehydrogenase.

4. Search for oligo-/polysaccharide structures of glycoproteins from human with following saccharide compositions:

(a) $Hex_1HexNAc_2dHex_1NeuAc_1$
(b) $Hex_2HexNAc_2dHex_2NeuAc_1$
(c) $Hex_3HexNAc_2dHex_1NeuAc_1$
(d) $Hex_4HexNAc_2dHex_2NeuAc_1$
(e) $Hex_5HexNAc_3dHex_2NeuAc_1$
(f) $Hex_6HexNAc_5dHex_1NeuAc_1$

Depict their linkage and antennary structures using abbreviations for monosaccharide units.

5. Physicochemical properties of amino acids are very useful descriptors for understanding the structures and properties of proteins. These properties are expressed numerically in indexes that can be retrieved from the AAindex database. Design an index database of physicochemical properties of amino acids with Microsoft Access that may facilitate the data retrieval according to their chemical similarities:

Chemical Characteristics	Residue Groups
Small	Ala, Gly, Pro
Small/relatively polar	Cys, Ser, Thr
Acidic/amide	Asp, Asn, Gln, Glu
Basic	Arg, His, Lys
Aliphatic/hydrophobic	Ile, Leu, Met, Val
Aromatic	Phe, Trp, Tyr

6. The carbonyl functionality at C1 and the primary hydroxyl group at C6 of D-glucose are readily oxidized/reduced to yield D-fucose, D-glucaric acid, D-glucitol,

D-gluconic acid, and D-glucuronic acid. Search Klotho and/or ChemFinder sites for their structures. How would you differentiate these redox derivatives by spectroscopic methods.

7. Mass spectrometry is one of the best techniques for identifying fatty acids, which are the major component of lipids. Its application (Benveniste and Davies, 1973) reveals the following:

(a) The molecular ion (M^+) of fatty acids and their esters are relatively abundant, thus it is easy to identify them and determine their molecular weight.

(b) The fragmentation process leading to intense peaks follows the simple C–C cleavage.

(c) The rearrangement event

leads to the intense peak.

(d) Most intense peaks are oxygen-containing fragments.

(e) The periodicity C–C cleavage of —$(CH_2)_4$— is favored.

Search Lipid Bank for structures and SDBS for mass spectra of common saturated fatty acids (Section 5.1.2), and identify the intense peaks with the above-mentioned characteristics.

8. Nuclear magnetic resonance spectroscopy is a powerful technique for investigating structure of biomolecules. The ^{1}H- and ^{13}C-NMR spectra of L-α-amino acids have been compiled (Wüthrich, 1986) and can be retrieved from SDBS. Design a database for ^{1}H- or ^{13}C-NMR data that can be used in the identification of amino acids.

9. The characteristic ^{13}C-NMR data for oligosaccharides of glucopyranosese have been compiled (Bok et al., 1984), and the chemical shifts (δ in ppm) for glucobioses are listed below:

Glcobiose	Nonred C1[a]	Red C1[a]	αGlucosid C2 to C4, C6[b]	βGlucosid C2 to C4, C6[b]	Other C2 to C5[c]	Other C6[c]
αGlc(1→1)-αGlc	94.0				72.0 ± 1.5	61.5
αGlc(1→1)-βGlc	103 ± 1				73.9 ± 3.5	62.0 ± 0.4
βGlc(1→1)-βGlc	100.7				74.2 ± 3.1	62.5
αGlc(1→2)-αGlc	97.1	90.4	76.7		72.4 ± 1.7	61.6
αGlc(1→2)-βGlc	98.6	97.1	79.5		73.7 ± 3.0	61.6
βGlc(1→2)-αGlc	104.4	92.4		81.4	73.5 ± 3.1	61.7
βGlc(1→2)-βGlc	103.2	95.1		82.1	73.5 ± 3.1	61.7
αGlc(1→3)-αGlc	99.8	93.1	80.8		72.4 ± 1.8	61.8
αGlc(1→3)-βGlc	99.8	97.0	83.2		73.6 ± 3.0	61.8
βGlc(1→3)-αGlc	103.2	92.7		83.5	72.7 ± 3.7	61.7

Glcobiose	Nonred C1[a]	Red C1[a]	αGlucosid C2 to C4, C6[b]	βGlucosid C2 to C4, C6[b]	Other C2 to C5[c]	Other C6[c]
βGlc(1→3)-βGlc	103.2	96.5		86.0	72.7 ± 3.7	61.7
αGlc(1→4)-αGlc	100.7	92.8	78.5		72.3 ± 1.9	61.6
αGlc(1→4)-βGlc	100.7	96.8	78.2		73.8 ± 3.4	61.7 ± 0.1
βGlc(1→4)-αGlc	103.6	92.9		79.9	73.8 ± 3.2	61.4 ± 0.4
βGlc(1→4)-βGlc	103.6	96.8		79.8	73.8 ± 3.2	61.5 ± 0.3
αGlc(1→6)-αGlc	98.5	92.9	66.5		72.3 ± 1.9	61.6
αGlc(1→6)-βGlc	98.5	96.8	66.5		73.3 ± 2.9	61.6
βGlc(1→6)-αGlc	103.0	92.5		69.4	73.3 ± 3.0	61.7
βGlc(1→6)-βGlc	103.0	96.4		69.4	73.3 ± 3.0	61.7

[a]Nonred C1 and Red C1 refer to C1 of nonreducing and reducing glucopyranose units.
[b]αGlucosid C2 to C4, C6 and βGlucosid C2 to C4, C6 refer to C2, C3, C4 or C6 of glucopyranose unit involved in the α- or β-(1→2), (1→3), (1→4) or (1→6) glucosidic linkages, respectively.
[c]Other refers to glucopyranose C's with free OH groups.
Source: Academic Press, reproduced with permission.

Design the ^{13}C-NMR database (Access) for glucobioses and create a query form to identify the structure of glucobiose with ^{13}C-NMR signals at δ103.65, 92.94, 79.88, 77.08, 76.66, 74.29, 72.47, 72.38, 71.20, 70.61, 61.74, and 61.08 ppm.

10. The anticodon of tRNA molecule interacts with codon of mRNA. The degeneracy of the genetic code means that several codons can specify a single amino acid. To reduce the number of tRNAs necessary to read the codons, the first nucleotide (5′ end) of the anticodon which pairs with the third base (3′ end) of the codon is wobbly such that it can pair with two or three different bases at the 3′ end of the codon. This reduces the specificity of interaction between the tRNA molecule and the transcript. It appears that one approach taken by nature to enhance the specific base pairing interaction between tRNA and mRNA is to use modified bases at positions flanking the anticodon in the tRNA molecule. Thus tRNA is noted for the number and variety of modified bases in its structure. The modification of purine and pyrimidine bases is required to generate mature tRNA (Piper and Clark, 1974). Examples of the modified bases are as follows:

Adenosine (A)	Cytidine (C)	Guanidine (G)	Uridine (U)
1-Methyl (m^{1}A)	3-Methyl (m^{3}C)	1-Methyl (m^{1}G)	Dihydro (D/hU)
N^6-Methyl (m^{6}A)	5-Methyl (m^{5}C)	7-Methyl (m^{7}G)	5-Methoxycarbonylmethyl(mcm^{5}U)
N^6-Isopentyl (t^{6}A)	N^4-Acetyl (ac^{4}C)	N^2-Methyl (m^{2}G)	4-Thio (s^{4}U)
2-Methylthio-t^{6}A	2-Thio (s^{2}C)	N^2, N^2-Dimethyl (m$_2^2$G)	5-Methylaminomethyl-s^{2}U

Search for the structures of methyl nucleosides in tRNA and suggest spectral characteristics that distinguish these minor bases from their corresponding parent (major) bases.

REFERENCES

Bax, A. (1989) *Annu. Rev. Biochem.* **58**:223–256.

Beechem, J. M., and Brand, L. (1985) *Annu. Rev. Biochem.* **54**:43–71.

Bell, J. E. Ed. (1981) *Spectroscopy in Biochemistry.* CRC Press, Boca Raton, FL.

Benveniste, R., and Davies, J. (1973) *Annu. Rev. Biochem.* **42**:471–506.

Bieman, K. (1992) *Annu. Rev. Biochem.* **61**:977–1010.

Blackburn, G. M., and Gait, M. G. (1995) *Nucleic Acids in Chemistry and Biology.* IRL Press/Oxford University Press, Oxford.

Bloomfield, V. A., Crothers, D. M., and Tinoco, I., Jr. (1999) *Nucleic Acids: Structure, Properties and Functions.* University Science Books, Sausalito, CA.

Blundell, T. L., and Johnson, L. N. (1976) *Protein Crystallography.* Academic Press, New York.

Bok, K., Pederson, C., and Pederson, H. (1984) *Adv. Carbohydr. Chem. Biochem.* **42**:193–225.

Boons, G.-J., Ed. (1998) *Carbohydrate Chemistry.* Blackie, New York.

Brandén, C., and Tooze, J. (1991) *Introduction to Protein Structure.* Garland, New York.

Brown, J. M. (1998) *Molecular Spectroscopy.* Oxford University Press, Oxford.

Campbell, I. D., and Dwek, R. A. (1984) *Biological Spectroscopy.* Benjamin/Cummings, Menlo Park, CA.

Collins, P. M. (1987) *Carbohydrates.* Chapman and Hall, London.

Cooper, C. A., Harrison, M. J., Wilkins, M. R., and Packer, N. H. (2001) *Nucleic Acids Res.* **29**:332–335.

Coscia, C. J. (1984) *Terpenoids.* CRC Press, Boca Raton, FL.

Creighton, T. E. (1993) *Proteins*, 2nd edition. W. H. Freeman, New York.

Deutscher, M. P., Ed. (1990) *Guide to Protein Purification, Methods in Enzymology*, Vol. 182. Academic Press, San Diego, CA.

Dyke, S. F. (1965) *The Chemistry of the Vitamins.* Wiley-Interscience, New York.

Edward, D. I. (1970) *Chromatography: Principles and Techniques.* Butterworths, London.

Fersht, A. (1999) *Structure and Mechanism in Protein Science.* W. H. Freeman, New York.

Gendreau, R. M., Ed. (1986) *Spectroscopy in Biomedical Sciences.* CRC Press, Boca Raton, FL.

Gurr, M. I. (1991) *Lipid Biochemistry: An Introduction.* Chapman and Hall, London.

Hames, B. D., and Rickwood, D., Eds. (1981) *Gel Electrophoresis of Proteins.* IRS Press, Oxford.

Heftmann, E., Ed. (1983) *Chromatography: Fundamentals and Applications of Chromatographic and Electrophoretic Methods.* Elsevier, New York.

Hobkirk, R., Ed. (1979) *Steroid Biochemistry.* CRC Press, Boca Raton, FL.

Jain, M. K. (1988) *Introduction to Biological Membranes*, 2nd edition. John Wiley & Sons, New York.

Kobata, A. (1993) *Acc. Chem. Res.* **26**:319–324.

Limbach, P. A., Crain, P. F., and McLoskey, J. A. (1994) *Nucleic Acids Res.* **22**:2183–2196.

Mann, M., and Pandey, A. (2001) *Trends Biochem. Sci.* **26**:54–61.

McHale, J. L. (1999) *Molecular Spectroscopy.* Prentice-Hall, Upper Saddle River, NJ.

Millner, P., Ed. (1999) *High Resolution Chromatography: A Practical Approach.* Oxford University Press, Oxford.

Moore, P. B. (1999) *Annu. Rev. Biochem.* **68**:287–300.

Murphy, K. P., Ed. (2001) *Protein Structure, Stability and Folding.* Humana Press, Totowa, NJ.

Nornam, A. W., and Litwack, G. (1997) *Hormones*, 2nd edition, Academic Press, San Diego, CA.

Piper, P. W., and Clark, B. F. C. (1974) *Nature* **247**:516–518.

Poole, C. F., and Poole, S. K. (1991) *Chromatography Today.* Elsevier, New York.

Rademacher, T. W., Parekh, R. B., and Dwek, R. A. (1988) *Annu. Rev. Biochem.* **57**:785–838.

Ramachandran, G. N., and Sasisekharan, V. (1968) *Adv. Protein Chem.* **23**:283–437.

Rickwood, D., and Hames, B. D., Eds. (1982) *Gel Electrophoresis of Nucleic Acids.* IRS Press, Oxford.

Saenger, W. (1983) *Principles of Nucleic Acid Structure.* Springer-Verlag, Berlin.

Schulz, G. E., and Schirmer, R. H. (1979) *Principles of Protein Structure.* Springer-Verlag, Berlin.

Sharon, N. (1984) *Trends Biochem. Sci.* **9**:198–202.

Siegel, L. M., and Monty, K. J. (1966) *Biophys. Biochim. Acta* **112**:346–362.

Simons, R. W., and Grunberg-Manago, M. (1997) *RNA Structure and Function.* Cold Spring Harbor Laboratory Press, Cold Spring Harbor, New York.

Sinden, R. R. (1994) *DNA Structure and Function.* Academic Press, San Diego, CA.

Singh, B. R. (2000) *Infrared Analysis of Peptides and Proteins.* Oxford University Press, Oxford.

Tomii, K., and Kaneshisa, M. (1996) *Protein Eng.* **9**:27–36.

Vance, D. E., and Vance, J. E., Eds. (1991) *Biochemistry of Lipids, Lipoproteins and Membranes.* Elsevier, New York.

Watson, J. D., and Crick, F. H. C. (1953) *Nature* **171**:737–738.

Weber, K., Pringle, J. R., and Osborn, M. (1972) *Meth. Enzymol.* **26**:3–27.

Wüthrich, K. (1986) *NMR of Proteins and Nucleic Acids.* John Wiley & Sons, New York.

Yamamoto, O., Someno, K., Wasada, N., Hiraishi, J., Hayamizu, K., Tanabe, K., Tamura, T., and Yanagisawa, M. (1988) *Anal. Sci.* **4**:233–239.

6

DYNAMIC BIOCHEMISTRY: BIOMOLECULAR INTERACTIONS

Understanding the structure–function relationship of biomacromolecules furnishes one of the strong motives for investigating biomacromolecule–ligand interactions that constitute an initial step in their biological functions. Various situations arising from multiple equilibria including cooperativity and allosterism are introduced. Binding equilibrium is analyzed with DynaFit. Concepts and databases for receptor biochemistry and signal transduction are presented.

6.1. BIOMACROMOLECULE–LIGAND INTERACTIONS

6.1.1. General Consideration

Most physiological processes are the consequences of an effector interaction with biomacromolecules (Harding and Chowdhry, 2001; Weber, 1992), such as interactions between enzymes and their substrates, between hormones and hormone receptors, between antigens and antibodies, between inducer and DNA, and so on. In addition, there are macromolecule-macromolecule interactions such as between proteins (Kleanthous, 2000), between protein and nucleic acid (Saenger and Heinemann, 1989), and between protein and cell-surface saccharide. The effector of small molecular weight is normally referred to as the ligand, and the macromolecular combinant is known as the receptor.

The biochemical interaction systems are characterized by general ligand interactions at equilibrium, site–site interactions, and cooperativity as well as linkage relationships regarding either (a) two different ligands binding to the same macro-molecule or (b) the same ligand binding to the different sites of the macromolecule (Steinhardt and Reynolds, 1969).

Consider a macromolecular receptor, R, which contains n sites for the ligand L. Each site has the microscopic ligand association constant K_i for the i-site.

$$R + L \rightleftharpoons RL \qquad\qquad K_1 = [RL]/[R][L]$$
$$RL + L \rightleftharpoons RL_2 \qquad\quad K_2 = [RL_2]/[RL][L]$$
$$\cdots \qquad\qquad\qquad\qquad \cdots$$
$$RL_{n-1} + L \rightleftharpoons RL_n \qquad K_n = [RL_n]/[RL_{n-1}][L]$$

Equilibrium measurement of ligand binding typically yields the moles of ligand bound per mole of macromolecule, v, which is given by

$$v = \frac{K(L)}{1 + K(L)} \qquad \text{for the single equilibrium}$$

and

$$v = \frac{\sum \{i \prod K_j\}(L)^i}{1 + \sum \{\prod K_j\}(L)^i} \qquad \text{for the multiple equilibrium}$$

The solution of v for n-sites gives different expressions under various situations, such as:

1. **n-Equivalent sites:** The macromolecular receptor contains n sites that are thermodynamically equivalent. The equilibrium treatment of the ligand binding to a receptor with n-equivalent sites which may be either independent (noninteracting) or interacting gives rise to

   a. Noninteracting n-equivalent sites

$$v = \frac{nK(L)}{1 + K(L)}$$

   The linear transformation of this expression gives

   Klotz's equation: $1/v = 1/n + 1\{nK(L)\}$

   or

   Scatchard's equation: $v/(L) = nK - vK$

   Both equations are used in the linear and graphical analysis of binding data.

b. Interacting n-equivalent sites

$$v = \frac{nK(L)^h}{1 + K(L)^h}$$

This expression is known as Hill's equation and undergoes linear transformation to yield

$$\log\{v/(n - v)\} = \log K + h \log(L)$$

The Hill's interaction coefficient, h, is a measure of the strength of interaction among n-sites. If the n-equivalent sites are noninteracting, then $h = 1$, whereas h approaches n (number of equivalent sites) for the strongly interacting n-equivalent sites.

2. **n-Nonequivalent sites:** The macromolecular receptor contains n sites that bind ligand with different equilibrium constants. For a macromolecule with m classes of independent sites, each class i has n_i sites with an intrinsic association constant, K_i:

$$v = \sum^{m} \frac{n_i K_i(L)}{1 + K_i(L)}$$

The receptors with n-nonequivalent sites are generally oligomeric, consisting of heterooligomers or homooligomers with different conformations. These receptors may display cooperativity in the ligand–receptor interactions.

6.1.2. Cooperativity and Allosterism

The ligand-induced conformational changes appear to be an important feature of receptors with regulatory functions. The concepts of cooperativity and allosterism (Koshland et al., 1966; Monad et al., 1965; Richard and Cornish-Bowden, 1987) in proteins have been applied to explain various ligand–receptor interaction phenomena. The observed changes in successive association constants for a multiple ligand binding system suggest the cooperativity in the interaction between binding processes. The interaction that causes an increase in successive association constants is called positive cooperativity, while a decrease in successive association constants is the consequence of negative cooperativity. The *cooperativity* refers to interaction between binding sites in which the binding of one ligand modifies the ability of a subsequent ligand molecule to bind to its binding site, whereas the *allosterism* refers to the binding of ligand molecules to different sites.

Because a single binding site cannot generate cooperativity, a cooperative binding system must consist of two or more binding sites on each molecule of protein. Although there is no need for such proteins to be oligomeric, nearly all known cases of cooperativity at equilibrium are found in proteins with separate binding sites on different subunits. Furthermore, the cooperativity can be observed with interactions of a single kind of ligand in homotropic (same ligand) interactions or those involving two (or more) different kinds of ligands in heterotropic interactions. Two models have been proposed for the cooperativity of homotropic interactions between ligand and proteins:

The *symmetry model* of Monond, Wyman, and Changeux (Monad et al., 1965). This model was originally termed the allosteric model. The model is based on three postulates about the structure of an oligomeric protein (allosteric protein) capable of binding ligands (allosteric effectors):

1. Each subunit (protomer) in the protein is capable of existing in either of two conformational states, namely, T (tight) and R (relaxed) states.

2. The conformational symmetry is maintained such that all protomers of the protein must be in the same conformation, either all T or all R at any instance. Therefore, a protein with n subunits is limited to only two conformational states, T_n and R_n. Thus only a single equilibrium constant $L = [T_n]/[R_n]$ is sufficient to express the equilibrium between them.

3. The dissociation constants, K_T and K_R for binding of ligand S to protomers in the T and R conformations, respectively, are different, with the ratio $K_R/K_T = C$ that is assumed to favor binding to the R conformation.

The fractional saturation has the following expression:

$$v = \frac{\alpha(1 + \alpha)^{n-1} + LC\alpha(1 + \alpha)^{n-1}}{(1 + \alpha)^n + L(1 + \alpha)^n}$$

where $\alpha = SK_R$ and $C\alpha = SK_T$.

The *sequential model* of Koshland, Némethy and Filmer (Koshland et al., 1966). The sequential model proposes that the conformational stability of each subunit is determined by the conformations of the subunits with which it is in contact. The model is based on three postulates:

1. Each subunit in the protein is capable of existing in either of two conformational states A and B.

2. There is no requirement for conformational symmetry; therefore mixed conformations are permitted. The stability of any particular state is determined by a product of equilibrium constants including a term K_t for each subunit in the B conformation expressing the free energy of the conformational transition from A to B of an isolated subunit, a term K_{AB} for each contact between a subunit in the A conformation with one in the B conformation, and a term K_{BB} for each contact between a pair of subunits in the B conformation.

3. Ligand binds only to conformation B, with association constant K_S. The binding to conformation A does not exist.

Because the sequential model deals with subunit interactions explicitly and refers to contacts between subunits, it is necessary to consider the geometry of the molecule; for example, a tetrameric protein may exist as square or tetrahedral of which tetrahedral is the preferred arrangement. Treatment of the tetrahedral tetrameric protein gives rise to the following expression:

$$v = \frac{K_S K_t K_{AB}^3[S] + 3K_S^2 K_t^2 K_{AB}^4 K_{BB}[S]^2 + 3K_S^3 K_t^3 K_{AB}^3 K_{BB}^3[S]^3 + K_S^4 K_t^4 K_{BB}^6[S]^4}{1 + 4K_S K_t K_{AB}^3[S] + 6K_S^2 K_t^2 K_{AB}^4 K_{BB}[S]^2 + 4K_S^3 K_t^3 K_{AB}^3 K_{BB}^3[S]^3 + K_S^4 K_t^4 K_{BB}^6[S]^4}$$

Treatment of other geometries leads to the same expressions for the concentrations except for the subunit interaction terms, which are different for each geometry.

The two models further differ in that the symmetry model predicts only positive cooperativity whereas both positive and negative cooperativities are possible according to the sequential model. It is noted that the application of the multiple equilibrium to cooperativity, though fundamentally important, is not practical because it requires knowledge of the number of binding sites and the values of the individual association constants. It is often convenient to define cooperativity with reference to the shape of saturation curve. Operationally, the Hill equation (or the Hill plot) provides a measure of cooperativity. The Hill coefficient, h (or the slope of the Hill plot), is larger than 1 for positive cooperativity, less than 1 for negative cooperativity, and equal to 1 for noncooperativity. Thus the ligand–biomacro-molecule interaction is positively cooperative, negatively cooperative, or non-cooperative according to the sign of $(h - 1)$.

6.2. RECEPTOR BIOCHEMISTRY AND SIGNAL TRANSDUCTION

A receptor (Hulme, 1990; Strader et al., 1994) is a molecule (commonly biomac-romolecule) in/on a cell that specifically recognizes and binds a ligand acting as a signal molecule. Ligand–receptor interactions constitute important initial steps in various cellular processes. The ligands such as hormones and neurotransmitters bind to plasma membrane receptors, which are transmembrane glycoproteins. These ligands include bioactive amines (acetylcholine, adrenaline, dopamine, histamine, serotonin), peptides (calcitonin, glucagon, secretin, angiotensin, bradykinin, inter-leukin, chemokine, endothelin, melanocortin, neuropeptide Y, neurotensin, somatos-tatin, thrombin, galanin, orexin), hormone proteins, prostaglandin, adenosine, and platelet activating factor. Other ligands such as steroids and thyroid hormones bind soluble DNA-binding proteins. The ligand–receptor interactions initiate various *signal transduction* (Heldin and Purton, 1996; Milligan, 1999) pathways that mobilize second messengers which activate/inhibit cascade of enzymes and proteins involved in specific cellular processes (Gilman, 1987; Hepher and Gilman, 1992; Hollenberg, 1991; Kaziro et al., 1991). Three membrane receptor groups that mediate eukaryotic transmembrane signaling processes are as follows:

1. Group 1, Single-transmembrane segment catalytic receptors: Proteins consist-ing of a single transmembrane segment with (a) a globular extracellular domain, which is the ligand recognition site, and (b) an intracellular catalytic domain, which is either tyrosine kinase or guanylyl cyclase.

2. Group 2, Seven-transmemebrane segment receptors: Integral membrane pro-teins consisting of seven transmembrane helical segments with extracellular recognition site for ligands and an intracellular recognition site for a GTP-binding protein.

3. Group 3, Oligomeric ion channels: Ligand-gated ion channels consisting of associated protein subunits that contain several transmembrane segments. Typically, the ligands are neurotransmitters that open the ion channels upon binding.

The binding of these receptors with many ligands stimulates a G-protein (GTP-binding protein), which in turn activates an effector enzyme (e.g., adenylyl cyclae/phospholipase C). Typically, G-proteins (Gilman, 1987; Kaziro et al., 1991; Strader et al., 1994) are heterotrimers ($G_{\alpha\beta\gamma}$) consisting of α-(G_α), β-(G_β), and γ-(G_γ) subunits. Binding of the ligand to receptor stimulates an exchange of GTP for bound GDP on G_α causing G_α to dissociate from $G_{\alpha\beta\gamma}$ and to associate with an effector enzyme which synthesizes the second messenger (Ross and Wilkie, 2000). G-Proteins are a universal means of signal transduction in higher organisms, activating many hormone-receptor-initiated cellular processes via activation of adenylyl cyclases, phospholipases A/C, phosphodiesterases, and ion (Ca^{2+}, Na^+, K^+) channels. Each hormone receptor protein interacts specifically with either a stimulatory G protein or an inhibitory G protein. Some of G proteins and their physiological effects (Clapnam, 1996) are given in Table 6.1.

Two stages of amplification occur in the G-protein-mediated signal response. First, a single ligand–receptor complex can activate many G proteins. Second, the G-protein-activated adenylyl cyclase or phospholipase synthesizes many second messenger molecules. Some of these intracellular second messengers and their effects are given in Table 6.2.

Cyclic AMP (cAMP) is produced from ATP by the action of an integral membrane enzyme, adenylyl cyclase. Various second messengers, such as inositol-1,4,5-triphosphate, diacylglycerol, and arachidonic acid, are generated via breakdown of membrane phospholipids by the action of phospholipases (Liskovich, 1992). Calcium ion is an important intracellular signal (Carafoli and Klee, 1999) that is affected by either cAMP or inositol-1,4,5-triphosphate. The Ca^{2+} signals are translated into the desired intracellular response by calcium binding proteins (e.g., calmodulin, parvalbumin), which in turn regulate cellular processes via protein kinase C (PKC) (Ferrell, 1997). PKC is a cellular transducer, translating the ligand

TABLE 6.1. Some G Proteins and Their Physiological Effects

G Protein	Ligand	Effector	Effect
G_s	Epinephrine, glucagon	Adenylyl cyclase	Glycogen/fat breakdown
G_s	Antidiuretic hormone	Adenylyl cyclase	Conservation of water
G_s	Luteinizing hormone	Adenylyl cyclase	Increase estrogen, progesterone synthesis
G_I	Acetyl choline	Potassium channel	Decrease heart rate
G_i/G_o	Enkephalins, endorphins, opioids	Adenylyl cyclase, ion channel	Changes in neuron electrical activity
G_q	Angiotensin	Phospholipase C	Muscle contraction, blood pressure elevation
G_{olf}	Odorant molecules	Adenylyl cyclase	Odorant detection

TABLE 6.2. Some Second Messengers and Their Effects

Second Messenger	Effect	Source
cAMP	Protein kinase activation	Adenylyl cyclase
cGMP	Protein kinase activation, ion channel regulation	Guanylyl cyclase
Ca^{2+}	Protein kinase and Ca^{2+}-regulatory protein activation	Membrane ion channel
Inositol-1,4,5-triP	Ca^{2+}-channel activation	Phospholipase C on P-inositol
Diacylglycerol	Protein kinase C activation	Phospholipase C on P-inositol
Phosphotidic acid	Ca^{2+}-channel activation, adenylyl cyclase inhibition	Phospholipase D products
Ceramide	Protein kinase activation	Phospholipase C on sphingomyelin
Nitric oxide	Cyclase activation, smooth muscle relaxation	Nitric oxide synthase
cADP-ribose	Ca^{2+}-channel activation	cADP-ribose synthase

message and the signals of second messengers to phosphorylate serine and threonine residue of a wide range of proteins that control growth and development. Nitric oxide (NO) (Bredt and Snyder, 1994), which is synthesized from arginine by nitric oxide synthetase, acts as a neurotransmitter and as a second messenger. Cyclic GMP generated by the NO-stimulated soluble guanylyl cyclase acts to regulate ion channel conductivity and to block gap junction conductivity. Steroid hormones and carcinogens exert their effect as transcription regulators to mediate gene expression.

6.3. FITTING OF BINDING DATA AND SEARCH FOR RECEPTOR DATABASES

6.3.1. Analysis of Binding Data

DynaFit (Kuzmic, 1996), which can be downloaded for academic use from http:// www.biokin.com/, is a versatile program for statistical analysis and fitting of the equilibrium binding data, the initial velocities, and the time course of enzymatic reactions according to the user defined mechanisms. The program performs non-linear least-squares regression of ligand–receptor binding and enzyme kinetic data. To run DynaFit, you must first create a user subdirectory into which you should place a script file and a data file or a data sub-subdirectory containing several data files. The script file is an ascii text file that contains all the information required for the program to perform analysis and simulation.

The general format of the script file for ligand (L) binding to two-site receptor (R) is

```
[task]
  data = equilibrium
  task = fit
  model = fixed?
[mechanism]
```

```
L + R⇔LR    :  K1 dissoc
L + LR⇔L2R:  K2 dissoc
[constants]
  K1 = 5.0  ?,    ; mM
  K2 = 20.0 ?     ; mM
[response]
  LR = 1.0
  L2R = 2.0
[concentrations]
  R = 20.00    ;enzyme concentration
[data]
  variable  L
  file              user/ligand/data.txt
[output]
  directory    user/ligand/output
[end]
```

The student should consult the script files in the example directory of the distribution program for the requisite format in order to run DynaFit successfully.

The data files are ascii text files arranged in columns separated by space, tabs, or commas with header lines. *Equilibrium files* contain total ligand concentrations in the first column and free ligand concentrations in the second column. *Initial velocity files* contain substrate concentrations in the first column and initial velocities of duplicate assays in the subsequent columns. *Progress curve files* contain time in the first column and the measured experimental signal in the second column.

Three-component windows open at the start of DynaFit (double click DynaFit.exe or DynaFit icon). Status window provides messages for the action to be taken or in progress. Text window gives mechanism, analytical, and fitted text results, while the fitted and simulated plots are displayed in Graphic window (Figure 6.1). From the File menu, load the script file which processes an analysis and fitting of input data in the data files. An initial plot based on the input data and parameters is displayed on the Graphic window, and the mechanism used is listed on the Text window. After the Status window requesting "Ready to run...,' choose Run from the File menu. The analytical and fitting results are displayed continuously on the Graphic and the Text windows with the Status window providing messages of progress. At the end of an analytical/fitting cycle, answer *Yes* to "Is the initial estimate good enough?" of the pop-up dialogue box to view results. Answer *Yes* to "Terminate the least square minimization?" of the pop-up dialogue box in order to terminate analysis. Otherwise, answer *No* to continue with an addition cycle of analysis and fitting. At the end of fitting (the Status window shows a message, "All tasks completed. Examine output files"), the final fitting is displayed in the Graphic window and the summary results are given in the Text window (Figure 6.2). The details of analysis and results can be viewed via Notepad by invoking Results from the Edit menu and saved as result.txt.

6.3.2. Retrieval of Receptor Information and Signal Transduction Pathways

The receptors are classified according to membrane receptors and nuclear receptors, which are also listed alphabetically for retrieval at Receptor database (http://impact.nihs.go.jp/RDB.html) (Nakata et al., 1999). A keyword query can be

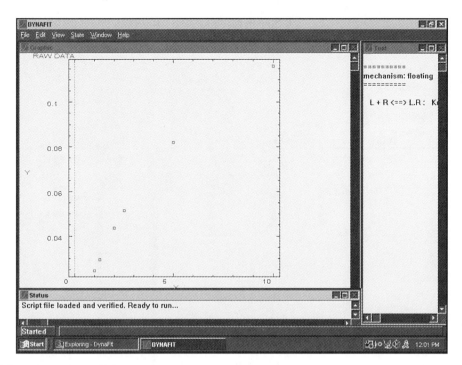

Figure 6.1. DynaFit start windows. After loading script file, the program automatically inputs data file and plots raw data in the graphic window. Start the analysis (File→Run!) upon receiving "Ready to run" instruction in the status window.

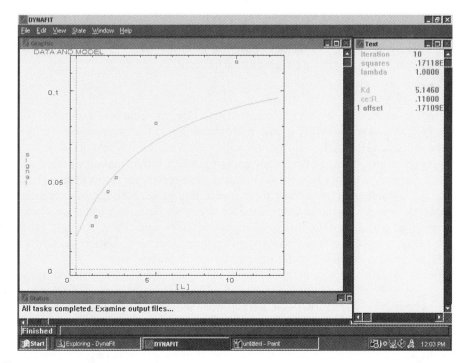

Figure 6.2. DynaFit result windows. After answering Yes to questions, "Is the initial estimate good enough?" and "Terminate the least-squares minimization?", the program terminates by plotting fitted results in the Graphic window and summary in the Text window.

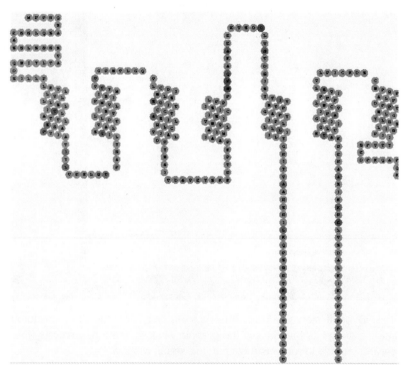

Figure 6.3. G Protein-coupled receptor retrieved from GPCRDB.Seven transmembrane segment receptor for serotonin retrieved from GPCRDB is visualized in a snake-like plot. The colored groups of residues show a correlated behavior as determined by correlated mutation analysis (CMA). The colored positions are hyperlinked to their corresponding residue locations in the multiple sequence alignments.

initiated to search for a specific receptor, its link to sequence databases (Gen-Bank,PIR, and SWISS-PROT), and multiple alignment of a receptor family via MView option. GPCRDB at http://www.gpcr.org/7tm/ and NucleaRDB at http://www.receptors.org/NR/ or http://www.gpcr.org/NR/ are Information Systems for G protein-coupled receptors (Figure 6.3) and nuclear receptors, respectively (Horn et al., 2001). Both sites provide links to sequence databases, bibliographic references, alignments, and phylogenetic trees based on sequence data. In addition, GPCRDB also provides mutation data, lists of available pdb files, and tables of ligand binding constants to receptors of acetylcholine, adrenaline, dopamine, histamine, serotonin, opioid, adenosine, cannabis, melatonin, and γ-aminobutyric acid.

The information for signaling pathways in human cells can be retrieved from the Cell Signaling Network Database (CSNDB) (Takai-Igarashi et al., 1998) at http://geo.nihs.go.jp/csndb/ (Figure 6.4). Under Global Search, enter the ligand name (e.g., ACTH, estradiol) as the keyword and click Send Request. The search returns list of information (signaling cell, function, and structure link). Click Find Pathways to view the sketch of signal pathway with clickable links. Alternatively, you may click Browse and select EndoMolecules (e.g., hormones) or ExoMolecules (e.g., drugs) to view signal pathways.

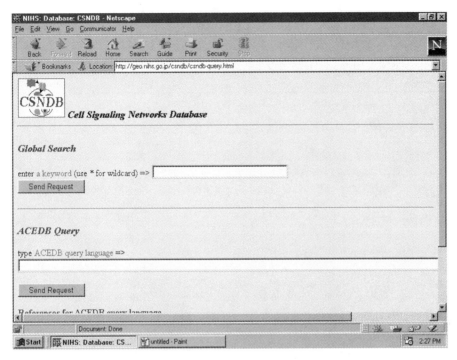

Figure 6.4. Cell signaling networks database. The cell signaling networks database (CSNDB) provides facilities for searching/retrieving transduction (signaling) pathways.

The Biomolecular Interaction Network Database (BIND) at http://www.binddb.org/ (Bader et al., 2001) provides a text query for information on biomolecular interactions relating to cellular communication, differentiation, and growth. Enter the name of receptor or second messenger but not ligand (e.g., the text query accepts dopamine receptor but not dopamine) and click Find. This returns a list of hits. Select the Interaction ID (full record) or desired information (Full record, PubMed abstract, BIND publication, View ASN report, or View XML report). Click Visualize Interaction to open the Interaction Viewer window showing the interaction between two components in the signaling pathway (Figure 6.5). Further interactions involving the pathway members can be viewed by double-clicking the highlighted boxes.

ReliBase at http://relibae.ebi.aci.uk/ is the resource site for receptor–ligand (Reli) interactions and provides facilities for similarity search/analysis of the Reli database for similar ligands, similar binding sites, and Reli complex (Figure 6.6). The query at ReliBase can be conducted via Text (e.g., chemical name of ligand), Sequence (of receptor), Smiles (strings), or 2D/3D (fragments of structure) of ligand by clicking respective menu button. The instructions on how to use Relibase can be obtained by clicking the Help button. A ligand and its receptor complex can be searched/retrieved by entering the ligand name after clicking Text button or entering SMILES string (by clicking the Smiles button). To save the atomic coordinate file (pdb format) of a ligand-binding complex, click Complex PDB file. To search for similar ligands with a receptor sequence, click Sequence on the ReliBase home page.

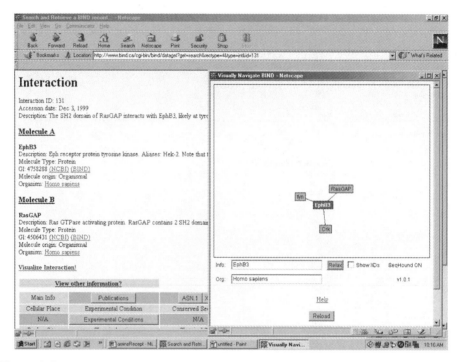

Figure 6.5. BIND Information and interaction viewer. The full record of BIND is shown for an interaction involving Ras. Clicking Visualize Interaction opens the Interaction viewer window. Activate (double-click) the highlighted molecule displays further interactions.

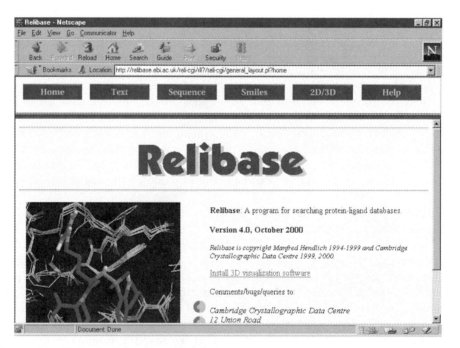

Figure 6.6. Relibase home page. Relibase provides resources for searching/retrieving structures, binding sites of receptor proteins, and chemical diagram of ligands. All nonprotein moieties (except water) in the PDB complexes are considered as ligands.

Copy and paste the amino acid sequence into the query box and click Show Ligands. Clicking the 2D/3D button opens the structure window for drawing a structure component used to search the database by substructure.

6.4. WORKSHOPS

1. The data below describe the binding of ligand L to its receptor R (100 μM). Build a script file and data file to solve for the association constant and the number of noninteracting equivalent sites.

Total Ligand (mM)	Bound Ligand (μM)
1.00	24.58
1.25	29.59
2.00	43.62
2.50	51.51
5.00	81.94
10.00	116.23

2. The data below describe the binding of ligand L to the oligomeric receptor R (100 μM). Create a script file and data file to evaluate the association constant and the number of equivalent sites. Calculate Hill's interaction coefficient by perform regression analysis.

Total Ligand (mM)	Bound Ligand (μM)
1.00	98.22
1.25	108.12
1.60	130.90
2.00	150.32
2.50	184.07
4.00	245.71
5.00	279.36
8.00	347.78
10.00	378.92

3. The binding data of ligand L with its receptor protein (100 μM) are given below. Estimate the association constants by assuming two classes of equivalent sites and deduce their cooperativity.

Total Ligand (mM)	Bound Ligand (μM)
1.000	10.30
1.250	15.90
1.425	17.78
1.600	19.55
2.000	22.94

Total Ligand (mM)	Bound Ligand (μM)
2.500	26.42
3.125	28.73
4.000	30.86
5.000	32.48
8.000	35.05
10.000	36.45
20.000	38.72
40.000	39.96

4. The binding data of ligand L with its receptor protein (100 μM) are given below. Estimate the association constants by assuming two classes of equivalent sites and deduce their cooperativity.

Total Ligand (mM)	Bound Ligand (μM)
1.000	11.42
1.250	13.07
1.425	14.05
1.600	15.06
2.000	16.47
2.500	18.07
3.125	19.58
4.000	21.58
5.000	23.69
8.000	19.80
10.000	33.19
14.000	38.40
20.000	41.61
40.000	48.74

5. Search receptor database for the classification of receptors according to the chemical nature of their ligands.

6. List hormones according to their receptors being membrane (G-protein coupled) proteins or soluble (nuclear bound) proteins.

7. Retrieve the sequence and plot of one of 7-transmembrane segment receptors and discuss its structural features.

8. Retrieve the sequence and 3D model of a neurotransmitter or hormone receptor that transduces nuclear signaling.

9. Depict one signal transduction pathway each for

   a. G-Protein-coupled, adenylylate-activating (via cAMP) pathway

   b. G-Protein-coupled, phospholipase C (via inositol triphosphate) pathway

   c. Nuclear protein mediating pathway

10. List candidate signal molecules for the receptor with the amino acid sequence given below.

```
MRTLNTSAMD  GTGLVVERDF  SVRILTACFL  SLLILSTLLG  NTLVCAAVIR  FRHLRSKVTN
FFVISLAVSD  LLVAVLVMPW  KAVAEIAGFW  PFGSFCNIWV  AFDIMCSTAS  ILNLCVISVD
RYWAISSPFR  YERKMTPKAA  FILISVAWTL  SVLISFIPVQ  LSWHKAKPTS  PSDGNATSLA
ETIDNCDSSL  SRTYAISSSV  ISFYIPVAIM  IVTYTRIYRI  AQKQIRRIAA  LERAAVHAKN
CQTTTGNGKP  VECSQPESSF  KMSFKRETKV  LKTLSVIMGV  FVCCWLPFFI  LNCILPFCGS
GETQPFCIDS  NTFDVFVWFG  WANSSLNPII  YAFNADFRKA  FSTLLGCYRL  CPATNNAIET
VSINNNGAAM  FSSHHEPRGS  ISKECNLVYL  IPHAVGSSED  LKKEEAAGIA  RPLEKLSPAL
SVILDYDTDV  SLEKIQPITQ  NGQHPT
```

REFERENCES

Bader, G. D., Donaldson, I., Wolting, C., Ouellette, B. F. F., Pawson, T., and Hogue, C. W. V. (2001) *Nucleic Acids Res.* **29**:242–245.

Bredt, D. S., and Snyder, S. H. (1994) *Annu. Rev. Biochem.* **63**:175–195.

Clapnam, D. E. (1996) *Nature* **379**:297–299.

Carafoli, E., and Klee, C. B., Eds. (1999) *Calcium As A Cellular Regulator.* Oxford University Press, Oxford.

Ferrell, J. E. (1997) *Trends Biochem. Sci.* **22**:288–289.

Gilman, A. (1987) *Annu. Rev. Biochem.* **56**:615–649.

Harding, S. E., and Chowdhry, B. Z., Eds. (2001) *Protein–Ligand Interactions: Structure and Spectroscopy.* Oxford University Press, Oxford.

Heldin, C.-H., and Purton, M., Eds. (1996) *Signal Transduction.* Chapman and Hall, New York.

Hepher, J., and Gilman, A. (1992) *Trends Biochem. Sci.* **17**:383–387.

Hollenberg, M. (1991) *FASEB J.* **5**:178–186.

Horn, F., Vriend, G., and Cohen, F. E. (2001) *Nucleic Acids Res.* **29**:346–349.

Hulme, E. C., Ed. (1990) *Receptor Biochemistry: A Practical Approach.* IRL Oxford University Press.

Kaziro, Y., Itoh, H., Kozasa, T., Nakafuku, M., and Satoh, T. (1991) *Annu. Rev. Biochem.* **60**:349–400.

Kleanthous, C., Ed. (2000) *Protein–Protein Recognition.* Oxford University Press.

Koshland, D. E., Némethy, G., and Filmer, D. (1966) *Biochemistry* **5**:365–385.

Kuzmic, P. (1996) *Anal. Biochem.* **237**:260–273.

Liskovich, M. (1992) *Trends Biochem. Sci.* **17**:393–399.

Milligan, G., Ed. (1999) *Signal Transduction: A Practical Approach*, 2nd edition. Oxford University Press.

Monad, J., Wyman, J., and Changeux, J. P. (1965) *J. Mol Biol.* **12**:88–118.

Nakata, K., Takai, T., and Kaminuma, T. (1999) *Bioinformatics* **15**:544–552.

Richard, J., and Cornish-Bowden, A. (1987) *Eur. J. Biochem.* **166**:255–272.

Ross, E. M., and Wilkie, T. M. (2000) *Annu. Rev. Biochem.* **69**:759–827.

Saenger, W., and Heinemann, U., Eds. (1989) *Protein–Nucleic Acid Interaction.* CRC Press, Boca Raton, FL.

Steinhardt, J., and Reynolds, J. A. (1969) *Multiple Equilibria in Proteins.* Academic Press, New York.

Strader, C. D., Fong, T. M., Tota, M. R., and Underwood, D. (1994) *Annu. Rev. Biochem.* **63**:101–132.

Takai-Igarashi, T., Nadaoka, Y., and Kaminuma, T. (1998) *J. Comput. Biol.* **5**:747–754.

Weber, G. (1992) *Protein Interactions.* Chapman and Hall, London.

7

DYNAMIC BIOCHEMISTRY:
ENZYME KINETICS

Enzymes are biocatalysts, as such they facilitate rates of biochemical reactions. Some of the important characteristics of enzymes are summarized. Enzyme kinetics is a detailed stepwise study of enzyme catalysis as affected by enzyme concentration, substrate concentrations, and environmental factors such as temperature, pH, and so on. Two general approaches to treat initial rate enzyme kinetics, quasi-equilibrium and steady-state, are discussed. Cleland's nomenclature is presented. Computer search for enzyme data via the Internet and analysis of kinetic data with Leonora are described.

7.1. CHARACTERISTICS OF ENZYMES

Enzymes are globular proteins whose sole function is to catalyze biochemical reactions. The most important properties of all enzymes are their catalytic power, specificity, and capacity to regulation. The characteristics of enzymes (Copeland, 2000; Fersht, 1985; Kuby, 1991; Price and Stevens, 2000) can be summarized as follows:

- All enzymes are proteins: The term enzymes refers to biological catalysts, which are proteins with molecular weights generally ranging from 1.5×10^4 to 10^8 daltons. The nonprotein biocatalysts such as catalytic RNA and DNA are known as *ribozymes* (Doherty and Doudna, 2000; Scott and Klug, 1996) and deoxyribozymes (Li and Breaker, 1999; Sheppard et al., 2000), respectively, while engineered catalytic antibodies are called *abzymes* (Benkovic, 1992; Hilvert, 2000).

- Enzymes increase the rate but do not influence the equilibrium of biochemical reactions: Enzymes are highly efficient in their catalytic power displaying rate enhancement of 10^6 to 10^{12} times those of uncatalyzed reactions without changing the equilibrium constants of the reactions.

- Enzymes exhibit a high degree of specificity for their substrates and reactions: Enzymes are highly specific both in the nature of the substrate(s) that they utilize and in the types of reactions that they catalyze. Enzymes may show *absolute specificity* by a catalyzing reaction with a single substrate. Enzymes may display *chemical (bond) specificity* by promoting transformation of a particular chemical functional group (e.g., hydrolysis of esteric bond by esterases or phosphorylation of primary hydroxy group of aldohexoses by hexokinase). The *stereospecificity* refers to the ability of enzymes to choose only one of the enantiomeric pair of a chiral substrate in *chiral stereospecificity*, such as D-lactate dehydrogenase versus L-lactate dehydrogenase. One of the most subtle stereospecificities of enzymes relates their ability to distinguish between two identical atoms/groups (proR versus proS) bonded to a carbon atom in *prochiral stereospecificity* (Hanson, 1966), such as proR glycerol-3-phosphate dehydrogenase versus proS glycerol-3-phosphate dehydrogenase.

- Some enzymes require cofactors for the activities: Enzymes that require covalent cofactors (prosthetic groups, e.g., heme in cytochromes) or non-covalent cofactors (coenzymes, e.g., $NAD(P)^+$ in dehydrogenases) for activities are called haloenzymes (or simply enzymes). The protein molecule of a haloenzyme is termed *proenzyme*. The prosthetic group/coenzyme dictates the reaction type catalyzed by the enzyme, and the proenzyme determines the substrate specificity.

- The active site of an enzyme (Koshland, 1960) is the region that specifically interacts with the substrate: A number of generalizations concerning the active site of an enzyme are as follows:

  1. The active site of an enzyme is the region that binds the substrates (and cofactors, if any) and contributes the catalytic residues that directly participate in the making and breaking of bonds.

  2. The active site takes up a relatively small part of the total dimension of an enzyme.

  3. The active site is a three-dimensional entity and dynamic.

  4. Active sites are generally clefts or crevices.

  5. Substrates are bound to the active site of enzymes by noncovalent bonds.

  6. The specificity of binding depends on the precisely defined arrangement of atoms in an active site.

  7. The induced fit model (Koshland, 1958) has been proposed to explain how the active site functions.

- Enzymatic catalysis involves formation of an intermediate enzyme–substrate complex.

- Enzymes lower the activation energies of reactions.

- The activity of some enzymes are regulated (Hammes, 1982) as follows:
  1. The enzyme concentration in cells is regulated at the synthetic level by genetic control (Gottesman, 1984), which may occur positively or negatively.
  2. Some enzymes are synthesized in an inactive precursor form known as proenzymes or zymogens, which are activated at a physiologically appropriate time and place.
  3. Another control mechanism is covalent modifications (Freeman and Hawkins, 1980, 1985) such as reversible phosphorylation, glycosylation, acylation, and so on. These modifications are normally reversible catalyzed by separate enzymes.
  4. The enzymatic activities are also subjected to allosteric/cooperative regulations (Monad et al., 1965; Perutz, 1990; Ricard and Cornish-Bowden, 1987). A regulatory molecule (effector) binds to the regulatory site (allosteric site) distinct from the catalytic site to affect one or more kinetic parameters of an enzymatic reaction.
  5. The compartmentation/solubility of an enzyme is another form of controlling the activity of an enzyme.
- Some enzymes exist as multienzyme or multifunctional complexes (Bisswanger and Schmincke-Ott, 1980; Perham, 2000; Reed, 1974; Srere, 1987).
- Some enzymes exist in isozymic forms: Within a single species, there may exist several different forms of enzyme catalyzing the same reaction; these are known as isozymes (Markert, 1975). Generally isozymes are derived from an association of different subunits in an oligomeric enzyme with different electrophoretic mobilities.

Enzymes are usually named with reference to the reaction they catalyze. It is customary to add the suffix -*ase* to the name of its major substrate. The Enzyme Commission (EC) has recommended nomenclature of enzymes based on the six major types of enzyme-catalyzed reactions (http://www.chem.qmw.ac.uk/iubmb/enzyme/):

EC 1: *Oxidoreductases* catalyze oxidation–reduction reactions.

EC 2: *Transferases* catalyze group transfer reactions.

EC 3: *Hydrolases* catalyze hydrolytic reactions.

EC 4: *Lyases* catalyze cleavage and elimination reactions.

EC 5: *Isomerases* catalyze isomerization reactions.

EC 6: *Ligases* catalyze synthetic reactions.

Thus, the EC numbers provide unique identifiers for enzyme functions and give us useful keyword entries in database searches.

The ENZYME database at http://www.expasy.ch/enzyme/ provides information on EC number, name, catalytic activity, and hyperlinks to sequence data of enzymes. The 3D structures of enzymes can be accessed via Enzyme Structures Database at http://www.biochem.ucl.ac.uk/bsm/enzyme/index.html. Some other enzyme databases are listed in Table 7.1.

TABLE 7.1. Enzyme Databases

Web Site	URL
ENZYME DB: General information	http://www.expasy.ch/enzyme/
Enzyme structure database: Structures	http://www.biochem.ucl.ac.uk/bsm/enzyme/index.html
LIGAND: Enzyme reactions	http://www.gebine.ad.jp/dbget/ligand.html
Brenda: General enzyme data	http://www.brenda.uni-koeln.de/
EMP: General, literature summary	http://wit.mcs.anl.gov/EMP/
Esther: Esterases	http://www.ensam.inra.fr/cholinesterase/
Merops: Peptidases	http://www.bi.bbsrc.ac.uk/Merops/Merops.htm
Protease	http://delphi.phys.univ-tours.fr/Prolysis
CAZy: Carbohydrate active enzymes	http://afmb.cnrs-mrs.fr/~pedro/CAZY/db.html
REBASE: Restriction enzymes	http://rebase.neh.com/rebase/rebase.html
Ribonuclease P database	http://www.mbio.ncsu.edu/RnaseP/home.html
PKR: Protein kinase	http://pkr.sdsc.edu
PlantsP: Plant protein kinases and phosphatase	http://PlantP.sdsc.edu
Aminoacyl-tRNA synthetases	http://rose.man.poznan.pl/aars/index.html
MDB: Metalloenzymes	http://metallo.scripps.edu/
Promise: Prosthetic group/Metal enzymes	http://bmbsgi11.leads.ac.uk/promise/
Aldehyde dehydrogenase	http://www.ucshc.edu/alcdbase/aldhcov.html
G6P dehydrogenase	http://www.nal.usda.gov/fnic/foodcomp/
2-Oxoacid dehydrogenase complex	http://qcg.tran.wau.nl/local/pdhc.htm

7.2. KINETICS OF ENZYMATIC REACTIONS

Enzyme kinetics (Ainsworth, 1977; Cornish-Bowden, 1995; Fromm, 1975; Plowman, 1972; Segel, 1975; Schulz, 1994), which investigates the rates of enzyme-catalyzed reactions as affected by various factors, offers an enormous potential to the study of enzyme reaction mechanisms and functions. Some important factors that affect the rates of enzymatic reactions are enzyme concentration, ligand (substrates, products, inhibitors, and activators) concentrations, solvent (solution, ionic strength, and pH), and temperature. When all these factors are properly analyzed, it is possible to learn a great deal about the nature of enzymes. The kinetic studies of an enzymatic reaction by varying ligand concentrations provide kinetic parameters that are essential for an understanding of the *kinetic mechanism* of the biochemical reaction. Operationally, the initial rate enzyme kinetics can be treated according to two assumptions:

7.2.1. Quasi-equilibrium Assumption

Quasi-equilibrium, also known as rapid equilibrium, assumes that an enzyme (E) reacts with substrate (S) rapidly to form an enzyme–substrate complex (ES) (with a rate constant, k_1) that breaks down to release the enzyme and product (P). The enzyme, substrate, and the enzyme–substrate complex are at equilibrium; that is the rate at which ES dissociates to E + S (rate constant of k_2) is much faster than the rate of forming E + P (rate constant of k_3). Because enzyme kinetic studies are carried out with excess concentrations of substrate — that is, $[S] \gg [E]$ — the conservation equations $[E]_0 = [E] + [ES]$ and $[S]_0 = [S] + [ES] \cong [S]_0$ apply. The overall rate of the reaction is limited by the breakdown of the enzyme–substrate complex, and the velocity is measured during the very early stage of the reaction. Because the quasi-equilibrium treatment expresses the enzyme–substrate complex in terms of $[E]$, $[S]$, and K_s — that is, $[ES] = [E][S]/(K_s + [S])$ — the kinetic expression is obtained if we insert the expression for the enzyme–substrate complex into the rate expression:

$$v = k_3[ES] = k_3[E][S]/(K_s + [S]) = V[S]/(K_s + [S])$$

This is known as the Michaelis–Menten equation, where there are two kinetic parameters, the maximum velocity $V = k_3[E]$ and the Michaelis constant $K_s(K_m) = k_2/k_1$.

The general rule for writing the rate equation according to the quasi-equilibrium treatment of enzyme kinetics can be exemplified for the random bisubstrate reaction with substrates A and B forming products P and Q (Figure 7.1), where $K_a K_{ab} = K_b K_{ba}$ and $K_p K_{pq} = K_q K_{qp}$.

1. Write the initial velocity expression: $v = k[EAB] - k'[EPQ]$, where the interconversion between the ternary complexes is associated with the rate constants k and k' in the forward and reverse directions, respectively.

2. Divide the velocity expression by the conservation equation for enzymes, $[E]_0 = [E] + [EA] + [EB] + [EAB] + [EPQ] + [EP] + [EQ]$, that is,

$$v/[E]_0 = (k[EAB] - k'[EPQ])/([E] + [EA] + [EB] + [EAB] + [EPQ] + [EP] + [EQ])$$

3. Set the $[E]$ term equal to unity, that is, $[E] = 1$.

4. The term for any enzyme-containing complex is composed of a numerator, which is the product of the concentrations of all ligands in the complex, and a denominator, which is the product of all dissociation constants between

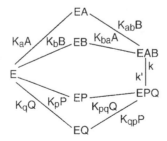

Figure 7.1. Diagram for random bi bi kinetic mechanism. The random addition of substrates, A and B to form binary (EA and EB) and ternary (EAB) complexes. The two ternary complexes EAB and EPQ interconvert with the rate constant of k and k'. The release of products P and Q also proceeds in a random manner. K_s are dissociation constants where $K_a K_{ab} = K_b K_{ba}$ and $K_p K_{pq} = K_q K_{qp}$.

the complex and free enzyme, that is, $[EA] = [A]/K_a$, $[EB] = [B]/K_b$, $[EP] = [P]/K_p$, $[EQ] = [Q]/K_q$, $[EAB] = ([A][B])/(K_a K_{ab})$, and $[EPQ] = ([P][Q])/(K_q K_{qp})$.

5. The substitution yields the rate expression:

$$v = \frac{\{k([A][B])/(K_a K_{ab}) - k'([P][Q])/(K_q K_{qp})\}[E]_0}{1 + [A]/K_a + [B]/K_b + ([A][B])/(K_a K_{ab}) + [P]/K_p + [Q]/K_q + ([P][Q])/(K_q K_{qp})}$$

The rate expression for the forward direction is simplified to

$$v = \frac{k([A][B])/(K_a K_{ab})[E]_0}{1 + [A]/K_a + [B]/K_b + ([A][B])/(K_a K_{ab})}$$

$$= \frac{V[A][B]}{K_a K_{ab} + K_{ab}[A] + K_{ba}[B] + [A][B]}$$

7.2.2. Steady-State Assumption

The steady-state treatment of enzyme kinetics assumes that concentrations of the enzyme-containing intermediates remain constant during the period over which an initial velocity of the reaction is measured. Thus, the rates of changes in the concentrations of the enzyme-containing species equal zero. Under the same experimental conditions (i.e., $[S]_0 \gg [E]_0$ and the velocity is measured during the very early stage of the reaction), the rate equation for one substrate reaction (uni uni reaction), if expressed in kinetic parameters (V and K_s), has the form identical to the Michaelis–Menten equation. However, it is important to note the differences in the Michaelis constant that is, $K_s = k_2/k_1$ for the quasi-equilibrium treatment whereas $K_s = (k_2 + k_3)/k_1$ for the steady-state treatment.

The general rule for writing the rate equation according to the steady-state treatment of enzyme kinetics by King and Altman method (King and Altman, 1956) can be illustrated for the sequential bisubstrate reaction. Figure 7.2 shows the ordered bi bi reaction with substrates A and B forming products P and Q. Each enzyme-containing species is associated with two rate constants, k_{odd} in the forward direction and k_{even} in the reverse direction.

The steady-state rate equation is obtained according to following rules (King and Altman method):

1. Write down all possible basic patterns with $n - 1$ lines (n = number of enzyme forms that is, enzyme-containing species), in which all lines are connected without closed loops—for example,

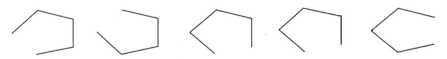

2. The total line patterns equals $m!\{(n - 1)!(m - n + 1)!\}$, where m is the number of lines in the patterns.

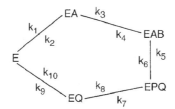

Figure 7.2. Diagram for ordered bi bi kinetic mechanism. The free enzyme, E, binds to A (first substrate) to form a binary complex, EA, which then interacts with B (second substrate) to form a ternary complex, EAB. The two ternary complexes EAB and EPQ interconvert. The release of P (first product) forms EQ, which then dissociates to E and Q (second product) in an ordered sequence. k_1 to k_{10} are rate constants.

3. Write a distribution equation for each enzyme form—for example, $E/E_0 = N_e/D$, where N_e and D are the numerator (for E) and denominator terms, respectively.

4. Numerator terms (e.g., N_e) are written:

   (a) Follow along the lines in the basic pattern in the direction from other enzyme forms (i.e., EA, EQ, EAB, and EPQ) leading toward the free enzyme form (i.e., E) for which the numerator is sought.

   (b) Multiply all the rate constants and concentration factors for this direction.

   (c) Repeat the process for all the basic patterns.

   (d) The numerator terms are the sum of all the products of rate constants and concentration factors.

5. Write the numerator terms for all the other enzyme forms by repeating the process—for example, N_{ea}, N_{eq}, N_{eab}, and N_{epq} for EA/E_0, EQ/E_0, EAB/E_0, and EPQ/E_0, respectively.

6. Denominator terms are the sum of all numerator terms, that is $D = N_e + N_{ea} + N_{eq} + N_{eab} + N_{epq}$.

7. Substitute appropriate distribution equations (e.g., EPQ/E_0 and EQ/E_0) into the initial velocity expression:

$$v = dP/dt = k_7[\text{EPQ}] - k_8[\text{EQ}][\text{P}]$$

$$= k_7\{\text{EPQ}/E_0\}[E]_0 - k_8\{\text{EQ}/E_0\}[E]_0[\text{P}]$$

$$= (k_1k_3k_5k_7k_9[\text{A}][\text{B}] - k_2k_4k_6k_8k_{10}[\text{P}][\text{Q}])[E]_0)/\{k_2(k_4k_6 + k_4k_7 + k_5k_7)k_9$$

$$+ k_1(k_4k_6 + k_4k_7 + k_5k_7)k_9[\text{A}] + k_3k_5k_7k_9[\text{B}] + k_2k_4k_6k_8[\text{P}]$$

$$+ k_1k_3(k_5k_7 + k_5k_9 + k_6k_9 + k_7k_9)[\text{A}][\text{B}]$$

$$+ (k_2k_4 + k_2k_5 + k_2k_6 + k_4k_6)k_8k_{10}[\text{P}][\text{Q}]$$

$$+ k_1k_4k_6k_8[\text{A}][\text{P}] + k_3k_5k_7k_{10}[\text{B}][\text{Q}]$$

$$+ k_1k_3k_8(k_5 + k_6)[\text{A}][\text{B}][\text{P}] + k_3k_8k_{10}(k_5 + k_6)[\text{B}][\text{P}][\text{Q}]$$

This steady-state equation expressed with rate constants can be converted into the rate equation expressed with kinetic parameters according to Clelend (Cleland, 1963a, 1963b):

$$v = \frac{V_1 V_2 (AB - PQ/K_{eq})}{(K_{ia}K_b + K_b A + K_a B + AB)V_2 + (K_p Q + K_q P + PQ)V_1/K_{eq} + (K_a BQ/K_{iq} + ABP/K_{ip})V_2 + (K_q AP/K_{ia} + BPQ/K_{ib})V_1/K_{eq}}$$

where the equilibrium constant, $K_{eq} = (V_1 K_p K_{iq})/(V_2 K_b K_{ia})$. V_1 and V_2 are maximum velocities for the forward and reverse reactions. K_a, K_b, K_p, and K_q are Michaelis constants, while K_{ia}, K_{ib}, K_{ip}, and K_{iq} are inhibition constants associated with substrates (A and B) and products (P and Q), respectively. The rate equation for the forward reaction can be simplified (Table 7.2) to

$$v = \frac{V_1 AB}{K_{ia}K_b + K_b A + K_a B + AB}$$

7.2.3. Cleland's Approach

The Cleland nomenclature (Cleland, 1963a) for enzyme reactions follows:

1. The number of kinetically important substrates or products is designated by the syllables *Uni*, *Bi*, *Ter*, *Quad*, *Pent*, and so on, as they appear in the mechanism.

TABLE 7.2. Cleland Nomenclature for Bisubstrate Reactions Exemplified[a]

Mechanism	Cleland Representation	Forward Rate Equation
Order bi bi	A B P Q E EA EXY EQ E	$v = \dfrac{V_1 AB}{K_{ia}K_b + K_b A + K_a B + AB}$
Random bi bi	A B P Q EA EQ E EAB EP E EB EP B A Q P	$v = \dfrac{V_1 AB}{K_{ia}K_b + K_b A + K_a B + AB}$
Ping pong bi bi or Uni uni uni uni ping pong	A P B Q E EA F EQ E	$v = \dfrac{V_1 AB}{K_b A + K_a B + AB}$

[a]Three common kinetic mechanisms for bisubstrate enzymatic reactions are exemplified. The forward rate equations for the order bi bi and ping pong bi bi are derived according to the steady-state assumption, whereas that of the random bi bi is based on the quasi-equilibrium assumption. These rate equations are first order in both A and B, and their double reciprocal plots ($1/v$ versus $1/A$ or $1/B$) are linear. They are convergent for the order bi bi and random bi bi but parallel for the ping pong bi bi due to the absence of the constant term ($K_{ia}K_b$) in the denominator. These three kinetic mechanisms can be further differentiated by their product inhibition patterns (Cleland, 1963b).

Note: V_1, K_a, K_b, and K_{ia} are maximum velocity, Michaelis constants for A and B, and inhibition constant for A, respectively.

2. A *sequential* mechanism will be one in which all the substrates must be present on the enzyme before any product can leave. Sequential mechanisms will be designated *ordered* or *random* depending on whether the substrate adds and the product releases in an obligatory sequence or in a nonobligatory sequence.

3. A *ping pong* mechanism will be designated if one or more products are released during the substrate addition sequence, thereby breaking the substrate addition sequence into two or more segments. Each segment is given an appropriate syllable corresponding to the number of substrate additions and product releases.

4. The letters A, B, C, D designate substrate in the order of their addition to the enzyme. Products are P, Q, R, S in the order of their release. Stable enzyme forms are designated by E, F, G, H, with E being free enzyme.

5. Isomerization of a stable enzyme form as a part of the reaction sequence is designated by the prefix *Iso*, such as *Iso ordered, Iso ping pong.*

6. For expressing enzymatic reactions, the sequence is written from left to right with a horizontal line or group of lines representing the enzyme in its various forms. Substrate additions and product releases are indicated by the downward ($\downarrow$) and upward ($\uparrow$) vertical arrows, respectively.

7. Each arrow is associated with the corresponding reversible step — that is, one rate constant each with the forward and reverse directions. Generally, odd-numbered rate constants are used for the forward reactions whereas even-numbered ones are used for the reverse direction.

Examples of the bisubstrate reactions according to Cleland nomenclature are listed in Table 7.2.

For multisubstrate enzymatic reactions, the rate equation can be expressed with respect to each substrate as an $n:m$ function, where n and m are the highest order of the substrate for the numerator and denominator terms respectively (Bardsley and Childs, 1975). Thus the forward rate equation for the random bi bi derived according to the quasi-equilibrium assumption is a 1:1 function in both A and B (i.e., first order in both A and B). However, the rate equation for the random bi bi based on the steady-state assumption yields a 2:2 function (i.e., second order in both A and B). The 2:2 function rate equation results in nonlinear kinetics that should be differentiated from other nonlinear kinetics such as allosteric/cooperative kinetics (Chapter 6, Bardsley and Waight, 1978) and formation of the abortive substrate complex (Dalziel and Dickinson, 1966; Tsai, 1978).

7.2.4. Environmental Effects

The rate of an enzymatic reaction is affected by a number of environmental factors, such as solvent, ionic strength, temperature, pH, and presence of inhibitor/activator. Some of these effects are described below.

Presence of Inhibitors: **inhibition Kinetics.** The kinetic study of an enzymatic reaction in the presence of inhibitors is one of the most important diagnostic procedures for enzymologists. The inhibition (reduction in the rate) of an enzyme

TABLE 7.3. Types of Enzyme Inhibitions

Type of Inhibition	Complex Formation	Forward Rate Equation
Control	$E + S = ES \rightarrow E + P$	$v = \dfrac{V_1 S}{K_s + S}$
Competitive	$E + I \overset{K_{is}}{=} EI$	$v = \dfrac{V_1 S}{K_s(1 + I/K_{is}) + S}$
Uncompetitive	$ES + I \overset{K_{ii}}{=} ESI$	$v = \dfrac{V_1 S}{K_a s + S(1 + I/K_{ii})}$
Noncompetitive or mixed competitive	$E + I \overset{K_{is}}{=} EI$ $ES + I \overset{K_{ii}}{=} ESI$	$v = \dfrac{V_1 S}{K_s(1 + I/K_{is}) + S(1 + I/K_{ii})}$

Note: K_{ii} and K_{is} are inhibition (dissociation) constants for the formation of the inhibitor complexes in which the subscripts denote the intercept effect and slope effect, respectively.

reaction is one of the major regulatory devices of living cells and offers great potentials for the development of pharmaceuticals. An irreversible inhibitor forms the stable enzyme complex or modifies the enzyme to abolish its activity, whereas a reversible inhibitor (I) forms dynamic complex(es) with the enzyme (E) or the enzyme substrate complex (ES) by reducing the rate of the enzymatic reaction (see Table 7.3).

Temperature Effect: Determination of Activation Energy. From the transition state theory of chemical reactions, an expression for the variation of the rate constant, k, with temperature known as the Arrhenius equation can be written

$$k = Ae^{-E_a/RT}$$

or

$$\ln k = \ln A - E_a/RT$$

where A, R, and T are preexponential factor (collision frequency), gas constant, and absolute temperature, respectively. E_a is the activation energy that is related to the enthalpy of formation of the transition state complex, $\Delta H^{\ddagger}$, of the reaction $E_a = \Delta H^{\ddagger} + RT$. The lowering of the activation energy of an enzymatic reaction is achieved by the introduction into the reaction pathway of a number of reaction intermediate(s).

pH Effect: Estimation of pKₐ Value(s). Some of the possible effects that are caused by a change in pH are:

1. Change in the ionization of groups involved in catalysis
2. Change in the ionization of groups involved in binding the substrate

3. Change in the ionization of substrate(s)

4. Change in the ionization of other groups in the enzyme

5. Denaturation of the enzyme

The pH effect on kinetic parameters (pH-rate/binding profile) may provide useful information on the ionizaing groups of the enzyme if the kinetic studies are carried out with nonionizable substrate in the pH region (pH 5–9) where enzyme denaturation is minimum. If the Michaelis constant (K) and/or the maximum velocity (V) vary with pH, the number and pK values of the ionizing group(s) can be inferred from the shape of pH-rate profile (pH versus pK and pH versus log V plots), namely, full bell shape for two ionizing groups and half bell shape for one ionizing group (see Table 7.4).

The initial rate enzyme kinetics uses very low enzyme concentrations (e.g., 0.1 μM to 0.1 pM) to investigate the steady-state region of enzyme-catalyzed reactions. To investigate an enzymatic reaction before the steady state (i.e., transient state), special techniques known as transient kinetics (Eigen and Hammes, 1963) are employed. The student should consult chapters of kinetic texts (Hammes, 1982; Robert, 1977) on the topics. KinTekSim (http://www.kintek-corp.com/kinteksim.htm) is the Windows version of KINSIM/FITSIM (Frieden, 1993) which analyzes and simulate enzyme-catalyzed reactions.

7.3. SEARCH AND ANALYSIS OF ENZYME DATA

7.3.1. Search for Enzyme Database

The ENZYME nomenclature database (Figure 7.3) of ExPASy (Expert Protein Analysis System) at http://www.expasy.ch/enzyme/ can be searched by entering EC number or enzyme names. The query returns information on EC number, enzyme name, catalytic activity, cofactors (if any) and pointers to Swiss-Prot sequence, ProSite, and human disease(s) of the enzyme deficiency.

LIGAND (Goto et al., 2002) at http://www.gebine.ad.jp/dbget/ligand.html is a composite database comprising

TABLE 7.4. pH Effects on Enzyme Kinetics

pH-Rate Profile	Rate Expression	Diagnostic	
		Low pH	High pH
Full bell	$K = K_m/(H/K_{1e} + 1 + K_{2e}/H)$	$\log K = \log K_m - pK_{1e} + pH$	$\log K = \log K_m + pK_{2e} - pH$
Left half bell	$K = K_m/(H/K_{1e} + 1)$	$\log K = \log K_m - pK_{1e} + pH$	
Right half bell	$K = K_m/(1 + K_{2e}/H)$		$\log K = \log K_m + pK_{2e} - pH$
Full bell	$V = V_m/(H/K_{1es} + 1 + K_{2es}/H)$	$\log V = \log V_m - pK_{1es} + pH$	$\log V = \log V_m + pK_{2es} - pH$
Left half bell	$V = V_m/(H/K_{1es} + 1)$	$\log V = \log V_m - pK_{1es} + pH$	
Right half bell	$V = V_m/(1 + K_{2es}/H)$		$\log V = \log V_m + pK_{2es} - pH$

Note: K_{1e} and K_{2e} or K_{1es} and K_{2es} are ionizing group(s) in the free enzyme or the enzyme–substrate complex, respectively.

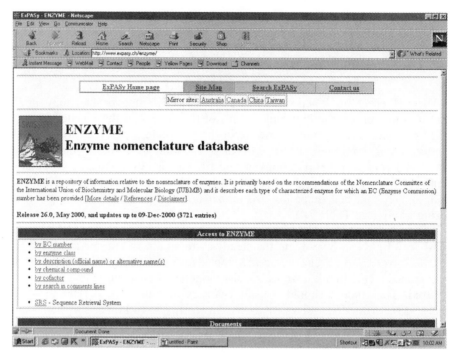

Figure 7.3. Enzyme nomenclature database of ExPASy.

- *Enzyme* for the information on enzyme molecules and enzymatic reactions
- *Compound* for the information on metabolites
- *Reaction* for the collection of substrate–product relationships

Select Search enzymes and compounds under DBGet/LinkDB Search to open query page (Figure 7.4). Enter the enzyme name or the substrate name in the *bfind* mode and click the Submit button. From the list of hits, select the desired entry by clicking the EC name. This returns information on name, class, reaction, pointers to structures of substrates/products/cofactor, links to pathway for which the selected enzyme is the member enzyme of the pathway, and related databases.

BRENDA (Schomburg et al., 2002) is the comprehensive enzyme information system that can be accessed at http://www.brenda.unikoeln.de/. Select New Query Forms to initiate Search by EC number, by Enzyme name or by Organism (Figure 7.5). Enter the EC number or the enzyme name (you can use * for partial name, e.g., *kinase) and click Query. From the list of hits (only one entry is returned with EC number), select the desired entry by clicking the EC number. The request returns the following information (the available information differs with each enzyme, e.g., for lysozyme): EC number, Organism, Systematic name, Recommended name, Synonyms, CAS registry number, Reaction, Reaction type, Substrates/products, Natural substrate, Turnover number [1/min], Specific activity [μmol/min/mg], K_m value [mM], pH optimum, pH range, Temperature optimum [$^\circ$C], Temperature range [$^\circ$C], Cofactors, Prosthetic groups, Activating substances, Metal/ions, Inhibitors, Source/tissue, Localization, Purification, Crystallization, Molecular weight, Subunits,

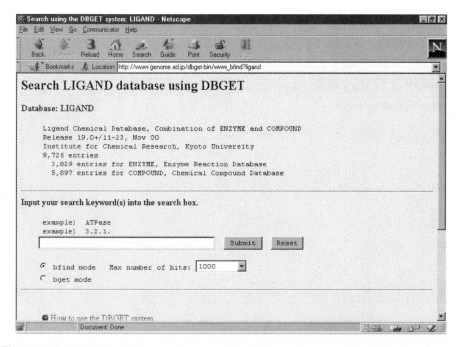

Figure 7.4. LIGAND database. A composite database for searching/retrieving enzyme information by enzyme name, EC number, substrates/products, and reactions.

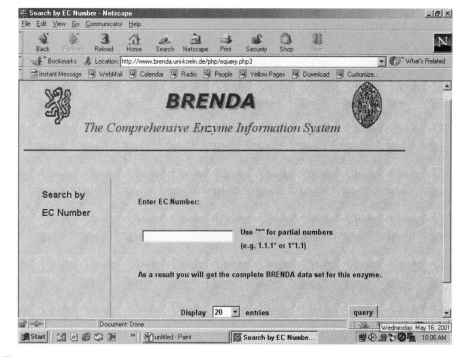

Figure 7.5. Search enzyme information at Brenda. Brenda is the comprehensive enzyme database for retrieving chemical, kinetic, and structural properties of enzymes via EC number, enzyme name, and organism (biological source). The search page by EC number is shown.

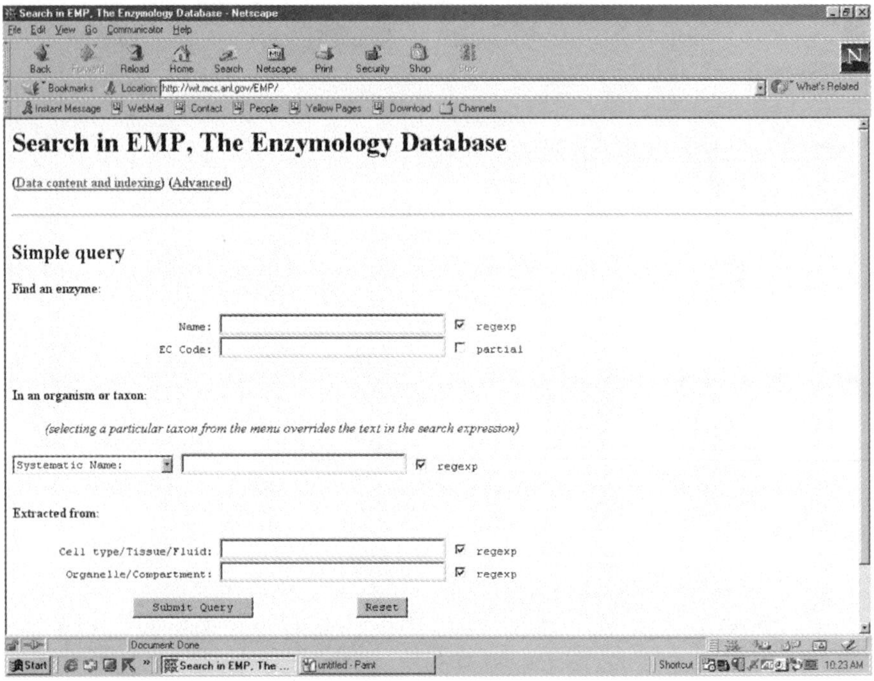

Figure 7.6. Enzymology database, EMP. The enzyme data published in literatures can be searched/retrieved from EMP via enzyme name/EC code and biological sources.

Cloned, pH stability, Temperature stability [°C], Organic solvent stability, Oxidation stability, General stability, Storage stability, Renatured, and Links to other databases and references.

EMP at http://wit.mcs.anl.gov/EMP/ is the resource site for summarized enzyme data that have been published in the literature. The site opens with the Simple query form (Figure 7.6). Enter the enzyme name into the name query box of 'Find an enzyme,' select 'Common name,' then enter the common organism name for 'In an organism or taxon,' and enter tissue name in response to 'Extracted from.' Clicking Submit Query returns an itemized summary of published enzyme data (data from one article may appear in more than one entries for different substrates) including concise assay and purification procedures, kinetic equations and kinetic parameters.

The Enzyme Structure Database (http://www.biochem.ucl.ac.uk/bsm/enzymes/ index.html), which contains the known enzyme structures of PDB, can be searched via EC number hierarchically (Figure 7.7). The search returns a list of individual pdb files with a link to CATH and pointers to PDBsum, ExPaSy, KEGG, and WIT. Clicking PDBsum opens the PDBsum page, which contains descriptive headings of enzyme, CATH classification, amino acid sequence with secondary structure designation, clickable PROMOTIF summary, TOPS (protein topology cartoon of related representative enzyme), PROSITE patterns, MolScript picture, as well as graphical presentations of ligand/ligand–active-site interactions. Click LIGPLOT of interactions (under Ligand) to display ligand–active site interactions (Figure 7.8). Pressing the RasMol button changes the view window into the RasMol window (if RasMol

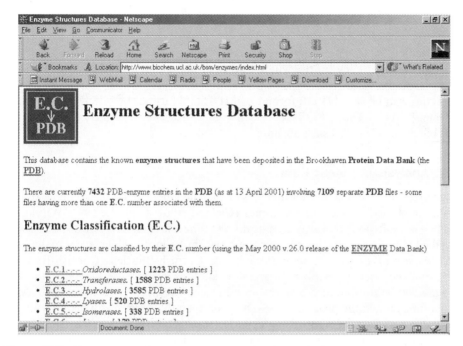

Figure 7.7. Home page of enzyme structure database. Links to enzyme structure and analysis servers are available at the Enzyme Structure Database which extracts and collects enzyme structures from pdb files.

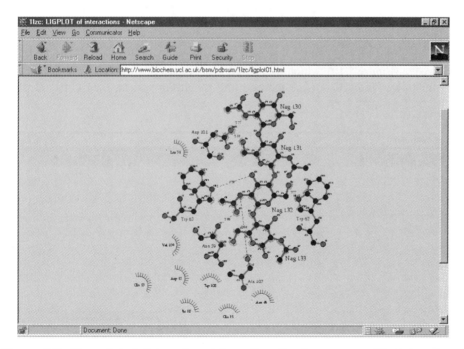

Figure 7.8. The substrate interaction at the active site of an enzyme. The interaction of tetra-*N,N,N,N*-acetylchitotetraose (NAG$_4$) with amino acid residues at the active site of lysozyme (1LZC.pdb) can be viewed/saved at PDBsum server (Enyme Structure Database→PDBsum→LIGPLOT of interactions under Ligand) linked to the Enzyme Structure Database.

is installed). Right click on the window to open the menu box (File, Edit, Display, Color, Options, and Rotation). The coordinate file of the ligand–active-site interaction can be saved (PDB format or MDL mol format) by selecting File→Save Molecule As. The PDBsum can be accessed directly at http:// www.biochem.ucl.ac.uk/bsm/pdbsum/.

7.3.2. Analysis of Kinetic Data

For the statistical and computer analysis of enzyme kinetic data, the students should consult published articles on the topics (Cleland, 1967; Crabble, 1992; Wilkinson, 1961). The software DynaFit, applicable to enzyme kinetic analysis, has been described (Chapter 6). In this chapter the program Leonora, which accompanies the text *Analysis of Enzyme Kinetic Data* by A. Cornish-Bowden (Cornish-Bowden, 1995), will be used to perform regression analysis of enzyme kinetic data. The software can be downloaded from http://ir2lcb.cnrs-mrs.fr/athel/leonora0.htm. After installation, launch the program (MS-DOS) to open the Main menu providing a list executable commands (Figure 7.9). Type **D** (select **D**ata) to bring the Data menu, and type **I** (select **I**nput new data) to enter kinetic data. Use Tab key to move across the columns and arrow keys to move up and down the rows. Enter label on the row 1 and data for the others. Press Esc to complete the data entry. Furnish short description for the Title, type **N** to enter filename, and save the data file (.mmd). Type **X** to exit the Data menu and return to the Main menu. Type **Q** (Equation) to select the appropriate rate equation from the pop-up Model menu (the list differs depending on kinetic data file). The menu entries of the equations are listed in Table 7.5.

To save the kinetic results, key in **O** for **O**utput requirement and then **R** for **R**esult page (Figure 7.10), which lists fitted kinetic parameters. Exit to the Calculations menu by typing **C**. Define **m**ethod and **w**eighting system, then **C** to calculate

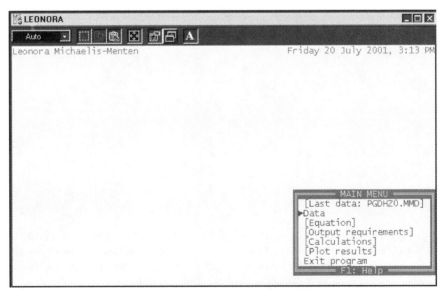

Figure 7.9. Starting page of Leonora.

TABLE 7.5. Representative Menu Entries of Kinetic Equations in Leonora

Equations by Name	Algebraical Equations
Michaelis–Menten	M: $v = V[S]/(K_m + [S])$
Substrate inhibition	S: $v = V[S]/(K_m + [S](1 + [S]/K_{si}))$
Michaelis-Menten (ignoring [I])	M: $v = V[S]/(K_m + [S])$
Primary Michaelis–Menten (at each [I])	P: $v = V_{app}[S]/(K_{m,app} + [S])$
Generic inhibition, (at each [S])	G: $v = v^0/(1 + [I]/K_i)$
Competitive inhibition	C: $v = V[S]/(K_m(1 + [I]/K_{ic}) + [S])$
Uncompetitive inhibition	U: $v = V[S]/(K_m + [S](1 + [I]/K_{iu}))$
Mixed inhibition	I: $v = V[S]/(K_m(1 + [I]/K_{ic}) + [S](1 + [I]/K_{iu}))$
Michaelis–Menten (ignoring [B])	M: $v = V[A]/(K_m + [A])$
Michaelis–Menten (ignoring [A])	I: $v = V[B]/(K_m + [B])$
Primary Michaelis–Menten (at each [B])	P: $v = V_{app}[A]/(K_{m,app} + [A])$
Primary Michaelis–Menten (at each [A])	R: $v = V_{app}[A]/(K_{m,app} + [A])$
Substituted enzyme mechanism	S: $v = V[A][B]/(K_{mB}[A] + K_{mA}[B] + [A][B])$
Ternary-complex mechanism	T: $v = V[A][B]/(K_{AB} + K_{mB}[A] + K_{mA}[B] + [A][B])$
Ordered equilibrium mechanism	O: $v = V[A][B]/(K_{AB} + K_{mB}[A] + [A][B])$
S-shaped pH profile	S: $k = K_{lim}/(1 + [H^+]/K_1)$
Z-shaped pH profile	Z: $k = K_{lim}/(1 + K_2/[H^+])$
Bell-shaped pH profile	B: $k = K_{lim}/(1 + [H^+]/K_1 + K_2/[H^+])$

Notes: Reprinted from table 9.1 (p. 156) from Analysis of Enzyme Kinetic Data by Athel Cornish-Bowden (1995) by permission of Oxford University Press.
1. The default Michaelis–Menten refers to uni uni or uni bi rate equation.
2. Mixed inhibition and noncompetitive inhibition can be used interchangeably.
3. Substituted-enzyme mechanism refers to ping pong bi bi mechanism.
4. Ternary-complex mechanism refers to order bi bi mechanism.
5. Michaelis–Menten, ignoring [X] refers to kinetic treatment of the bi bi reaction, A + B by ignoring the X (either A or B).
6. S-Shaped, Z-shaped and bell-shaped pH profiles refer to right-half, left-half and full bell profiles respectively.

best fit. To view the graphical results, type **P** for **P**lot results (which becomes active after calculations). Define **A**xes and **S**cale ranges. Use Tab key to move between *abscissa* and *ordinate*, and use arrow keys to define plotting parameters (e.g., 1/v). Type **P** to Plot.

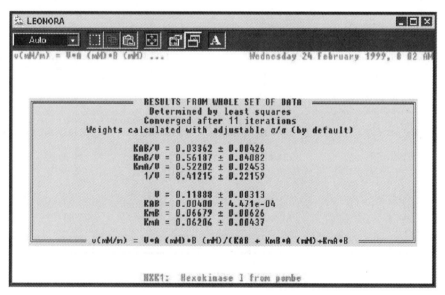

Figure 7.10. Result page of Leonora.

7.4. WORKSHOPS

1. The active-site-directed inhibition of enzymes has been an important research topic in pharmaceutical drug design (Sandler, 1980). An early development of anti-cancer agents involved inhibitions of dihydrofolate reductase and thymidylate synthetase. Search enzyme resource sites for kinetic data (turnover number, K_m and K_i) of these two enzymes.

2. Enzymes are highly selective of the substrates with which they interact and in the reactions that they catalyze. This selective nature of enzymes collectively known as enzyme specificity can be best illustrated with oxidoreductases (dehydrogenases), which display substrate and bond specificities (e.g., acting on —CHOH—, versus —CHO versus —CH—CH— versus —CHNH$_2$, and cis versus $trans$ for unsaturated substrates), coenzyme specificity (e.g., NAD(H) versus NADP(H)), chiral stereospecificity (D- versus L- or R- versus S-stereoisomers), and prochiral stereospecificity (A versus B corresponding to proR- versus proS isomers and re face versus si face, respectively). The table lists some dehydrogenases and their coenzyme, substrate, product and stereospecificities (You, 1982):

Oxidoreductase	Source	Substrate	Product	Coenzy	Stereo
Alcohol DH	Horse liver	Ethanol	Acetaldehyde	NAD	A
Alcohol DH	Yeast	Ethanol	Acetaldehyde	NAD	A
Alcohol DH(Aldehyde, RD)	Fruit fly	Glycerol	D-Glyceraldehyde	NADP	A
Alcohol DH(Aldehyde, RD)	Human liver	Glycerol	D-Glyceraldehyde	NADP	A
Homoserine DH	Pea	L-Homoserine	L-Adpβsemiald	NADP	B
Glycerol DH	*Aerobacter aerogenes*	Glycerol	DiOHacetone	NAD	A
Glycerol DH	Rabbit muscle	Glycerol	D-Glyceraldehyde	NADP	A
Glycerol-3P DH	*E. coli*	SnGlycerol-3P	DiOHacetone P	NADP	B
Glycerol-3P DH	Rabbit muscle	SnGlycerol-3P	DiOHacetone P	NAD	B
XylitolDH (Xylu RD)	Yeast	Xylitol	D-Xylulose	NAD	B
Xylitol DH(Xylu RD)	Pigeon liver	Xylitol	L-Xylulose	NADP	B
Mannitol-1P DH	*E. coli*	D-Mannitol-1P	D-Fructose-6P	NAD	B
Polyol DH (Aldose RD)	Human placenta	D-Sorbitol	D-Glucose	NADP	A
UDPGlucose DH	Beef liver	UDPGlucose	UDPGlucuronate	NAD	B
Shikimate DH	Pea	Shikimate	5-DeHshikimate	NADP	A
L-Lactate DH	*Lactobacillus arabinosus*	L-Lactate	Pyruvate	NAD	A
D-Lactate DH	*Lactobacillus arabinosus*	D-Lactate	Pyruvate	NAD	A
Glycerate DH	Spinach	D-Glycerate	Hydroxyacetone	NAD	A
3-Hydroxybutyrate DH	Beef heart	Dβ-OHbutyrate	Acetoacetate	NAD	B
Malate DH	Pig heart	L-Malate	Oxaloacetate	NAD	A
Malate DH (decarboxyl)	Pigeon liver	L-Malate	Pyruvate	NADP	A
Isocitrate DH	Pea	*threo*-Isocitrate	α-Ketoglutarate	NAD	A
Isocitrate DH	Pea	*threo*-Isocitrate	α-Ketoglutarate	NADP	A
P-Gluconate DH	Yeast	6P-D-Gluconate	D-Ribulose-5P	NADP	B
Glucose DH	Beef liver	β-D-Glucose	DGluc-δ-lactone	NAD	B
Galactose DH	*Pseud. fluorescens*	β-D-Galactose	DGal-γ-lactone	NAD	B
Glucose-6P DH	*L. mesenteroides*	D-Glucose-6P	DGlu-δlactone6P	NAD	B
Glucose-6P DH	Yeast	D-Glucose-6P	DGlu-δlactone6P	NADP	B
Aryl alcohol DH	Rabbit kid cortex	Benzyl alcohol	Benzaldehyde	NADP	B
P-Glycerate DH	*E. coli*	P-OHPyruvate	Glycerate-3P	NAD	A
Carnitine DH	*Pseud. aeruginosa*	Carnitine	3-DeHCarnitine	NAD	B

Oxidoreductase	Source	Substrate	Product	Coenzy	Stereo
L-Fucose DH	Sheep liver	L-Fucose	L-Fuc-δ-lactone	NAD	B
Sorbose DH	Yeast	L-Sorbose	5Keto-Dfructose	NADP	A
Aldehyde DH	Yeast	Acetaldehyde	Acetate	NAD	A
Acetaldehyde DH (acyl)	*Clost. kluyveri*	Acetaldehyde	Acetyl CoA	NAD	A
Asp-semialdehyde DH	*E. coli*	L-Asp-βsemaldehyde	L-β-Asp-P	NADP	B
Glyceraldehyde-P DH	Rabbit muscle	D-Glycerald-3P	D-3-P Glycerate	NAD	B
Glyceraldehyde-P DH	Pea	D-Glycerald-3P	1,3-diPGlycerate	NAD	B
Succinate semiald DH	*Psudomonas*	Succ semialdehyde	Succinate	NAD	A
Inosine monoP DH	*Aerob. aerogenes*	5'-IMP	5'-XMP	NAD	B
Xanthine DH	Chicken liver	Xanthine	Urate	NAD	B
Cortisone β-RD	Rat liver, soluble	Testosterone	5βHtestosterone	NADPH	A
Cortisone β-RD	Rat liver, micros	Progesterone	5βPregnandione	NADPH	A
Cortisone α-RD	Rat liver, soluble	Testosterone	5αHtestosterone	NADPH	B
Cortisone α-RD	Rat liver, micros	Progesterone	5αPregnandione	NADPH	B
Meso-Tartrate DH	*Pseud. putida*	*meso*-Tartrate	DiOH Fumarate	NAD	A
Acyl CoA DH	Rat liver, mictochondria	Octanoyl CoA	Oct-2-enoyl CoA	NADP	A
Alanine DH	*Bacillus subtilis*	L-Alanine	Pyruvate	NAD	A
Glutamate DH	Pea mitochondria	L-Glutamate	α-Ketoglutarate	NAD	B
Glutamate DH	Yeast	L-Glutamate	α-Ketoglutarate	NADP	B
Dihydrofolate RD	Chicken liver	Tetrahydrofolate	Dihydrofolate	NADP	A
Glutathione RD	Yeast	ox Glutathione	red Glutathione	NADPH	B
Lipoamide DH	Pig heart	DiHlipoamide	α(+)-Lipoamide	NAD	B
Nitrate RD	Spinach	NO_3^-	NO_2^-	NADH	A
Nitrite RD	Yeast	NO_2^-	NH_2OH	NADPH	B
Hydroxylamine RD	Yeast	NH_2OH	NH_3	NADPH	B

Note: Abbreviations used are: DH, dehydrogenase; RD, reductase; coenzy, coenzyme; stereo, stereospecificity; ox, oxidized; red, reduced; H, hydro-; OH, hydroxy-; and P, phospho-/phosphate.

Source: Partially reproduced with the permission from Academic Press.

Apply Microsoft Access to design a database appended with queries for retrieving groups of dehydrogenases according to their coenzyme specificity and stereospecificity.

3. Search enzyme databases for information to construct a database of glucosidases (EC 3.2.1.x) with retrievable fields on substrate (anomeric) specificity, catalytic mechanism (stereochemical, e.g., inversion versus retention) and kinetic constants (e.g., K_m and V).

4. The catalytic residues of serine proteases such as chymotrypsin generally involve catalytic triad of Asp, His, and Ser residues, for example,

Search the Enzyme Structure Database for γ-chymotrypsin active site (by the aid of the active-site-modified enzyme or active-site-specific inhibitor–enzyme complex) to identify and depict (save pdb file) the catalytic triad of γ-chymotrypsin.

5. Initial rates of a hydrolase (10.0 nM)-catalyzed reaction are measured and tabulated overleaf. Evaluate the kinetic parameters of this Uni-substrate reaction. Calculate the turnover number of the enzyme.

Substrate (mM)	Rates (μM/min)
1.00×10^{-2}	0.179
2.00×10^{-2}	0.305
5.00×10^{-2}	0.534
1.00×10^{-1}	0.718
2.00×10^{-1}	0.870
5.00×10^{-1}	0.981

6. Initial rates of an esterase-catalyzed reactions in the absence and presence of 0.10 mM each of inhibitors I and J are measured and tabulated below. Evaluate the kinetic and inhibition parameters of this Uni-substrate reaction.

Ester (mM)	Rates (μM/min)				
	No Inhibitor	I (mM)		J (mM)	
		0.10	0.25	0.10	0.25
1.00×10^{-2}	0.114	0.099	0.072	0.097	0.084
2.00×10^{-2}	0.190	0.169	0.128	0.163	0.141
5.00×10^{-2}	0.321	0.291	0.238	0.272	0.239
1.00×10^{-1}	0.414	0.385	0.340	0.361	0.309
2.00×10^{-1}	0.485	0.465	0.424	0.417	0.362
5.00×10^{-1}	0.542	0.526	0.483	0.467	0.392

7. The steady-state kinetic studies of liver alcohol dehydrogenase (12.5 nM) are performed. The initial rates (v in μM/min) with varying substrate concentrations in both directions (forward for ethanol oxidation and reverse for ethanal reduction) are given below. Evaluate their kinetic parameters and equilibrium constant.

NAD$^+$ (mM) Ethanal (mM)	0.10	0.20	0.50	0.80
0.50	0.16	0.19	0.22	0.26
1.0	0.24	0.28	0.30	0.34
2.0	0.31	0.37	0.42	0.49
5.0	0.38	0.45	0.52	0.60
10	0.45	0.52	0.63	0.72

NADH (mM) Ethanal (mM)	0.010	0.020	0.040	0.050
0.50	0.042	0.049	0.056	0.062
1.0	0.050	0.059	0.067	0.074
2.0	0.054	0.063	0.073	0.084
5.0	0.063	0.074	0.086	0.098

8. The initial rates (v in $\mu M/min$) of liver alcohol dehydrogenase-catalyzed ethanal reduction are measured in the presence of pyrazole as an inhibitor at the constant concentration of NADH (0.02 M) and the constant concentration of ethanal (2.0 mM), respectively. Propose respective inhibition types and estimate their inhibition constants.

Pyrazole (μM)	Ethanal (mM) 0.50	1.0	2.0	5.0
0	0.034	0.040	0.044	0.050
1.0	0.027	0.030	0.033	0.036
2.0	0.022	0.024	0.026	0.028
5.0	0.008	0.010	0.011	0.013

Pyrazole (μM)	NADH (mM) 0.010	0.020	0.050	0.10
0	0.032	0.040	0.046	0.052
1.0	0.022	0.024	0.027	0.030
2.0	0.016	0.018	0.020	0.022
5.0	0.006	0.008	0.010	0.011

9. Kinetic studies of an esterase catalysis are carried out at various pH values and initial rates (v in $\mu M/sec$) of hydrolysis are tabulated below:

pH	Ester (mM) 0.01	0.02	0.05	0.10	0.20
5.5	8.04×10^{-3}	1.29×10^{-2}	1.84×10^{-2}	2.36×10^{-2}	2.48×10^{-2}
6.0	2.95×10^{-2}	4.40×10^{-2}	6.13×10^{-2}	7.27×10^{-2}	7.91×10^{-2}
6.5	0.105	0.158	0.225	0.263	0.279
7.0	0.227	0.342	0.498	0.559	0.625
7.5	0.296	0.455	0.625	0.735	0.826
8.0	0.256	0.481	0.690	0.794	0.847
8.5	0.190	0.282	0.412	0.485	0.505
9.0	9.62×10^{-2}	0.144	0.205	0.238	0.256
9.5	3.02×10^{-2}	4.53×10^{-2}	6.46×10^{-2}	7.54×10^{-2}	8.14×10^{-2}

Perform data analysis and evaluate pK value(s) of ionizing group(s).

10. Quantitative structure–activity relationship (QSAR) (Hansch and Klein, 1986; Hansch and Leo, 1995) represents an attempt to correlate structural descriptors of compounds with activities. The physicochemical descriptors include numerical parameters to account for electronic properties, steric effect, topology, and hydrophobicity of analogous compounds. In its simplest form, the biochemical activities are correlated to the numerical substituent descriptors of analogous compounds tested by a linear equation such as

$$\log \mathrm{k} \text{ (or } K \text{ or } 1/C) = \rho\sigma + \delta E_s + \phi\pi + d$$

where k, K, and C are rate constant, binding constant, and molar concentration producing a standard biological response in a constant time interval by the compound. σ (σ for aromatic compounds and σ^* for aliphatic compounds), E_s, and π are the most widely used descriptors for electronic property, steric hindrance, and hydrophobicity associated with the substituent of congeners under investigation. The correlation equation is solved by the regression analysis to yield correlation coefficients for the electronic (ρ), steric (δ), and hydrophobic (ϕ) effects of the compounds on the measured biochemical activities.

Kinetic studies of α-chymotrypsin catalyzed hydrolysis of p-nitrophenyl esters are carried out (Duprix et al., 1970):

$$R-\overset{\overset{\textstyle O}{\|}}{O}-O-\!\!\!\left\langle\!\!\!\bigcirc\!\!\!\right\rangle\!\!\!-NO_2 + H_2O \rightleftharpoons R-\overset{\overset{\textstyle O}{\|}}{C}-OH + HO-\!\!\!\left\langle\!\!\!\bigcirc\!\!\!\right\rangle\!\!\!-NO_2$$

The rate constants for deacylation step (k_{da}) and common descriptors associated with substituents (R) are summarized below:

Substituent, R	σ	E_s	π	$\log k_{da}$
H	0.49	1.10	0.00	0.18
CH_3	0.00	0.00	0.50	-2.00
$(CH_3)_2CH$	-0.19	-0.47	1.30	-2.47
$(CH_3)_3C$	-0.30	-1.54	1.68	-3.74
CH_3OCH_2	0.64	-0.19	0.03	-0.47
ICH_2	0.85	-0.37	1.50	-0.24
$ClCH_2$	1.05	-0.24	0.89	-0.42
$Cl(CH_2)_2$	0.38	-0.90	1.39	-1.68
$Cl(CH_2)_3$	0.14	-0.40	1.89	-1.29
$Cl(CH_2)_4$	0.05	-0.40	2.39	-1.35
$C_6H_5CH_2$	0.21	-0.38	2.63	-1.73
$C_6H_5(CH_2)_2$	0.08	-0.38	3.13	-0.75
$C_6H_5(CH_2)_3$	0.02	-0.45	3.63	-0.92
$C_6H_5(CH_2)_4$	0.02	-0.45	4.13	-1.73

Perform the regression analyses for the descriptors to assess the contribution of substituent effect(s) on the rate of α-chymotrypsin-catalyzed hydrolysis of p-nitrophenyl esters. Referring to the catalytic triad of chymotrypsin, rationalize your results for the plausible reaction mechanism.

REFERENCES

Ainsworth, S. (1977) *Steady-State Enzyme Kinetics.* University Park Press. Baltimore, MD.

Bardsley, W. G., and Childs, R. E. (1975) *Biochem. J.* **149**:313–328.

Bardsley, W. G., and Waight, R. D. (1978) *J. Theor. Biol.* **70**:135–156.

Benkovic, S. J. (1992) *Annu. Rev. Biochem.* **61**:29–54.

Bisswanger, H., and Schmincke-Ott, E., Eds. (1980) *Multifunctional Proteins.* John Wiley & Sons, New York.

Cleland, W. W. (1963a) *Biochim. Biophys. Acta* **67**:104–172.

Cleland, W. W. (1963b) *Biochim. Biophys. Acta* **67**:188–196.

Cleland, W. W. (1967) *Adv. Enzymol.* **29**:1–32.

Copeland, R. A. (2000) *Enzymes: A Practical Introduction to Structure, Mechanism and Data Analysis*, 2nd edition. John Wiley & Sons, New York.

Cornish-Bowden, A. (1995) *Analysis of Enzyme Kinetic Data.* Oxford University Press, Oxford.

Crabble, M. J. (1992) in *Microcomputer in Biochemistry* (Bryce, C. F. A., Ed.). IRL Press, pp. 107–150.

Dalziel, K., and Dickinson, F. M. (1966) *Biochem. J.* **100**:34–46.

Doherty, E. A., and Doudna, J. A. (2000) *Annu. Rev. Biochem.* **69**:597–615.

Duprix, A., Béchet, J. J., and Roucous, C. (1970) *Biochem. Biophys. Res. Commun.* **41**:464–470.

Eigen, M., and Hammes, G. (1963) *Adv. Enzymol.* **25**:1–38.

Fersht, A. (1985) *Enzyme Structure and Mechanism*, 2nd edition. W. H. Freeman, New York.

Freeman, R. B., and Hawkins, H. C., Eds. (1980, 1985) *The Enzymology of Post-translational Modifications of Proteins*, Vols. 1 and 2. Academic Press, New York.

Frieden, C. (1993) *Trends Biochem. Sci.* **18**:58–60.

Fromm, H. J. (1975) *Initial Rate Enzyme Kinetics.* Springer-Verlag, New York.

Goto, S., Okuno, Y., Hattori, M., Nishioka, T., and Kanehisa, M. (2002) *Nucleic Acids Res.* **30**:402–404.

Gottesman, S. (1984) *Annu. Rev. Genet.* **18**:415–441.

Hammes, G. G. (1982) *Enzyme Catalysis and Regulation.* Academic Press, New York.

Hansch, C., and Klein, T. E. (1986) *Acc. Chem. Res.* **19**:396–400.

Hansch, C., and Leo, A. (1995) *Exploring QSAR: Fundamental and Applications in Chemistry and Biochemistry.* American Chemical Society, Washington, D.C.

Hanson, K. R. (1966) *J. Am. Chem. Soc.* **88**:2731–2742.

Hilvert, D. (2000) *Annu. Rev. Biochem.* **69**:751–793.

King, E. L., and Altman, C. (1956) *J.Phys. Chem.* **60**:1375–1381.

Koshland, D. E. Jr. (1958) *Proc. Natl. Acad. Sci. U.S.A.* **44**:98–104.

Koshland, D. E. Jr. (1960) *Adv. Enzymol.* **22**:45–97.

Kuby, S. (1991) *Enzyme Catalysis, Kinetics and Substrate Binding.* CRC Press, Boca Raton, FL.

Li, Y., and Breaker, R. R. (1999) *Curr. Opin. Struct. Biol.* **9**:315–323.

Markert, C. L., Ed. (1975) *Isozymes*, Vols 1–4. Academic Press, New York.

Monod, J., Wyman, J., and Changeux, J. P. (1965) *J. Mol. Biol.* **12**:88–118.

Perham, R. N. (2000) *Annu. Rev. Biochem.* **69**:961–1004.

Perutz, M. (1990) *Mechanisms of Cooperativity and Allosteric Regulation in Proteins.* Cambridge University Press, Cambridge.

Plowman, K. M. (1972) *Enzyme Kinetics.* McGraw-Hill Book New York.

Price, N. C., and Stevens, L. (1982) *Fundamentals of Enzymology*, 3rd edition. Oxford University Press, Oxford.

Reed, L. J. (1974) *Acc. Chem. Res.* **7**:40–46.

Ricard, J., and Cornish-Bowden, A. (1987) *Eur. J. Biochem.* **166**:255–272.

Robert, D. V. (1977) *Enzyme Kinetics.* Cambridge University Press, Cambridge.

Sandler, M. (1980) *Enzyme Inhibitors as Drugs.* University Park Press, Baltimore.

Schomburg, I., Chang, A., and Schomburg, D. (2002) *Nucleic Acids Res.* **30**:47–49.

Schulz, A. R. (1994) *Enzyme Kinetics*. Cambridge University Press, Cambridge.

Scott, W. G., and Klug, A. (1996) *Trends Biochem. Sci.* **21**:220–224.

Segel, I. H. (1975) *Enzyme Kinetics*. Wiley–Interscience. New York.

Sheppard, T. L., Ordoukhanian, P., and Joyce, G. F. (2000) *Proc. Natl. Acad. Sci. U.S.A.* **97**:7802–7807.

Srere, P. A. (1987) *Annu. Rev. Biochem.* **56**:89–124.

Tsai, C. S. (1978) *Biochem. J.* **173**:483–496.

Wilkinson, G. N. (1961) *Biochem. J.* **80**:324–332.

You, K. (1982) *Meth. Enzymol.* **87**:101–126.

8

DYNAMIC BIOCHEMISTRY: METABOLIC SIMULATION

Almost the same biochemical reactions occur in all living organisms. These reactions, known as primary metabolism, are organized into catabolic pathways and anabolic pathways. In addition, specialized biochemical transformations, known as secondary metabolism, are found in specialized tissues of some organisms. Xenometabolism refers to biochemical reactions that transform foreign compounds. Internet resources for retrieving metabolic information are presented. Biochemical reactions are subject to metabolic regulations. Metabolic control analysis and metabolic simulation using Gepasi are introduced.

8.1. INTRODUCTION TO METABOLISM

8.1.1. Primary Metabolism

Metabolism represents the sum of the chemical changes that convert nutrients, the raw materials necessary to nourish living organisms, into energy and the chemically complex finished products of cells. Metabolism consists of a large number of enzymatic reactions organized into discrete reaction sequences/pathways (Dagley and Nicholson, 1970; Saier, 1987). The metabolic pathways that are common to all living organisms are sometimes referred to as *primary metabolism* or simply metabolism. The synthesis and degradation of small molecules, termed intermediary metabolism, comprises all reactions concerned with storing and generating metabolic energy and with using that energy in biosynthesis of low-molecular-weight compounds and energy storage compounds. Energy metabolism is that part of intermediary metabolism consisting of pathways that stores or generates metabolic energy. The Boehringer Mannheim chart of metabolic pathways is available as a series of images at http://expasy.houge.ch/cgi-bin/search-biochem-index. The information and links related to metabolic pathways can be obtained from KEGG (*Kyoto Encyclopedia of Genes and Genomes*) Pathway site at http://www.genome.ad.jp/dbget/ (Kanehisa et al., 2002).

The metabolism serves two fundamentally different purposes: the generation of energy and the synthesis of biological molecules. Thus metabolism can be subdivided into *catabolism*, which produces energy and reducing power, and *anabolism*, which consumes these products in the biosynthetic processes. Both catabolic and anabolic pathways occur in three stages of complexity: stage 1, the interconversion of biopolymers and complex lipids with monomeric intermediates; stage 2, the interconversion of monomeric sugars, amino acids, nucleotides and fatty acids with still simpler organic compounds; and stage 3, the ultimate degradation to, or synthesis from, raw materials including CO_2, H_2O and NH_3. Catabolism involves the oxidative degradation of complex nutrient molecules such as carbohydrates, lipids, and proteins. The breakdown of these molecules by catabolism leads to the formation of simpler molecules and generates the chemical energy that is captured in the form of ATP. Because catabolism is mostly oxidative, part of the chemical energy may be conserved in the coenzyme forms, NADH and NADPH. These two nicotinamide coenzymes have very different metabolic roles. NAD^+ reduction is part of catabolism, and NADPH oxidation is an important aspect of anabolism. The energy released upon oxidation of NADH is coupled to the phosphorylation of ADP to ATP. In contrast, NADPH is the source of the reducing equivalent needed for reductive biosynthetic reactions. Anabolism is a synthetic process in which the varied and complex biomolecules such as proteins, nucleic acids, polysacchrides, and lipids are assembled from simpler precursors. The ATP and NADPH provide the energy and reducing power, respectively, to drive endergonic and reductive anabolic reactions. Despite their divergent roles, anabolism and catabolism are interrelated in that the products of one provide the substrates of the other. Many metabolic intermediates are shared between the two processes that occur simultaneously in the cell. However, biosynthetic and degradative pathways are rarely, if ever, simple reversals of one another, even though they often begin and end with the same metabolites. The existence of separate pathways is important for energetic and regulatory reasons.

The living cell uses a marvelous array of regulatory devices to control its functions (Denton and Pogson, 1976; Martin, 1987; Stadtman, 1966). The mechanisms include those that act primarily to control enzyme concentrations at synthetic level via induction and repression, or protein degradation. The enzyme activity can be further controlled by the concentrations of metabolite/inhibitor, allosteric regulation, and protein modifications. In eukaryotic cells, compartmentation represents another regulatory mechanism with the fate of a metabolite being controlled by its flow through a membrane. Overlying all of these mechanisms are the actions of hormones, chemical messengers that act at all levels of regulation.

The flow of genetic information involves biosyntheses of DNA, RNA, and proteins, known as replication, transcription, and translation, respectively. DNA replication in prokaryotes (Nossal, 1983) and eukaryotes (Campbell, 1985) are very similar, though eukaryotic replication is more complex (DePamphilis, 1996). A number of enzymes mediate DNA synthesis (Kornberg, 1988), such as helicase and topoisomerase (unwinding and provision of the template), primase (synthesis of primer), DNA-dependent DNA polymerase (synthesis of polynucleotide chain), and ligase (joining of Okazaki fragments). DNA replication has the following general features (Kornberg and Baker, 1992):

1. The double helix is unwound so that the single strand of DNA molecule may serve as the template.
2. An oligonucleotide complementary to 3'-sequence of the template is required as the primer.
3. The incoming nucleotide is selected as directed by Watson–Crick base pairing with the template.
4. The polynucleotide chain elongates from the $5' \rightarrow 3'$ direction.
5. The replication is bidirectional from the origin of replication.
6. The replication is semiconservative — that is, continuous for the leading strand and fragmentational (Okazaki fragments) for the lagging strand.

Cells contain three major classes of RNA (mRNA, rRNA, and tRNA), all of which are synthesized from DNA templates by DNA-dependent RNA polymerase (Moldave, 1981), which binds to the promoters (typically ~ 40-bp region upstream of the transcription start site containing a hexameric TATA box). Prokaryotic and eukaryotic RNA transriptions show strong parallels, though there are several important differences.

1. Only one RNA polymerase catalyzes the synthesis of all three classes of RNA in prokaryotes. While three RNA polymerases mediate the synthesis of eukaryotic RNA, namely, RNA polymerase I for major rRNA, RNA polymerase II for mRNA, and RNA polymerase III for tRNA (Palmer and Folk, 1990; Woychik and Young, 1990).
2. Various regulatory proteins are involved in the eukaryotic transcription (Johnson and McKnight, 1989). The classification of transcription factors can be found at TRANFAC (http://transfac.gbf.de/TRANFAC/cl/cl.html) (Wingender et al., 2001).
3. Transcription and translation occur concomitantly in prokaryotes. In eukaryotes, the two processes are spatially and temporally disconnected. Eukaryotic transcription occurs on DNA in the nucleus, and translation takes place on ribosomes in the cytoplasm.
4. The eukaryotic transcript undergoes post-transcriptional modifications such as capping (5'-methyl-GTP cap) and 3'-polyadenylation (3'-polyA tail).
5. Prokaryotic mRNA is polycitronic, whereas eukaryotic mRNA tends to be monocitronic.
6. The eukaryotic transcript is the precursor of mRNA (pre-mRNA) consisting of coding regions (exons) and noncoding regions (introns). The mature mRNA is formed after splicing of pre-mRNA (Sharp, 1986).

Translation converts the genetic information embodied in the base sequence (codon) of mRNA into the amino acid sequence of a polypeptide chain on ribosomes. Protein biosynthesis (Arnstein and Cox, 1992; Moldave, 1981) is characterized by three distinct phases:

1. Initiation in prokaryotes involves binding of mRNA by small ribosomal subunit (30S), followed by association of the fMet-tRNA^{met} (initiator formyl-methionyl-tRNA^{met}) that recognizes the initiation codon. Large ribosomal subunit (50S) then joins to form the 70S initiation complex. In eukaryotes, the initiator aminoacyl-tRNA is not formylated. Instead Met-tRNA$_f^{met}$ forms 40S preinitiation complex with small ribosomal subunit (40S) in the absence of mRNA. The association of mRNA results in a 40S preinitiation complex, which forms an 80S initiation complex after large ribosomal subunit (60S) joins.

2. Elongation includes the synthesis of all peptide bonds of a polypeptide chain. This is accomplished by a repetitive cycle of events in which successive aminoacyl-tRNA add to the A site (acceptor site) of the ribosome–mRNA complex and the growing peptidyl-tRNA occupying the P site (peptidyl site). The elongation reaction transfers the peptide chain from the peptidyl-tRNA on the P site to the aminoacyl-tRNA on the A site by forming a new peptide bond. The E site (exit site) is transiently occupied by the deacylated tRNA as it exits the P site. The new longer peptidyl-tRNA moves from the A site into the P site as the ribosome moves one codon further along the mRNA. This translocation vacates the A site, which can accept the new incoming aminoacyl-tRNA.

3. Termination is triggered when the ribosome reaches a stop codon on the mRNA. At this stage, the polypeptide chain is released and the ribosomal subunits dissociate from the mRNA. Various protein factors are involved in all three phases of protein biosynthesis.

8.1.2. Secondary Metabolism

In addition to the primary metabolic reactions, which are similar in all living organisms, a vast number of metabolic pathways lead to the formation of compounds peculiar to a few species or even to a single chemical race only. These reactions are summed up under the term *secondary metabolism*, and their products are called secondary metabolites (Grierson, 1993; Herbert, 1989; Porter and Spurgeon, 1981, 1983; Stafford, 1990).

The wide variety of secondary products formed in nature includes such well-known groups as alkaloids, antibiotics, cardiac glycosides, tannins, saponins, volatile oils, and others. A considerable number of them are of economic importance in therapeutics or technology. Although secondary products are produced by microorganisms, plants, and animals, most of the substances are found in the plant kingdom. The lack of mechanisms for true excretion in higher plants may result in this unequal distribution, with the "waste products" of metabolism in plants instead being accumulated in the vacuoles, the cell walls, or in special excretory cells or spaces of the organism.

Secondary metabolism is characterized by the following:

1. It may be considered as an expression of the chemical individuality of organism.

2. It is usually restricted to specific developmental stage of the organism in the specialized cells.

3. It is regulated by the concerted action of the ability of the cell to respond to certain internal or external stimuli or the presence of the suitable effectors at the particular time.

4. Some of its products (secondary products) may be eliminated from primary metabolism and are of no importance to the living organisms producing them.

5. All secondary products are derived ultimately from compounds originated during primary metabolism such as sugars, acetate, isoprene, amino acids, and so on.

6. The biosynthetic mechanisms of secondary products are probably of physio-logical significance based on ecological demands.

7. Secondary metabolic reactions may be catalyzed by enzymes of primary metabolism, enzymes specific to secondary metabolism or enzymes that occur spontaneously.

Many secondary substances have, however, a direct biological function. They can be regulatory effectors (e.g., hormones or signal transducers) or they can be ecologically significant substances (e.g., pigments, substances involved in defense mechanisms, or factors enabling life in special ecological niches). The KEGG Pathway site at http://www.genome.ad.jp/dbget/ also provides information on me-tabolisms of terpenoids, flavonoids, alkaloids, and antibiotics.

8.1.3. Xenometabolism

Living organisms are exposed to a steadily increasing number of synthetic chemicals that are without nutritive value but nevertheless ingested, inhaled, or absorbed by the organisms. These compounds are foreign to the organisms and are called *xenobiotics*. These xenobiotics are detoxified by xenobiotic metabolizing enzymes in the processes known as *xenometabolism* (Copley, 2000; Gorrod et al., 1988). In animals, a main site of xenometabolism is the liver, where the normally lipophilic xenobiotics are metabolized to more soluble forms and then excreted in three phases. Phase 1 enzymes oxidize, reduce, or hydrolyze the xenobiotics, introducing a reactive group for the subsequent conjugation catalyzed by phase 2 transferases. In phase 3, the hydrophilic conjugates are excreted. Xenometabolism in plants can also be divided into three phases, although plants have no effective excretion pathways. Instead transformation (phase 1) and conjugation (phase 2) reactions are coupled to internal compartmentation and storage processes (phase 3). Cellular storage sites are vacuoles for soluble conjugates and are cell wall for insoluble conjugates.

Table 8.1 represents some of the enzymes (Jakoby and Ziegler, 1990) that participate in altering xenobiotics so as to make them either more readily excretable or less toxic for the storage. These enzymes have a preference for lipophilic compounds, although all hydrophilic compounds, are not excluded. Each enzyme has a phenomenally broad substrate range, and many appear to be inducible. The information for xenometabolism can be obtained from the University of Minnesota Biocatalysis/Biodegradation Database (UM-BBD) at http://umbbd.ahc.umn.edu/index.html (Ellis et al., 2000).

TABLE 8.1. Some Xenometabolic Enzymes[a]

Transformation Enzymes	Conjugation Enzymes
Flavin-containing monooxygenases	Glutathione transferase
P450-dependent monooxygenases	UDP-Glucuronyl transferase
Alcohol dehydogenase	Phenol sulfotransferase
Aldehyde dehydrogenase	Tyrosine ester sulfotransferase
Dihydrodiol dehydrogenase	Alcohol sulfotransferase
Glutathione peroxidase	Amine N-sulfotransferase
Monoamine oxidase	Cysteine conjugate N-acetyltransferase
Xanthine oxidase	Catechol O-methyltransferase
Carbonyl reductase	Amine N-methyltransferase
Quinone reductase	Thiol S-methyltransferase
Epoxide hydrolase	Thioltransferase
Amidases	Acetyltransferase
Esterases	Rhodanase

[a]The phase 1 transformation involves oxidoreductases and hydrolases, while the phase 2 conjugation is mediated by transferases.

8.2. METABOLIC CONTROL ANALYSIS

The computer simulation is one of the essential means to investigate dynamic and steady-state behavior as well as control of metabolic pathways. A metabolic simulator is a computer program that performs one or several of the tasks including solving the steady state of a metabolic pathway, dynamically simulating a metabolic pathway, or calculating the control coefficient of a metabolic pathway. Its mathematical model generally consists of a set of differential equations derived from rate equations of the enzymatic reactions of the pathway.

Metabolic control analysis (MCA) is the application of steady-state enzyme networks to the problem of the control of metabolic flux (Fell, 1992; Kacser and Burns, 1995). Consider a pathway:

$$S_0 \xrightarrow{E_1} S_1 --- \xrightarrow{E_i} S_i --- \xrightarrow{E_n} P$$

The importance of each step in the pathway ($E_1 -- E_i -- E_n$) in controlling the flow through that pathway (J) in a steady state is given by $(\delta J/J)/(\delta E_i/E_i) = C_i^J$, and a measure of the effect of any parameter (P) on J would be $(\delta J/J)/(\delta P/P) = R_P^J \cdot C_i^J$ is the *flux control coefficient*, which is defined as the fractional change in a pathway flux ($\delta J/J$) for an infinitestimal fractional change in the activity of the particular enzyme ($\delta E_i/E_i$) under study. Thus C_i^J are properties of the defined system. The change in the rate caused by the presence of inhibitor/activator can be thought of as equivalent to some change in the concentration (activity) of enzyme. R_P^J is called the *response coefficient*, and its value may be thought of as an overall measure of the control exerted by P at its value. Kinetic constants of enzymes such as turnover number or maximum velocity, Michaelis constants, inhibition constants, and so on, are parameters. The *response coefficient* can be usually separated into two parts: the *flow control coefficient* and *elasticity coefficient* (ε_S^i), which is concerned with the response of an isolated enzyme to a variable such as metabolite or effector whose concentration is set by the metabolic system, that is, $(\delta v_i/v_i)/(\delta S/S) = \varepsilon_S^i$. The *elasticity coefficient* is defined as the fractional change in the flux through the enzyme $(\delta v_i/v_i)$ caused by an infinitestimal fractional change in the concentration of the metabolite or effector $(\delta S/S)$. It is a measure of the extent to which the substrate or the effector has the potential to influence the flux.

The following relationships hold in MCA:

1. The *Flux Summation Theorem* states that the sum of all the flux control coefficients of any pathway is equal to unity:

$$\sum_{i=1}^{n} = C_i^J = 1$$

In linear pathways, individual flux control coefficient will normally lie between zero (no control) and 1 (full control). But in branched pathways, negative flux control coefficients arise where the stimulation of an enzyme in one branch may decrease the flux through a competing branch. This gives rise to values greater than 1 occurring in that pathway.

2. The *Connectivity Theorem* states that the flux control coefficient and elasticity are related, that is,

$$C_1^J/C_2^J = -\varepsilon_S^2/\varepsilon_S^1 \qquad \text{for the adjacent enzymes}$$

or

$$C_1^J : C_2^J : C_3^1 \cdots = \cdots \varepsilon_S^3 : \varepsilon_S^2 : \varepsilon_S^1$$

and therefore,

$$\sum C_i^J \varepsilon_S^i = 0$$

8.3. METABOLIC DATABASES AND SIMULATION

8.3.1. Search for Metabolic Pathways and Information

Metabolic databases serve as online reference sources making metabolic information readily accessible via the Internet. These databases typically describe collections of enzymes, reactions, and biochemical pathways with pointers to genetic, sequence, and structural servers. Table 8.2 lists some of metabolic databases.

In addition to the metabolic databases listed above, some of the enzyme databases described in the previous chapter (Chapter 6) also serve as useful metabolic resources. All of the enzyme and metabolic databases make use of EC (Enzyme Commission) numbers which are available at the International Union of Biochemistry and Molecular Biology (IUBMB) site (http://www.chem.qmw.ac.uk/iubmb/enzyme/).

KEGG Metabolic Pathways is a comprehensive metabolic database that can be accessed from Kegg home page at http://www.genome.ad.jp/kegg/ or DBGet at http://www.genome.ad.jp/dbget/. From the Kegg home page, select Open KEGG to

TABLE 8.2. Some Metabolic Databases

Database	URL
PathDB	http://www.ncgr.org/software/pathdb/
KEGG	http://www.genome.ad.jp/kegg/kegg.html
EcoCyc	http://ecocyc.panbio.com/ecocyc/
Soybase	http://cgsc.biology.yale.edu/
Biocatalysis/Biodegradation DB	http://www.labmed.umn.edu/umbbd/index.html
Boehringer Mannheim	http://expasy.houge.ch/cgi-bin/search-biochem-index

view the Kegg table of contents. Select either Metabolic Pathways or Regulatory Pathways to open the list of Pathways. Choosing the desired pathway returns the reference pathway in which enzymes (boxes) are represented by their EC numbers and metabolites (circles) are named (Figure 8.1). Some of these boxed EC numbers become highlighted (green) after selecting an organism from the pull-down list and clicking Exec button.

Clicking the nonhighlighted EC number returns general information of the enzyme (name, reaction, substrates and products). Clicking the highlighted EC number provides, in addition to general information, nucleotide and amino acid sequences of the enzyme. Clicking the metabolite returns molecular formula and structure of the compound, which can be viewed by Launching ISIS/Draw. Right click on the structure and open the pop-up menu. Select File→Save Molecule As to save the structure as cpdname.mol, which can be viewed with RasMol or WebLab Viewer. Select Edit→Transfer to ISIS/Draw to display the structure on the ISIS window (if ISIS/Draw is installed) from which the structure can be saved as cpdname.skc.

You may search metabolic pathways between the two metabolites at KEGG. On the KEGG home page, select Search and Compute KEGG then Generate possible reaction pathways between two compounds under Prediction Tools to open the search form (Figure 8.2). Choose organism from a list of "Search against." Enter the initial substrate and the final product and click the Exec button. The same form reappears with an addition of the compound ID to the requested substrate/product (a change in the organism name to Standard Dataset indicates an unavailability of the requested pathway for the specified organism). Clicking the Exec button again returns a list of linked entries of enzymes (EC number) and metabolites (Compound ID). Choose Show as diagram to view the pathways with connected clickable Compound ID for metabolites and EC number for enzymes:

$$C00469 \xrightarrow{2.3.1.152b} C00293 \xrightarrow{2.3.1.152} C00132 \xrightarrow{1.1.1.244} C00067 \xrightarrow{1.8.3.4} C00409 \xrightarrow{4.2.99.10} C00033$$

$$\xrightarrow{2.4.1.9} C00089 \xrightarrow{2.7.1.69} C00615 \xrightarrow{2.7.3.9} C00022 \xrightarrow{1.2.2.2}$$

The primary metabolic information can also be obtained from PathDB of National Center for Genome Resources (NCGR) at http://www.ncgr.org/software/pathdb/. Click Simple Web Query to open the query page. Choose Pathway/Step/Catalyst or Compound from the Retrieve pop-up list (search by compound is the

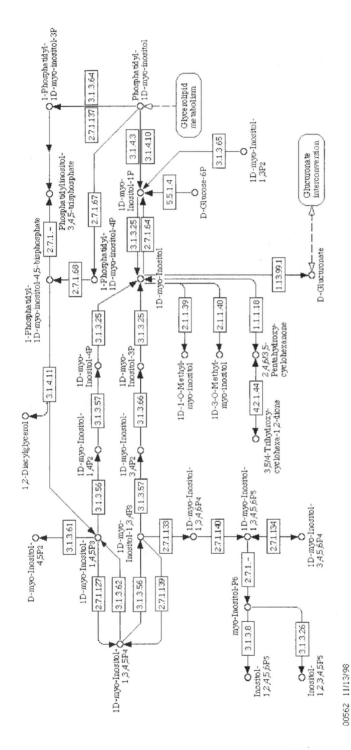

Figure 8.1. Retrieval of metabolic pathway from KEGG. The diagram of metabolic pathways retrieved from KEGG server is exemplified for inositol phosphate metabolism. Clicking small circles (metabolites) opens information pages providing the molecular structures that can be saved as cpdname.skc (ISIS Draw) and links to connected pathways/enzymes. Clicking the highlighted boxes (enzyme corresponding to the EC number) opens information pages providing enzyme nomenclature, catalyzed reaction, substrates/products/cofactors, and links to the connected pathways, genes, structure, and disease of its deficiency.

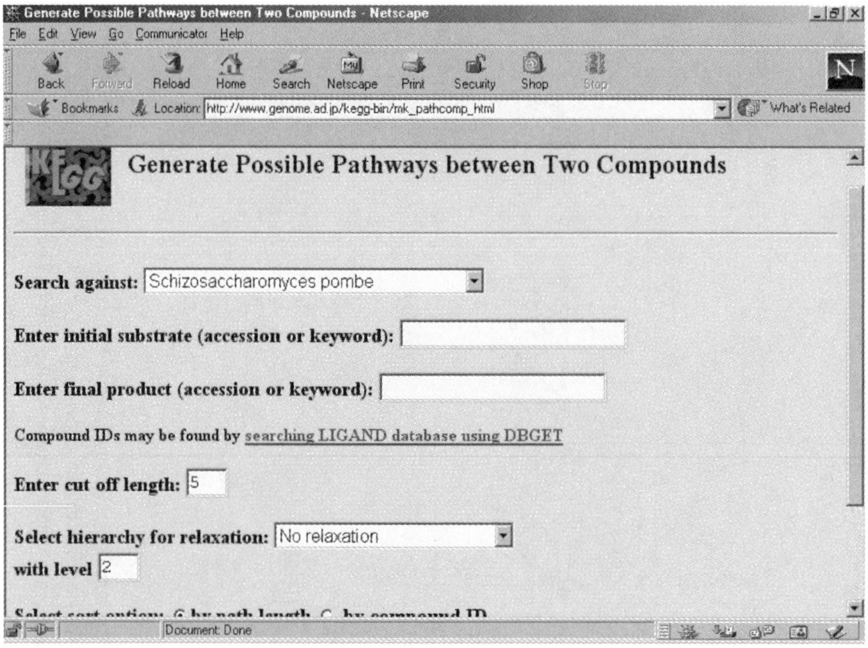

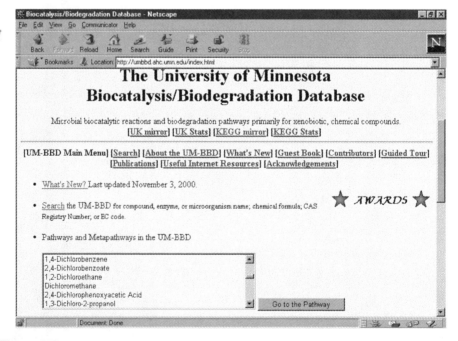

Figure 8.2. Query page for searching metabolic pathway between two compounds. A search for metabolic pathway(s) between two compounds in a specific organism can be performed at the KEGG server.

Figure 8.3. Biocatalysis/biodegradation database for xenometabolism. The BBD is the primary resource for searching/retrieving information on xenobiotics and their metabolic pathways.

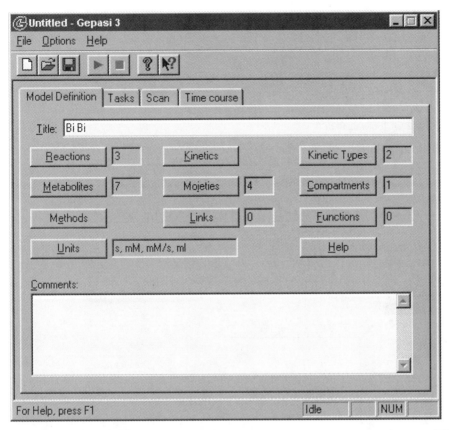

Figure 8.4. Main window of Gepasi. The main window of Gepasi consists of menus (File, Options, and Help), icons, and four tabs (Model definition, Tasks, Scan, and Time course). Activation of any of the tab opens an indexed page. At the start of Gepasi, the Model definition page is opened. Enter name of the metabolic pathway to the Title box. Click Reactions button to define enzymatic reactions (e.g., $E + A + B = EAB$ for R1, $EAB = EPQ$ for R2, and $EPQ = E + P + Q$ for R3 shows 3 reactions and 7 metabolites), and then click Kinetics button to select kinetic type. Activate Tasks tab to assign Time course (end time, points, simufile.dyn), Steady state (simufile.ss) and Report request. Activate Scan tab to select scan parameters. Activate Time course tab to select data to be recorded and then initiate the time course run.

simplest), enter the pathway/enzyme/compound name in the Value box, and click the Submit Query button. Selecting the desired entry from the hit list returns a table of pathway information.

Secondary metabolic pathways and some xenometabolic pathways are likewise accessible from the KEGG Pathway server (mirror site of BBD). However, the biocatalysis/biodegradation database at the University of Minnesota (UM-BBD) at http://umbbd.ahc.umn.edu/index.html is the main resource site for xenometabolisms (Figure 8.3). On the UM-BBD home page, select (highlight) xenobiotic from the list of compounds and click Go to the Pathway. The request returns descriptions of the xenobiotic and pathway with organisms involved and pointers to enzymes as well as

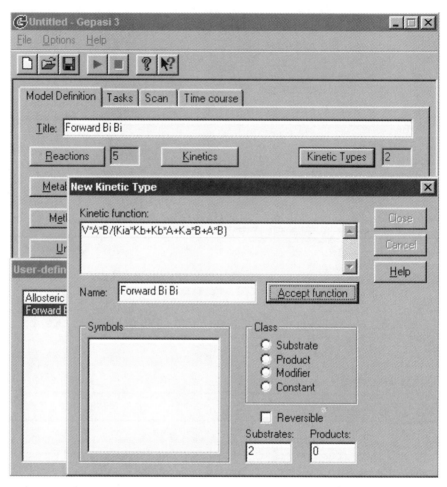

Figure 8.5. Definition of kinetic type for Gepasi. To define new kinetic type, open User-defined kinetic types dialog box by clicking Kinetic types button (Model definition page). Click Add button to open New kinetic type dialog box. Enter kinetic equation as shown in the Kinetic function box (forward ordered Bi Bi) of the inset and click Accept function button.

metabolites. Setting the pointer to the enzyme returns enzyme links and the catalyzed reaction. Setting the pointer to the metabolite returns the molecular structure, Smiles string of the compound and the links to ChemFinder, toxicological resources, and UM-BBD reaction(s).

8.3.2. Metabolic Simulation

The metabolic simulation can be performed with a software program, Gepasi (*General pathway simulator*) (Mendes, 1993), which can be downloaded from http://www.gepasi.org/. Gepasi is a metabolic modeling program that generates data for time courses and analyzes the steady-state using MCA from the input model pathway, kinetic parameters, and variables. The main window of Gepasi consists of four indexed pages corresponding to four tabs (Model definition, Tasks, Scan, and Time course). The program starts with the Model definition page (Figure 8.4). Enter the title and click buttons labeled Reactions, Metabolites, Units, Kinetic types, and so on, to activate dialog boxes in order to enter appropriate information for the simulation. Initially select the predefined kinetic type such as Henri–Michaelis–Menten. Exit the Model definition page and open the Tasks page (via activating the Tasks tab). Click to activate appropriate dialog box and enter the required information regarding the tasks of simulation such as time duration of simulation, data points required, simulation analysis, and so on.

For kinetic mechanisms that are not predefined by the program, such as "order bi bi," the new kinetic function has to be defined. This can be accomplished through the User-defined kinetic type (via Medel definition) as shown in Figure 8.5.

To record concentrations of metabolites/intermediates during the simulation, click Edit to activate a pop-up dialog box, and use arrow keys to select (transfer)

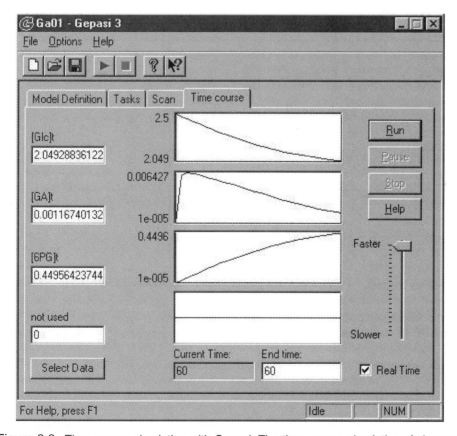

Figure 8.6. Time course simulation with Gepasi. The time course simulation of changes in metabolite concentrations can be viewed after clicking the Run button from the Time course page. The simulation is exemplified for gluconate pathway (D-Glucose→Gluconate→6-Phosphogluconate) by glucose dehydrogenase and gluconate kinase from *Schizosaccharomyces pombe*.

the desired metabolites/intermediates with time or at steady state. Exit the Tasks page and open the Time course page. Select data for the metabolites/intermediates to view and click Run to start the simulation that terminates at the specified time (Figure 8.6). Save the file (simufile.gps) before you enter new data or leave the program. The time course of the simulation can be exported as a data file (simufile.dyn) to Excel or can be plotted by activating the accompanying program, Gnuplot.

8.4. WORKSHOPS

1. The free energy changes at 25°C for glycolytic and gluconeogenetic reactions in rat liver are given below:

Enzyme	$\Delta G^{0\prime}$ (KJ mol^{-1})	$\Delta G^{\prime}$ (KJ mol^{-1})
Glucokinase	−15.6	−25.3
G6Phosphatase	−13.9	−2.0
Phosphoglucoisomerase	+1.67	+4.6
Phosphofructokinase	−14.2	−8.2
Fructose-1,6-bisphosphatase	−16.7	−9.4
Aldolase	+23.9	−22.7
Triosephosphate isomerase	+7.56	+2.41
Glycerald3P dehydrogenase	+6.3	−1.29
Phosphoglycerate kinase	−18.9	+0.1
Phosphoglyceromutase	+4.4	0.1
Enolase	+1.8	2.9
Pyruvate kinase	−31.7	0.7
Pyruvate carboxylase + PEP caqrboxykinase	−4.82	−21.9

Apply spreadsheet, Excel, to compute the equilibrium constant (K_{eq}) and the mass-action ratio (Q) for each reaction. Identify and rationalize the futile cycles (Koshland, 1984) as possible sites of regulation in the glucose metabolism, noting that a regulatiory enzyme will catalyze a "nonequilibrium reaction" under intracellular conditions. This can be accomplished by comparing the established equilibrium constant for the reaction with the mass-action ratio as it exists within the cells.

2. A cellular redox regulator, glutathione, which undergoes NADP(H)-linked interconversion between the oxidized and reduced forms, also interconverts with its constituent amino acids (Glu, Gly, and Cys). Construct (search metabolic databases) the annotated glutathione metabolic cycle including its redox and anabolic/catabolic interconversions.

3. Depict the biosynthetic pathway of tetracycline by showing structural transformation of each metabolic step.

4. Depict biosynthetic pathways for eukaryotic and prokaryotic tRNA and highlight the differences between the two systems.

5. Prostaglandins (PG) denoted by letters A through F have shown to be involved in inflammation and are responsible for pharmacological effect of non-

steroidal antiinflammatory agents such as aspirin. Search for prostaglandin metabolic pathway and depict the interconversion of prostaglandins (with reference to their molecular structures by furnishing their 2D and 3D structural files).

6. Catechol is an important biodegradative intermediate of aromatic xenobiotics. List biochemical reactions that involve catechol.

7. A second messenger, 1,2-diacylglycerol, can be formed from phosphatidylcholine (lecithin) by hydrolytic (Uni-substrate) reactions catalyzed by phospholipase C or by the combination of 3-*sn*-phosphatidate phosphatase and phospholipase D:

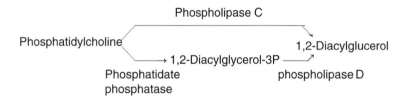

The kinetic parameters for these hydrolases are given below:

	V (mM/min/mg)	K_a (mM)
Phospholipase C	1010	4.5
Phospholipase D	153	0.7
3-*sn*-Phosphatidate phosphatase	1.4	0.38

Compare the time course for 1,2-diacylglycerol formation from 1.0 mM of phosphatidylcholine by the two pathways.

8. A trisaccharide, raffinose, is hydrolyzed to its component monosaccharides by the actions of glycosidases:

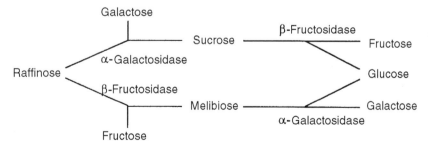

Search the Internet for kinetic parameters of β-fructosidase and α-galactosidase. Perform metabolic simulation to compare the effectiveness of the two reaction sequences to produce D-glucose.

9. Perform metabolic simulation for ethanol metabolism using kinetic parameters from Exercise 7 of Chapter 7, assuming ethanol and NAD$^+$ concentrations of 10 mM and 1.0 mM, respectively.

10. D-Glucose may enter pentose phosphate pathway in *Schizosaccharomyces pombe* via the oxidative phase of the pentose phosphate pathway:

D-Glucose $\xrightarrow{\text{Hexokinase}}$ D-Glucose-6-phosphate $\xrightarrow{\text{G6P Dehydrogenase}}$ 6-Phospho-D-gluconate

or gluconate pathway:

D-Glucose $\xrightarrow{\text{D-Glucose dehydrogenase}}$ D-Gluconate $\xrightarrow{\text{Gluconate kinase}}$ 6-Phospho-D-gluconate

Their kinetic parameters are given below:

	HxK	G6PDH	GlcDH	GaK
Enzyme (μg)	2.0	1.0	1.0	2.0
V (μM/min)	0.1229	0.2154	0.4181	0.1666
K_a (mM)	0.0569	0.0680	0.0173	0.0898
K_b (mM)	0.0619	0.6626	0.2637	1.9071
K_{ia} (mM)	0.0679	0.0443	0.6459	0.0551

Note: HxK, G6PDH, GlcDH, and GaK are hexokinase, D-glucose-6-phosphate dehydrogenase, glucose dehydrogenase, and gluconate kinase.

Perform metabolic simulation to compare preferred pathways under different experimental conditions:

REFERENCES

Arnstein, H. R. V., and Cox, R. A. (1992) *Protein Biosynthesis*. IRL Press/ Oxford University Press, Oxford.

Campbell, J. L. (1985) *Annu. Rev. Biochem.* **55**:733–771.

Copley, S. D. (2000). *Trends Biochem. Sci.* **25**:261–265.

Dagley, S., and Nicholson, D. E. (1970) *An Introduction to Metabolic Pathways*. John Wiley & Sons, New York.

Denton, P. M., and Pogson, C. I. (1976) *Metabolic Regulation*. John Wiley & Sons, New York.

DePamphilis, M. L., Ed. (1996) *DNA Replication in Eukaryotic Cells*. Cold Spring Harbor Laboratory Press, Cold Spring Harbor, New York.

Ellis, L., Hershberger, C. D., and Wackett, L. P. (2000) *Nucleic Acids Res.* **28**:377–379.

Fell, D. A. (1992) *Biochem. J.* **286**:313–330.

Gorrod, J. W., Oelschläger, H., and Caldwell, J., Ed. (1988) *Metabolism of Xenobiotics*. Taylor & Francis, London.

Grierson, D., Ed. (1993) *Biosynthesis and Manipulation of Plant Products*. Chapman and Hall, London.

Herbert, R. B. (1989) *The Biosynthesis of Secondary Metabolites*, 2nd edition. Chapman and Hall, New York.

Jakoby, W. B., and Ziegler, D. M. (1990). *J. Biol. Chem.* **265**:20715–20718.

Johnson, P. F., and McKnight, S. L. (1989) *Annu. Rev. Biochem.* **58**:799–839.

Kacser, H., and Burns, J. A. (1995) *Biochem. Soc. Trans.* **23**:341–366.

Kanehisa, M., Goto, S., Kawashima, S. and Nakaya, A. (2002) *Nucleic Acids Res.* **30**:42–46.

Kornberg, A. (1988) *J. Biol. Chem.* **263**:1–4.

Kornberg, A., and Baker, T. A. (1992) *DNA Replication*, 2nd edition. W. H. Freeman, New York.

Koshland, D. E., Jr. (1984) *Trends Biochem. Sci.* **9**:155–159.

Martin, B. R. (1987) *Metabolic Regulation: A Molecular Approach.* Blackwell Scientific, Oxford.

Mendes, P. (1993) *Comput. Appl. Biosci.* **9**:563–571.

Moldave, K., Ed. (1981) *RNA and Protein Synthesis.* Academic Press, New York.

Nossal, N. G. (1983) *Annu. Rev. Biochem.* **52**:581–615.

Palmer, J. M., and Folk, W. F. (1990) *Trends Biochem. Sci.* **15**:300–304.

Porter, J. W., and Spurgeon, S. L. (1981, 1983) *Biosynthesis of Isoprenoid Compounds*, Vols. 1 and 2. John Wiley & Sons, New York.

Saier, M., Jr. (1987) *Enzymes in Metabolic Pathways.* Harper & Row, New York.

Sharp, P. A. (1986) *Annu. Rev. Biochem.* **55**:1119–1150.

Stadtman, D. E. (1966) *Adv. Enzymol.* **28**:41–154.

Stafford, H. A. (1990) *Flavonoid Metabolism.* CRC Press, Boca Raton, FL.

Wingender, E., Chen, X., Fricke, E., Geffer, R., Hehl, R., Liebich, I., Krull, M., Matys, V., Michael, H., Ohnhäuser, R., Prüß, M., Schacherer, F., Thiele, S., and Urbach, S. (2001) *Nucleic Acids Res.* **29**:281–283.

Woychik, N. A., and Young, R. A. (1990) *Trends Biochem. Sci.* **15**:347–351.

9

GENOMICS: NUCLEOTIDE SEQUENCES AND RECOMBINANT DNA

Genomics refers to systematic investigation of genomes including nuclear and extranuclear genes of an organism. The investigation includes DNA sequencing, management and analysis of sequence data, discovery of new genes, gene mapping, and new genetic technologies. This chapter deals with the Internet search and retrieval of genome databases, recombinant DNA technology including polymerase chain reaction, and application of a program package BioEdit to analyze DNA sequences.

9.1. GENOME, DNA SEQUENCE, AND TRANSMISSION OF GENETIC INFORMATION

Genomes of living organisms have a profound diversity (Lewin, 1987; Miklos and Rubin, 1996). This diversity related not only to genome size but also to the storage principle as either single- or double-stranded DNA or RNA. Moreover, some genomes are linear (e.g., mammals) whereas others are closed and circular (e.g., most bacteria). Cellular genomes are always made of DNA, while phage and viral genomes may consist of either DNA or RNA. In single-stranded genomes, the information is read in the positive sense, the negative sense, or in both directions in which case one speaks of an ambisense genome. The positive direction is defined as going from the 5′ to the 3′ end of the molecule. In double-stranded genomes the information is read only in the positive direction (5′ to 3′ on either strand). The smallest genomes are found in non-self-replicating bacteriophages and viruses. Such very small genomes normally come in one continuous piece of sequence. But other larger genomes may have several chromosomal components. For example, the approximately 3-billion-base-pair (Bbp) human genome with fewer than 30,000 protein-coding genes (Claverie, 2001) is organized into 22 chromosomes plus the two that determine sex. Viral genomes have sizes in the interval from 3.5 to 280 killobase pairs (Kbp), bacteria range from 0.5 to 10 million base pairs (Mbp), fungi range from around 10 to 50 Mbp, plants start at around 50 Mbp, and mammals are found to be around 1 Gbp (Miklos and Rubin, 1996).

There are two basic protocols for sequencing DNA molecules (^{32}P- or chemiluminescent-labeled). Maxim–Gilbert sequencing (also called chemical degradation method) uses chemicals to cleave DNA at specific bases, resulting in fragments of different lengths (Maxim and Gilbert, 1980). Sanger sequencing (also called chain termination or dideoxy method) involves using DNA polymerase, in the presence of all four nucleoside triphosphates and one of the dideoxynucleoside triphosphate, to synthesize DNA chains of varying lengths in four different reactions. DNA replication would be stopped at positions occupied by the dideoxynucleotides (Smith, 1980). The resulting fragments with different lengths corresponding to the nucleotide sequence are analyzed by polyacrylamide gel electrophoresis, which is capable of resolving fragment lengths differing in one nucleotide.

In 1995 the first complete genome of a free-living organism, the prokaryote *Haemophilus influenzae*, was published and made available for analysis (Fleischmann et al., 1995). This circular genome contains 1,830,137 bp with 1743 predicted protein coding regions and 76 genes encoding RNA molecules. The human genome consisting 2.91 billion base pairs of DNA has been completely sequenced (Venter et al., 2001); however, only a small fraction (1.1%) of the DNA is a coding sequence (exons). Having complete sequences and knowing what they mean are two distinct stages in understanding any genome. Genomes On Line Database (GOLD) at http://igweb.integratedgenomics.com/GOLD/ monitors worldwide genome projects listed according to the published complete genomes, prokaryotic ongoing genomes, and eukaryotic ongoing genomes (Bernal et al., 2001). The published complete genomes are listed according to organism, tree, information, size (ORF number) with map, data search, institution, funding, genome database, and publication (underlines indicate linked for further information).

The important DNA sequence data repositories as the primary resources known as International Nucleotide Sequence Database Collaboration are:

GenBank of the National Center for Biotechnology Information: http://www.ncbi.nlm.nih.gov/Genbank/

EMBL (European Molecular Biology Laboratory): http://www.ebi.ac.uk/

DDBJ (DNA Data Bank of Japan): http://www.ddbj.nig.ac.jp/.

All three centers are separate points of data submission, but they all exchange this information and make the same database available to the international communities.

GeneBank (Benson et al., 2002), the DNA database from the National Center for Biotechnology Information (NCBI), incorporates sequences from publicly available sources, primarily from direct author submissions and genome sequence projects. The resource exchanges data with EMBL and DDBJ on a daily basis to ensure comprehensive coverage worldwide. To facilitate fast and specific searches, the GenBank database is split into several subsets such as PRI (primate), ROD (rodent), MAM (other mammalian), VRT (other vertebrate), IVT (invertebrate), PLN (plant, fungus, algae), BCT (bacterial), RNA (structural RNA), VRL (viral), PHG (bacteriophage), SYN (synthetic), UNA (unannotated), EST (Expressed sequence tags), PAT (patent), STS (sequence tagged sites), GSS (genome survey sequences), and GSS (genome survey sequences).

Each GenBank entry consists of a number of keywords. The LOCUS keyword introduces a short label and other relevant facts including the number of base pairs, source of sequence data, division of database, and date of submission. A concise description of the sequence is given after the DEFINITION and a unique accession number after the ACCESSION. The KEYWORDS line includes short phrases assigned by author, describing gene products and other relevant information. The SOURCE keyword and ORGANISM sub-keyword indicate the biological origin of the entry. The REFERENCE and its sub-keywords record bibliographic citation with the MEDLINE line pointing to an online link for viewing the abstract of the given article. The FEATURES keyword introduces the feature table describing properties of the sequence in detail and relevant cross-linked information. The BASE COUNT line provides the frequency count of different bases in the sequence. The ORIGIN line records the location of the first base of the sequence within the genome, if known. The nucleotide sequence of the genome (in GenBank format, Chapter 4) follows, and the entry is terminated by the // marker.

EMBL (Stoesser et al., 2002), the nucleotide sequence database maintained by the Europiean Bioinformatics Institute (EBI) (Emmert et al., 1994), produces sequences from direct author submissions and genome sequencing groups and from the scientific literature and patent applications. The database is produced in collaboration with GenBank and DDJB. All new and updated entries are exchanged between the groups. Information can be retrieved from EBI using the SRS Retrieval System (http://srs.ebi.ac.uk/). Each EMBL entry consists of the following lines with headings such as ID (identifier), AC (accession number), DE (description), KW (keyword/name), OS/OC (biological origin), RX/RA/RT/RL (reference pointer to MEDLINE, authors, title and literature), DR (pointer to amino acid sequence), FT (features), and SQ (the nucleotide sequence in Embl format is preceded by the base count and terminated with the // marker (Chapter 4).

DDBJ (Tateno et al., 2002) is the DNA Data Bank of Japan in collaboration with GenBank and EMBL. The database is produced, maintained, and distributed at the National Institute of Genetics. The entry of DDBJ follows the keywords adapted by the GenBank.

Using sequence analysis techniques such as BLAST, it is feasible to identify similarities between novel query sequences and database sequences whose structures and functions have been elucidated. This is straightforward at high levels of sequence identity (above 50% identity), where relationships are clear, but at low levels (below 50% identity) it becomes difficult to establish relationships reliably. Alignment of random sequences can produce around 20% identity; below that the alignments are no longer statistically significant. The essence of sequence analysis is the detection of homologous sequences by means of routine database searches, usually with unknown or uncharacterized query sequences. The identification of such relationships is relatively easy when levels of similarity remain high. But if two

TABLE 9.1. Standard Codons

| 5′-Terminal Base | U(T) | Middle Base | | | 3′-Terminal Base |
		C	A	G	
U(T)	Phe	Ser	Tyr	Cys	U(T)
	Phe	Ser	Tyr	Cys	C
	Leu	Ser	Stop	Stop	A
	Leu	Ser	Stop	Trp	G
C	Leu	Pro	His	Arg	U(T)
	Leu	Pro	His	Arg	C
	Leu	Pro	Gln	Arg	A
	Leu	Pro	Gln	Arg	G
A	Ile	Thr	Asn	Ser	U(T)
	Ile	Thr	Asn	Ser	C
	Ile	Thr	Lys	Arg	A
	Met(Init)	Thr	Lys	Arg	G
G	Val	Ala	Asp	Gly	U(T)
	Val	Ala	Asp	Gly	C
	Val	Ala	Glu	Gly	A
	Val	Ala	Glu	Gly	G

sequences share less than 20% identity, it becomes difficult or impossible to establish whether they might have arisen through the evolutionary processes. Sequences are homologous if they are related by divergence from a common ancestor. Therefore, homology is a statement that sequences have a divergent rather than a convergent relationship.

The genetic information flows from DNA to RNA and hence to proteins (Crick, 1970). The replication copies the parent DNA to form daughter DNA molecules having identical nucleotide sequences. The transcription rewrites parts of the genetic messages in DNA into the form of RNA (mRNA). The translation is the process in which the genetic message coded by mRNA is translated (via carriers, tRNA) into the 20-letter amino acid sequence of protein. The genetic code is read in continuum with stepwise groups of three bases. It is a nonoverlapping, comma-free, degenerate triple code. The elucidation of the triplet *codon* for all amino acids permits the translation of nucleotide sequences of DNA into amino acid sequences of proteins or vice versa in the reverse translation. The compilation of the codon usage is available from http://www3.ncbi.nlm.nih.gov/htbin-post/Taxonomy/wprintge?/mode = t. The 64 codons of the standard genetic code are listed in Table 9.1.

It is noted that three codons (T(U)AA, T(U)AG, and T(U)GA) are assigned as stop signals (stop codons) that code for the chain termination. The initiation codon, AT(U)G, is shared with that for methionine. The observation that one kind of organism can accurately translate the genes from quite different organisms based on the standard genetic code is the basis of genetic engineering. However, the genetic codes of certain mitochondria that contain their own genes and protein synthesizing systems are variants of the standard genetic code (Fox, 1987).

In the translation process, codons in mRNA and anticodons in tRNA interact in an antiparallel manner such that the two 5′ bases of the codon interact with two

3′ bases of the anticodon according to the AU-pairing and GC-pairing. However, the 3′ base of the codon may interact with the 5′ base of the anticodon with a certain pattern of redundancy known as the wobble hypothesis:

3′ Base of Codon	5′ Base of Anticodon
G	C
U	A
A or G	U
C or U	G
U, C, or A	I

The anticodon can be found in the middle of the unpaired anticodon loop (7 nucleotides) of tRNA ($\sim$76 nucleotides) between positions 30 and 40 (at around position 35). The anitcodon is bordered by an unpaired pyrimidine, U, on the 5′ side and often an unpaired alkylated purine on the 3′ side.

9.2. RECOMBINANT DNA TECHNOLOGY

The elucidation of molecular events, enzymology of genetic duplicative process (Eun, 1996; Wu, 1993), and knowledge of dissecting nucleotide sequences in DNA (Brown, 1994) provides impetus for the development of recombinant DNA technology (Greene and Rao, 1998; Watson et al., 1992; Wu, 1993). The main objective of recombinant DNA technology or molecular cloning is to insert a DNA segment (gene) of interest into an autonomously replicating DNA molecule, known as a *cloning vector*, so that the DNA segment is replicated with the vector. The result is a selective amplification of that particular DNA segment (gene). The technology entails the following general procedures (Berger and Kimmel, 1987; Brown, 2001; Lu and Weiner, 2001; Rapley, 2000).

1. Cutting DNA at precise locations to yield the DNA segment (gene) of interest by use of sequence specific restriction endonuclease (restriction enzyme).
2. Selecting a proper vector and cutting the vector preferably with the identical restriction enzyme.
3. Construction of the recombinant DNA by inserting the DNA segment of interest into the vector by joining the two DNA fragments with DNA ligase.
4. Introduction of the recombinant DNA into a host cell that can provide the enzymatic machinery for DNA replication.
5. Selecting and identifying those host cells that contain recombinant DNA.
6. Identification of the recombinant DNA.

Various bacterial plasmids, bacteriophages, and yeast artificial chromosomes are used as cloning vectors (Brown, 2001). At the heart of the general approach to generating and propagating a recombinant DNA molecule is a set of enzymes which synthesize, modify, cut, and join DNA molecules.

Some restriction enzymes cleave both strands of DNA so as to leave no unpaired bases on both ends known as blunt ends (subsequent joining requires linkers).

Others make staggered cuts on two DNA strands, leaving two to four nucleotides of one strand unpaired at each resulting end referred to as cohesive ends which can be joined by DNA ligase if both the DNA segment and vector are cleaved by the same restriction enzyme or an isoschizomer. The type II restriction enzymes that cleave the DNA within the recognition sequences are most widely used for cutting DNA molecules at specific sequences. These sequences are normally short (four to six base pairs) and palindromic. A sampling of sequences recognized by some common type II restriction enzymes are listed in Table 9.2. REBASE at http://rebase.neb.com/rebase/rebase.html (Roberts and Macelis, 2001) is a resource site for restriction enzymes.

The average size of the DNA fragments produced with a restriction enzyme depends on the frequency with which a particular restriction site occurs in the original DNA molecule and on the size of the recognition sequence of the enzyme. In a DNA molecule with a random sequence, the restriction enzyme with six-base recognition sequence produces larger fragments than the enzyme with four-base recognition sequence. Another important consideration is the preservation of the original reading frames in the recombinant DNA.

Polymerase chain reaction (PCR) offers a convenient method of amplifying the amount of a DNA segment (Mullis et al., 1994; Newton and Graham, 1997; Saiki et al., 1988; Timmer and Villalobos, 1993). In this technique, the DNA synthesizing preparation contains denatured DNA with the segment of interest that serves as template for DNA polymerase (DNA-directed DNA polymerase), two oligonucleot-

TABLE 9.2. Some Type II Restriction Endonucleases[a]

Name	Sequence	Name	Sequence[b]
Acc I	GT↓MKAC	Mlu I	A↓CGCGT
Alu I	AG↓CT	Msp I	C↓CGG
Apy I	CC↓WGG	Nco I	C↓CATGG
Bal I	TGG↓CCA	Nde I	CA↓TATG
BamH I	G↓GATCC	Nru I	TCG↓CGA
Ban III	AT↓CGAT	Pst I	CTGCA↓G
Bgl I	GCCN$_4$↓NGGC	Pvu II	CAG↓CTG
Bgl II	A↓GATCA	Rsa I	GT↓AC
Cfo I	GCG↓C	Sal I	G↓TCGAC
Cla I	AT↓CGAT	Sca I	AGT↓ACT
Dra I	TTT↓AAA	Sph I	GCATG↓C
EcoR I	G↓AATTC	Sse I	CCTGCA↓GG
EcoR II	↓CCWGG	Ssp I	AAT↓ATT
Fsp I	TGC↓GCA	Sun I	C↓GTACG
Hae III	GG↓CC	Taq I	T↓CGA
Hha I	GC↓GC	Tha I	CG↓CG
Hinc II	GTY↓RAC	Vsp I	AT↓TAAT
Hind III	A↓AGCTT	Xba I	T↓CTAGA
Kpn2 II	T↓CCGGA	Xho I	C↓TCGAG

[a] The recognition sequences of type II restriction endonucleases are shown with the arrow (↓) indicating the cleavage site.
[b] R for A or G; M for C or A; K for T(U) or G; W for T(U) or A; N for any base.

ide primers that direct the polymerase, and all of the four deoxyribonucleoside triphosphates. These primers, designed to be complementary to the two 3′-ends of the specific DNA segment to be amplified, are added in excess amounts to prime the DNA polymerase-catalyzed synthesis of the two complementary strands of the desired segment, effectively doubling its concentration. The DNA is heated to dissociate the duplexes and then cooled so that primers bind to both the newly formed and the old strands. Another cycle of DNA synthesis ensues. The reaction is limited by the efficiency of DNA polymerase to segments 10 kbp or less in size. The protocol has been automated through the invention of thermal cyclers that alternately heat the reaction mixture to 95°C to dissociate the DNA, followed by cooling, annealing of primers, and another round of DNA synthesis. The use of thermal stable *Taq* DNA polymerase has improved the effectiveness of the process. The amplification efficiency varies significantly during the course of amplification, and the amounts of product produced at each cycle eventually level off. The average efficiency (E) of a series of PCR cycles can be described by

$$N = n(1 + E)^c$$

where N, n, and c are the final amount of product, the initial amount of target DNA, and the number of PCR cycles, respectively. The reported average efficiencies are in the range of 0.60–0.95. The PCR amplification is an effective cloning strategy if sequence information for the design of appropriate primers is available.

Complementary DNAs (cDNA) are DNA copies of mRNAs. Reverse transcriptase is the RNA-directed DNA polymerase that synthesizes DNA strand, using purified mRNA as the template. DNA polymerase is then used to copy the DNA strand forming a double-stranded cDNA, which is cloned into a suitable vector. Once a cDNA derived from a particular gene has been identified, the cDNA becomes an effective probe for screening genomic libraries (Cowell and Austin, 1997). Annotated human cDNA sequences can be accessed from HUNT at http://www.hri.co.jp/HUNT (Yudate, 2001).

Perhaps no scientific research other than gene cloning (and perhaps stem cell research) has generated lively discussions concerning its social, moral, and ethical impacts. These topics are beyond the scope of this text. The contemporary issues on these aspects of gene cloning can be found in many publications (Lauritzen, 2001; Rantak and Milgram, 1999; Thompson and Chadwick, 1999; Yount, 2000).

9.3. NUCLEOTIDE SEQUENCE ANALYSIS

9.3.1. Sequence Databases

Sequence data can be analyzed for (a) Sequence characteristics by knowledge-based sequence analysis, (b) Similarity search by pairwise sequence comparison, (c) Multiple sequence alignment, (d) Sequence motif discovery in multiple alignment, and (e) phylogenetic inference.

The nucleotide sequences can be retrieved from one of the three IC (International Collaboration) nucleotide sequence repositories/databases: GenBank, EMBL Nucleotide Sequence Database, and DNA Data Bank of Japan (DDBJ). The retrieval can be conducted via accession numbers or keywords. Keynet (http://www.ba.cnr.it/keynet.html) is a tree browsing database of keywords extracted from

EMBL and GenBank aimed at assisting the user in biosequence searching. GenBank nucleotide sequence database (http://www.ncbi.nlm.nih.gov/GenBabk/) can be accessed from the integrated database retrieval system of NCBI, Entrez (http://www.ncbi.nlm.nih.gov), by selecting the Nucleotide menu. Entering the text keyword displays the summary of hits. Pick the desired records and view them in the summary text or graphics that display the sequence, CDS, and protein product. Save the sequences in GenBank format or fasta format (Display = fasta and View = plain text) for subsequent sequence analyses. Access to EMBL Nucleotide Sequence Database (http://www.ebi.ac.uk/embl.html) is accomplished via the European Bioinformatics Institute (EBI) at http://www.ebi.ac.uk/embl/Access/index.html. The database search at EBI can be performed by the Sequence Retrieval System (SRS). EMBL incorporates sequence data produced by a number of genome projects and maintains Genome MOT (genome monitoring table). DNA Data Bank of Japan (DDBJ) can be found at http://srs.ddbj.nig.ac.jp/index.html. Genome Information Broker, GIB (http://mol.genes.nig.ac.jp/gib/), can be used to retrieve the complete genome data. Chapter 3 describes the procedures for database retrieval from GenBank, EMBL, and DDBJ.

9.3.2. Similarity Search

All of the three IC centers also provide facilities for sequence similarity search and alignment. The widely used database search algorithms are FASTA (Lipman and Pearson, 1985) at http://www.nbrf.georgetown.edu/pirwww/search/fasta.html and BLAST (Altschul et al., 1990) at http://www.ncbi.nlm.nih.gov/BLAST/. For BLAST

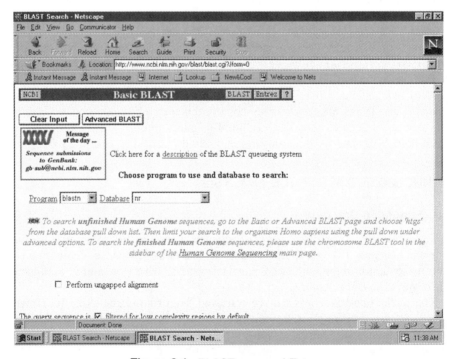

Figure 9.1. BLAST server of Entrez.

nucleotide sequence analysis at NCBI (Figure 9.1), paste the sequence, and select *blastn* (for nucleotides) followed by choosing the basic BLAST and alignment view. Click the Search button to submit the query sequence. After successful submission of the query sequence as indicated by an assignment of the Request ID, click Format results to display the search results. The similarity searches using FASTA (http://www2.ebi.aci.uk/fasta3/) and BLAST (http://www2.ebi.ac.uk/blastall/) are also available at EBI and DDBJ (http://spiral.genes.nig.ac.jp/homology/top-e.html). A *P*-value refers to the probability of obtaining, by chance, a pairwise sequence comparison of the observed similarity given the length of the query sequence and the size of the database searched. Thus, low *P*-values indicate sequence similarities of high significance.

9.3.3. Recombinant DNA

To search for a vector at Riken Gene Bank (http://www.rtc.riken.go.jp/), click DNA Database Search and then select Vector Database to open the vector search page. Enter the keyword (e.g., pBR322, cosmid, or using wild card in pBR*, p*), and click the Start button. Choose the desired vector from the hit list by clicking detail-id #. The search returns with description (name, classification, size of vector DNA. restriction sites, cloning site, genetic markers, host organism, growth condition, GenBank accession, and reference) and the restriction map of the vector (Figure 9.2). A catalog of vectors is available from American Type Culture Collection (ATCC) at http://www.atcc.org/. From the list of Search a Collection, select Molecular Biology and then Vectors. A tabulated list of name, map and/or sequence, hosts, and brief description of vectors is returned. Select map or sequence to view/save the restriction map or the nucleotide sequence (with references) of the vector. The nucleotide sequence of the known vector can be retrieved from the Nucleotide tool of Entrez

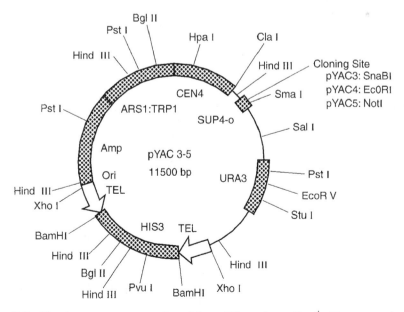

Figure 9.2. Shuttle vector map retrieved from Riken Gene Bank. The map of shuttle vector pYAC 3/4/5 shows major restriction sites.

```
        RarI
        Eam1104I                                                  AlwNI
tatctctcttcaagcattaaaaaaatctctttagagtcagtggatcaatagacagttcctgttttccacacaact base pairs
atagagagaagttcgtaatttttttagagaaatctcagtcacctagttatctgtcaaggacaaaaggtgtgttga 151 to 225
        Ksp632I


        BanII
        Eco24I
gaaagggtggagcccccaaaccacaaggggaagaaggaagttaaaagatgttaaatactggggccagctcaccct base pairs
ctttcccacctcggggttttggtgttcccttcttccttcaattttctacaatttatgaccccggtcgagtggga 226 to 300
        FriOI


                        HincII
ggtcagcctagcactctgacctagcagtcaacatgaaggctctcattgttctggggcttgtcctcctttctgtta base pairs
ccagtcggatcgtgagactggatcgtcagttgtacttccgagagtaacaagaccccgaacaggaggaaagacaat 301 to 375
                                                            HindII
```

Figure 9.3. Restriction map produced by Webcutter. The partial restriction map shows the nucleotide sequence of human lysozyme gene submitted to Webcutter using options for all restriction endonucleases with recognition sites equal to or greater than six nucleotides long and cutting the sequence 2–6 times (at least 2 times and at most 6 times). The restriction profile (map) is returned if "Map of restriction sites" is selected for display. The tables by enzyme name and by base pair number can be also returned if displays for "Table of sites, sorted alphabetically by enzyme name" "Table of sites, sorted sequentially by base pair number" are chosen.

(http://www.ncbi.nlm.nih.gov). Select and save the plasmid with circular DNA (check the header of GenBank format for circular DNA).

To search for an appropriate restriction enzyme and its restriction profile, subject the query DNA to Webcutter at http://www.firstmarket.com/cutter/cut2.html. Upload the sequence file (enter drive:directoryseqfilename) or paste the sequence into the query box. Indicate your preferences with respect to the type of analysis, site display, and restriction enzymes to include in the analysis. After clicking the Analyze Sequence button, the restriction map (duplex sequence with restriction enzymes at the cleavage sites), as shown in Figure 9.3, is returned if Map of restriction sites is selected for display. You may also select Table of sites, sorted alphabetically by enzyme name for display which lists number of cuts, positions of sites, and recognition sequences.

The primer selection for PCR can be accessed via Primer3 server of Whitehead Institute/MIT Center for Genome Research at http://genome.wi.mit.edu/cgi-bin/primer/primer3www.cgi (Figure 9.4). Paste the nucleotide template into the query box on the Primer3 home page. Key in the desired specifications—for example, included targets, excluded regions if any, product size, and primer picking conditions as desired. Click the Pick primers button. The returned output lists Oligo (left primer and right primer) with their start position, length, Tm, GC%, and sequences ($5' \rightarrow 3'$). The corresponding primers ($\gg\gg$for left primer and $\ll\ll$ for right primer) are also shown with the source (template) sequence (Figure 9.5).

The Web Primer (http://genome-www2.stanford.edu/cgi-bin/SGD/web-pirmer) searches 35 base pairs upstream and 35 base pairs downstream of the coding sequence to locate primers. On the entry page of Web Primer (Figure 9.6), paste the query sequence, select Sequencing [info] and click Submit button. The parameters page return with options for information on location of primer (length of DNA in which to search for valid primer, choice of DNA strand, distance between sequencing primers and primer length), primer composition (expressed in %GC content), and

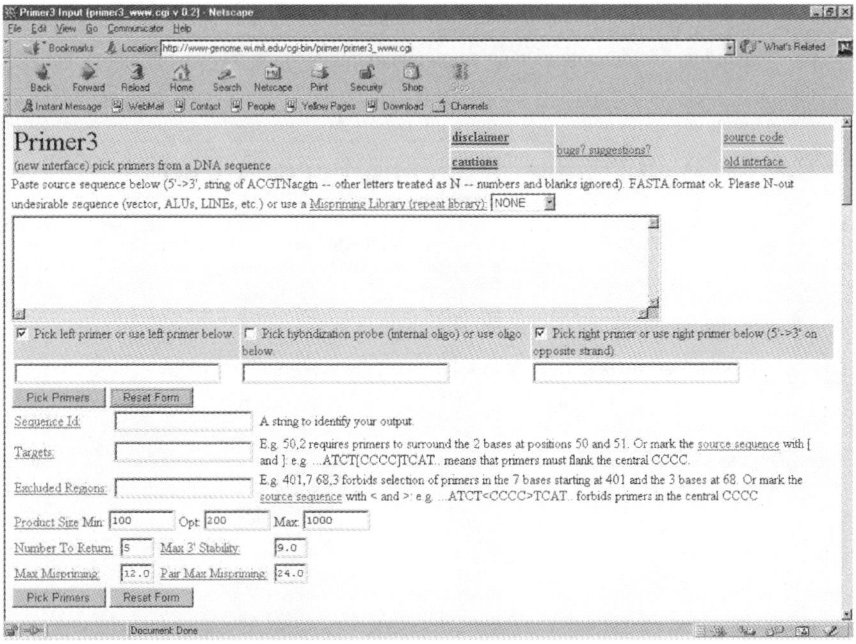

Figure 9.4. Request form for primer selection. The nucleotide sequence of a target DNA for polymerase chain reaction can be submitted for primer selection at Primer3 server.

primer annealing. Accept or modify the default options and click the Submit button. The user is instructed to click here for the list of primers. This returns data for primer-pairs listing the starting position and sequence of octadodecanucleotide primers of the coding strand.

The catalog of synthetic oligonucleotides which are proven useful as PCR primers or gene probes can be downloaded from National Cancer Research Institute

```
OLIGO             start  len      tm    gc%    any    3' seq
LEFT PRIMER         632   20   60.52  55.00   4.00   0.00 TTCAAGAGGGAGCTGACTGG
RIGHT PRIMER        840   20   60.05  50.00   6.00   2.00 TTCTCGGAGGTCATGAAACC
SEQUENCE SIZE: 1071
. . . . .

 601  GGACGAGTCGGCCGAGGCCTTCCCCCTGGAGTTCAAGAGGGAGCTGACTGGCCAGCGACT
                                    >>>>>>>>>>>>>>>>>>>>

 661  CCGGGAGGGAGATGGCCCCGACGGCCCTGCCGATGACGGCGCAGGGGCCCAGGCCGACCT

 721  GGAGCACAGCCTGCTGGTGGCGGCCGAGAAGAAGGACGAGGGCCCCTACAGGATGGAGCA

 781  CTTCCGCTGGGGCAGCCCGCCCAAGGACAAGCGCTACGGCGGTTTCATGACCTCCGAGAA
                                    <<<<<<<<<<<<<<<<<<<<
```

Figure 9.5. Output of primer selection for PCR. The abbreviated output of the primer selection for PCR by Primer3 server shows the input nucleotide sequence with appended primer oligonucleotide segments (>>> and <<<).

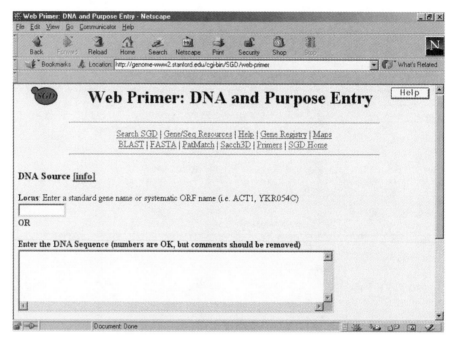

Figure 9.6. Home page of Web Primer.

in Genova, Italy by selecting molprob.gz (PC version) from ftp://ftp.biotech.ist.unige.it/pub/MPDB. For example, the following information describe the primer for tumor protein p53 gene:

ID: MP04028

Name: VNTRa

Type: primer

Sequence: 5′ CGAAGAGTGAAGTGCACAGG 3′

DataSource: Literature

Methods: PCR primers

Applications: Loss Of Heterozygosity

Species: human

TargetGene: TP53

GeneDescription: Tumor Protein p53

ComplementaryPrimer: VNTRb

Bibliography: Cancer Genet Cytogenet 1995;82:106-115)PMID: 7664239]

9.3.4. Application of BioEdit

BioEdit is a software program for nucleic acid/protein sequence editing, alignment, manipulation, and analysis. It can be downloaded from http://www.mbio.ncsu.edu/RnaseP/info/program/BIOEDIT/bioedit.html as BioEdit.zip. After installation, click BioEdit icon to open the main window. Select Open (to open new file in fasta format) or New from clipboard (to copy sequence) from the File menu. The input of sequence(s) changes the menu bar (with File, Edit, Sequence, Alignment, View, WWW, Accessory application, RNA, Option, Window, and Help menus) of the window. The Edit menu provides tools for manipulating nucleotide sequences. The Sequence menu provides tools for global alignment/calculation of identity/similarity of two sequences, creating plasmid from nucleotide sequence, analyses of nucleic acid, and protein sequences. The Alignment menu provides tools for multiple alignment, creating consensus sequence, entropy plot, positional nucleotide numerical summary, and finding conserved regions. The version 4.7.8 supports up to 20,000 sequences per document.

To analyze a nucleotide sequence for base composition, complement sequence, RNA transcription, protein translation (choice of Frames), creating plasmid, and restriction map from the sequence menu, for example to construct restriction map:

- Choose Nucleic acid tool of the Sequence menu (i.e., SequenceNucleic Acid-Restriction Map) to open a dialog box.
- Select output display and desired restriction enzymes.
- Click Generate map to return the restriction map (Figure 9.7).

To construct recombinant DNA:

- Input a desired plasmid sequence via File → Open.
- Copy the desired insertion DNA sequence on the clipboard.
- Identify cloning site by viewing restriction map of the plasmid.

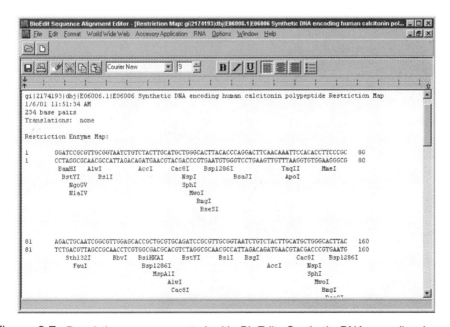

Figure 9.7. Restriction map generated with BioEdit. Synthetic DNA encoding human calcitonin is subjected to restriction with all REBASE restriction endonucleases to generate restriction map.

- Return to the plasmid window, place cursor at the cloning site, and paste the insertion DNA sequence from the clipboard (i.e., Edit → Paste).
- Rename the vector name if desired (click the highlighted name and type in the new name).
- Highlight the vector and select Sequence → Create Plasmid from Sequence to display the circular vector.
- Select Add Feature tool of the new Vector menu to open the dialog box.
- Enter the name of the insertion DNA as Feature name, identify the region of the insertion sequence by entering the start position and the end position, select type of display for the insertion sequence (color in box or arrow), and click Apply & Close.
- Display the reference restriction sites by selecting Vector → Restriction Sites to open the selection box. Transfer the desired enzyme sites to the Show box and click Apply & Close to display the recombinant DNA (Figure 9.8).
- Save the file as recdna.pmd:

To perform multiple alignment:

- Input a file containing multiple sequences by choosing File → Open.
- Highlight the headings (ID) of all sequences to be aligned.
- Select Alignment → "Plot identities to first sequence with a dot" to align all sequences with reference to the first sequence.

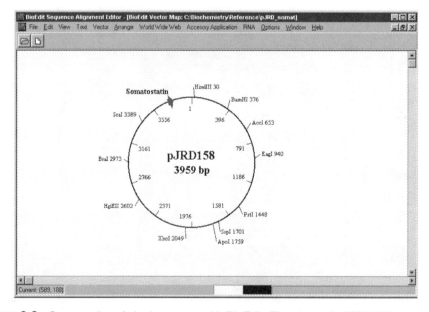

Figure 9.8. Construction of cloning vector with BioEdit. The plasmid pJRD158 is retrieved from Entrez and used to construct vector for cloning DNA encoding human somatostatin with BioEdit. The cloning vector with somatostatin gene (arrow) is displayed.

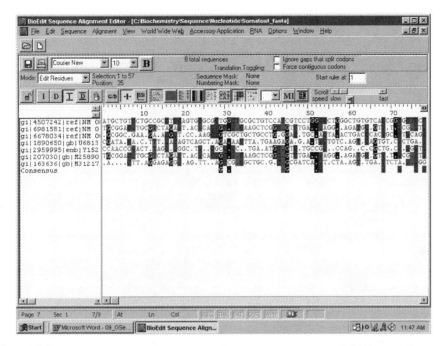

Figure 9.9. Sequence alignment with BioEdit. The sequences of DNA encoding prep-rosomatostatin mRNA are aligned to identify the consensus sequence.

- Shade the identity/similarity by clicking the Shade identities and similarity button.
- Display the consensus sequence by choosing Alignment → Create Consensus Sequence (Figure 9.9).
- Save the file (File → Save As) as conseq.bio.

9.4. WORKSHOPS

1. Mitochondria of various organisms use different genetic code. Search the Internet site to obtain information for the mitochondrial codons from either vertebrates or invertebrates. Discuss usage differences between the standard codons and mitochondrial codons.

2. Retrieve one each of the cytosolic tRNA specific for the following amino acids from *Saccharomyces cerevisiae*:

(a) Aspartic acid

(b) Phenylalanine

(c) Serine

(d) Tyrosine

Identify their anticodons.

3. Retrieve nucleotide sequences (in fasta format) and restriction maps for one each of bacterial plasmid, cosmid and shuttle vector.

4. Retrieve DNA sequence encoding for human glucagon mRNA. Subject the sequence to Webcutter and construct the restriction map using REBASE restriction enzymes.

5. Search the candidate primers of PCR for the human pro-optiomelanocortin gene with the following sequence:

```
>gi|4505948|ref|NM 000939.1|Homo sapiens proopiomelanocortin (POMC)
AGCGGCGGCGAAGGAGGGGAAGAAGAGCCGCGACCGAGAGAGGCCGCCGAGCGTCCCCGCCCTCAG
AGAGCAGCCTCCCGAGACAGAGCCTCAGCCTGCCTGGAAGATGCCGAGATCGTGCTGCAGCCGCTC
GGGGGCCCTGTTGCTGGCCTTGCTGCTTCAGGCCTCCATGGAAGTGCGTGGCTGGTGCCTGGAGAG
CAGCCAGTGTCAGGACCTCACCACGGAAAGCAACCTGCTGGAGTGCATCCGGGCCTGCAAGCCCGA
CCTCTCGGCCGAGACTCCCATGTTCCCGGGAAATGGCGACGAGCAGCCTCTGACCGAGAACCCCCG
GAAGTACGTCATGGGCCACTTCCGCTGGGACCGATTCGGCCGCCGCAACAGCAGCAGCAGCGGCAG
CAGCGGCGCAGGGCAGAAGCGCGAGGACGTCTCAGCGGGCGAAGACTGCGGCCCGCTGCCTGAGGG
CGGCCCCGAGCCCCGCAGCGATGGTGCCAAGCCGGGCCCGCGCGAGGGCAAGCGCTCCTACTCCAT
GGAGCACTTCCGCTGGGGCAAGCCGGTGGGCAAGAAGCGGCGCCCAGTGAAGGTGTACCCTAACGG
CGCCGAGGACGAGTCGGCCGAGGCCTTCCCCCTGGAGTTCAAGAGGGAGCTGACTGGCCAGCGACT
CCGGGAGGGAGATGGCCCCGACGGCCCTGCCGATGACGGCGCAGGGGCCCAGGCCGACCTGGAGCA
CAGCCTGCTGGTGGCGGCCGAGAAGAAGGACGAGGGCCCCTACAGGATGGAGCACTTCCGCTGGGG
CAGCCCGCCCAAGGACAAGCGCTACGGCGGTTTCATGACCTCCGAGAAGAGCCAGACGCCCCTGGT
GACGCTGTTCAAAAACGCCATCATCAAGAACGCCTACAAGAAGGGCGAGTGAGGGCACAGCGGGCC
CCAGGGCTACCCTCCCCCAGGAGGTCGACCCCAAAGCCCCTTGCTCTCCCCTGCCCTGCTGCCGCC
TCCCAGCCTGGGGGGTCGTGGCAGATAATCAGCCTCTTAAAGCTGCCTGTAGTTAGGAAATAAAAC
CTTTCAAATTTCACA
```

Suggest which of the candidate pairs lead to the synthesis of the constituent hormone(s) (adrenocorticotropin, β-lipotropin, α-melanocyte stimulating hormone, β-melanocyte stimulating hormone, and/or β-endorphin).

6. Use BioEdit to translate the human pro-optomelanocortin. Deduce the frame from which the constituent hormones are likely to be translated.

7. Retrieve complete CDS of human alcohol dehydrogenase (ADH) isozymes from UniGene and record the comparison with other organisms (UniGene listings).

8. Submit the following nucleotide sequence to homology search with the BLAST tool at one of the three major nucleotide sequence databases.

```
CCCAGCGCACCCGCACCATGGCCGGCCCCAGCCTCGCTTGCTGTCTGCTCGGCCTCCTGGCGCTGA
CCTCCGCCTGCTACATCCAGAACTGCCCCCTGGGAGGCAAGAGGGCCGCGCCGGACCTCGACGTGC
GCAAGTGCCTCCCCTGCGGCCCCGGGGGCAAAGGCCGCTGCTTCGGGCCCAATATCTGCTGCGCGG
AAGAGCTGGGCTGCTTCGTGGGCACCGCCGAAGCGCTGCGCTGCCAGGAGGAGAACTACCTGCCGT
CGCCCTGCCAGTCCGGCCAGAAGGCGTGCGGGAGCGGGGGCCGCTGCGCGGTCTTGGGCCTCTGCT
GCAGCCCGGACGGCTGCCACGCCGACCCTGCCTGCGACGCGGAAGCCACCTTCTCCCAGCGCTGAA
ACTTGATGGCTCCGAACACCCTCGAAGCGCGCCACTCGCTTCCCCCATAGCCACCCCAGAAATGGT
GAAAATAAAATAAAGCAGGTTTTTCTCCTCT
```

9. Use BioEdit to identify the consensus sequence for the following somatostatin sequences:

```
>gi|6678034|ref|NM_09215.1|Mouse somatostatin (Smst), mRNA
AGCGGCTGAAGGAGACGCTACCGAAGCCGTCGCTGCTGCCTGAGGACCTGCGACTAGACTGACCCA
CCGCGCTCCAGCTTGGCTGCCTGAGGCAAGGAAGATGCTGTCCTGCCGTCTCCAGTGCGCCCTGGC
TGCGCTCTGCATCGTCCTGGCTTTGGGCGGTGTCACCGGCGCGCCCTCGGACCCCAGACTCCGTCA
GTTTCTGCAGAAGTCTCTGGCGGCTGCCACCGGGAAACAGGAACTGGCCAAGTACTTCTTGGCAGA
GCTGCTGTCCGAGCCCAACCAGACAGAGAATGATGCCCTGGAGCCCGAGGATTTGCCCCAGGCAGC
TGAGCAGGACGAGATGAGGCTGGAGCTGCAGAGGTCTGCCAACTCGAACCCAGCAATGGCACCCCG
```

```
GGAACGCAAAGCTGGCTGCAAGAACTTCTTCTGGAAGACATTCACATCCTGTTAGCTTTAATATTG
TTGTCCTAGCCAGACCTCTGATCCCTCTCCCCCAAACCCCATATCTCTTCCTTAACTCCTGGCCCC
CGATGCTCAACTTGACCCTGCATTAGAAATTGAAGACTGTAAATACAAAATAAAATTATGGTGAGA
TTATG
```

>gi|207030|gb|M25890.1|RATSOMX Rat somatostatin mRNA, complete cds
```
TGCGGACCTGCGTCTAGACTGACCCACCGCGCTCAAGCTCGGCTGTCTGAGGCAGGGGAGATGCTG
TCCTGCCGTCTCCAGTGCGCGCTGGCCGCGCTCTGCATCGTCCTGGCTTTGGGCGGTGTCACCGGG
GCGCCCTCGGACCCCAGACTCCGTCAGTTTCTGCAGAAGTCTCTGGCGGCTGCCACCGGGAAACAG
GAACTGGCCAAGTACTTCTTGGCAGAACTGCTGTCTGAGCCCAACCAGACAGAGAACGATGCCCTG
GAGCCTGAGGATTTGCCCCAGGCAGCTGAGCAGGACGAGATGAGGCTGGAGCTGCAGAGGTCTGCC
AACTCGAACCCAGCCATGGCACCCCGGGAACGCAAAGCTGGCTGCAAGAACTTCTTCTGGAAGACA
TTCACATCCTGTTAGCTTTAATATTGTTGTCTCAGCCAGACCTCTGATCCCTCTCCTCCAAATCCC
ATATCTCTTCCTTAACTCCCAGCCCCCCCCCCAATGCTCAACTAGACCCTGCGTTAGAAATTGAAG
ACTGTAAATACAAAATAAAATTATGGTGAAATTATG
```

>gi|163636|gb|M31217.1|BOVPSOMA Bovine somatostatin mRNA, complete cds
```
AAGCTGCTTTAGGAGAGGCAAGGTTCGAGCCGTCGCTGCTGCCTGCGATCAGCTCCTAGAGTTTGA
ACTCTAGCTCGGCTTCGCCGCCGCCGCCGAGATGCTGTCCTGCCGCCTCCAGTGCGCGCTGGCCGC
GCTCTCCATCGTCCTGGCTCTTGGCGGTGTCACCGGCGCGCCCTCGGATCCCCGGCTCCGTCAGTT
TCTGCAGAAATCCCTGGCTGCTGCCGCGTGGCAAGCAGGAACTGGCCAAGTACTTCTTGGCAGAGCT
GCTGTCTGAACCCAACCAGACAGAGATTGATGCCCTGGAGCCTGAAGATTTGTCCCAGGCTGCTGA
GCAGGATGAAATGAGGCTGGAGCTGCAGAGATCTGCTAACTCAAACCCGGCCATGGCACCCCGAGA
ACGCAAAGCTGGCTGCAAGAATTTCTTCTGGAAGACTTTCACATCCTGTTAACTTTATTAATGATT
GTTGCCCATATAAGACCTCTGATTCCTCTTCTCCAAACCCCTTCTCACCTCCCTAATCCCTCCAAT
CCTCAATAAGACCCTCGTGTTAGAAATTGAAGACTGTAAATACAAAATAAAATTATGGGAAATTAT
G
```

10. Construct recombinant DNA by inserting the synthetic DNA encoding calcitonin polypeptide with the following sequence into the circular plasmid which you have retrieved from Exercise 3.

>gi|2174193|dbj|E06006.1|E06006 Synthetic DNA encoding human calcitonin
```
GGATCCGCGTTGCGGTAATCTGTCTACTTGCATGCTGGGCACTTACACCCAGGACTTCAACAAATT
CCACACCTTCCCGCAGACTGCAATCGGCGTTGGAGCACCGCTGCGTGCAGATCCGCGTTGCGGTAA
TCTGTCTACTTGCATGCTGGGCACTTACACCCAGGACTTCAACAAATTCCACACCTTCCCGCAGAC
TGCAATCGGCGTTGGAGCACCGCTGCGTGCAGATCT
```

REFERENCES

Altschul, S. F., Gish, W., Miller, W., Myers, E. W., and Lipman, D. J. (1990) *J. Mol. Biol.* **215**:403–410.

Benson, D. A., Karsch-Mizrachi, I., Lipman, D. J., Ostell, J. Rapp, B. A., and Wheeler, D. L. (2002). *Nucleic Acids Res.* **30**:17–20.

Berger, S. L., and Kimmel, A. R., Eds. (1987) *Guide to Molecular Cloning Techniques. Methods in Enzymology*, Vol. 152. Academic Press, New York.

Bernal, A., Ear, U., and Kyrpides, N. (2001) *Nucleic Acids Res.* **29**:126–127.

Brown, T. A. (1994) *DNA Sequencing: A Practical Approach.* IRL Press. Oxford, U.K.

Brown, T. A. (2001) *Gene Cloning and DNA Analysis*, 4th edition. Blackwell Science, Malden, MA.

Claverie, J.-M. (2001) *Science* **219**:1255–1257.

Cowell, I. G. and Austin, C. A., Eds. (1997) *cDNA Library Protocols.* Humana Press, Totowa, NJ.

Crick, F. H. C. (1970) *Nature* **227**:561–563.

Emmert, D. B., Stoehr, P. J., Stoesser, G., and Cameron, G. N. (1994) *Nucleic Acids Res.* **22**:3445–3449.

Eun, H.-M. (1996) *Enzymology Primer for Recombinant DNA Technology.* Academic Press, San Diego, CA.

Fleischmann, R. D. *et al.* (1995) *Science* **269**:496–512.

Fox, T. D. (1987) *Annu. Rev. Genet.* **21**:67–91.

Greene, J. J., and Rao, V. B. (1998) *Recombinant DNA: Principles and Methodologies.* Marcel Dekker, New York.

Lauritzen, P., Ed. (2001) *Cloning and the Future of Human Embryo Research.* Oxford University Press, Oxford.

Lewin, B. (1987) *Genes*, 3rd edition. John Wiley & Sons, New York.

Lipman, D. J., and Pearson, W. R. (1985) *Science* **227**:1435–1441.

Lu, Q., and Weiner, M. P. Eds. (2001) *Cloning and Expression Vectors for Gene Function Analysis.* Eaton Publishers, Natick, MA.

Maxim, A. M. and Gilbert, W. (1980) *Meth. Enzymol.* **65**:499–560.

Miklos, G. L. G., and Rubin, G. M. (1996). *Cell* **86**:521–529.

Mullis, K. B., Ferré, F., and Gibbs, R. A., Eds. (1994) *The Polymerase Chain Reaction.* Birkhauser, Boston.

Newton, C. R., and Graham, A. (1997) *PCR*, 2nd edition. Springer, New York.

Rantak, M. L., and Milgram, A. J., Eds. (1999) *Cloning.* Open Court, Chicago.

Rapley, R., Ed. (2000) *The Nucleic Acid Protocols Handbook.* Humana Press, Totowa, NJ.

Roberts, R. J., and Macelis, D. (2001) *Nucleic Acids Res.* **29**:268–269.

Saiki, R. K., Gelfrand, D. H., Stoffel, S., Scharf, S. J., Higuchi, R., Horn, G. T., Mullins, K. B., and Erlich, H. A. (1988) *Science* **239**:487–494.

Smith, A. J. H. (1980) *Meth. Enzymol.* **65**:560–580.

Stoesser, G., Baker, W., van den Broek, A., Camon, E., Garcia-Pastor, M., Kanz, C., Kulikowa, T., Leinonen, R., Lin, Q., Lombard, V., Lopez, R., Redaschi, N., Stoehr, P., Tuli, M. A., Tzouvara, K., and Vaughan, R. (2002) *Nucleic Acids Res.* **30**:21–26.

Tateno, Y., Imanishi, T., Miyazaki, S., Fukami-Kobayashi, K., Saitou, N., Sugawara, H., and Gojobori, T. (2002) *Nucleic Acids Res.* **30**:27–30.

Thompson, A. K., and Chadwick, R. F., Eds. (1999) *Genetic Information: Acquisition, Access and Control.* Kluwer Academic/Plenum Publishers, New York.

Timmer, W. C., and Villalobos, J. M. (1993) *J. Chem. Ed.* **70**:273–280.

Venter, J. C., *et al.* (2001) *Science* **291**:1304–1351.

Watson, J. D., Gilman, M., Witkowski, J., and Zoller, M. (1992) *Recombinant DNA*, 2nd edition. Scientific American/W.H. Freeman, New York.

Wu, R. (1993) *Meth. Enzymol.* **67**:431–468.

Yount, L., Ed. (2000) *Cloning.* Greenhaven Press, San Diego, CA.

Yudate, H., Suwa, M., Irie, R., Matsui, H., Nishikawa, T., Nakamura, Y., Yamaguchi, D., Peng, Z. Z., Yamamoto, T., Nagai, K., Hayashi, K., Otsuki, T., Sugiyama, T., Ota, T., Suzuki, Y., Sugano, S., Isogai, T., and Masuko, Y. (2001) *Nucleic Acids Res.* **29**:185–188.

10

GENOMICS: GENE IDENTIFICATION

Computational genome annotation is the process of genome database mining, which involves the gene identification and the assignment of gene functionality. Approaches to the gene identification are discussed. Online genome databases are surveyed, and the Internet resources for the gene identification are described.

10.1. GENOME INFORMATION AND FEATURES

The amount of information in biological sequences is related to their compressibility. Conventional text compression schemes are so constructed that the original data can be recovered perfectly without losing a single bit. Text compression algorithms are designed to provide a shorter description in the form of a less redundant representation, normally called a code, which may be interpreted and converted back into the uncompressed message in a reversible manner (Rival et al., 1996). Biochemistry is full of such code words (e.g., A, T, G, C) for nucleotides of DNA sequences which provide genome information. There are three levels of genome information to be considered:

1. The chromosomal genome, or simply genome referring to the genetic information common to every cell in the organism.
2. The expressed genome or transcriptome referring to the part of the genome that is expressed in a cell at a specific stage in its development.

3. The proteome referring to the protein molecules that interact to give the cell its individual character.

Online bibliographies of publications relevant to analysis of nucleotide sequences are accessible at SEQANALREF (http://expasy.hcuge.ch).

DNA sequence databases typically contain genomic sequence data, which includes information at the level of the untranslated sequence, introns and exons (eukaryotics), mRNA, cDNA (complementary DNA), and translations. Sequence segments of DNA which encode protein products or RNA molecules are called coding regions, whereas those segments that do not directly give rise to gene products are normally called noncoding regions. A coding sequence (CDS, cds) is a subsequence of a DNA sequence that is surmised to encode a gene. A CDS begins with an initiation codon (ATG) and ends with a stop codon. In the cases of spliced genes, all exons and introns should be within the same CDS. Noncoding regions can be parts of genes, either as regulatory elements or as intervening sequences. Untranslated regions (UTRs) occur in both DNA and RNA. They are portions of the sequence flanking the final coding sequence. The 5' UTR at the 5' end contains promoter site, and the 3' UTR at the 3' end is highly specific both to the gene and to the species from which the sequence is derived.

It is possible to translate a piece of DNA sequence into protein by reading successive codons with reference to a genetic code table. This is termed the *conceptual translation*, which has no biological validation or significance. Because it is not known whether the first base marks the start of the CDS, it is always essential to perform a *six-frame translation*. This includes three *forward frames* that are accomplished by beginning to translate at the first, second, and third bases, respectively, and three *reverse frames* that are achieved by reversing DNA sequence (the complementary strand) and again beginning on the first, second, and third bases. Thus for any DNA genome, the result of a six-frame translation is six potential protein sequences of which only one is biologically functional.

The correct reading frame is normally taken to be the longest frame uninterrupted by a stop codon (TGA, TAA, or TAG). Such a frame is known as an *open reading frame* (ORF). An ORF corresponds to a stretch of DNA that could potentially be translated into a polypeptide. For an ORF to be considered as a good candidate for coding a cellular protein, a minimum size requirement is often set — for example, a stretch of DNA that would code for a protein of at least 100 amino acids. An ORF is not usually considered equivalent to a gene until there has been shown to be a phenotype associated with a mutation in the ORF, and/or an mRNA transcript or a gene product generated from the DNA molecule of ORF has been detected.

The initial codon in the CDS is that for methionine (ATG), but methionine is also a common residue within the CDS. Therefore its presence is not an absolute indicator of ORF initiation. Several features may be used as indicators of poetential protein coding regions in DNA such as sufficient ORF length, flanking Kozak sequence (CCGCCATGG) (Kozak, 1996), species-specific codon usage bias, and detection of ribosome binding sites upstream of the start codon of prokaryotic genes. Ultimately, the surest way of predicting a gene is by alignment with a homologous protein sequence.

The eukaryotic genes are characterized by (a) regions that contribute to the CDS, known as exons, and (b) those that do not contribute, known as introns. The presence of exons and introns in eukaryotic genes results in potential gene products with different lengths because not all exons are jointed in the final transcribed mRNA. The proteins resulting from the mRNA editing process with different translated polypeptide chains are known as *splice variants* or *alternatively spliced forms.* Therefore, database searches with cDNA show that the sequences with substantial deletions in matches to the query sequences could be the result of alternative splicing.

To prepare a cDNA library (Cowell and Austin, 1997) suitable for use in rapid sequence experiments, a sample of cells is obtained, then RNA is extracted from the cells and is employed as the template for reverse transcriptase to synthesize cDNA which is transformed into a library. A sample of clones, each between 200 and 500 bases representing part of the genome, is selected from the library at random for sequence analysis. The sequences that emerge from this process are called *expressed sequence tags* (EST). Good libraries contain at least 1 million clones and probably substantially more, though the actual number of distinct genes expressed in a cell may be a few thousand. The number varies according to cell type—for example, human brain cells with $\sim$ 15,000 genes and gut cells with $\sim$ 2000 genes. Thus, a large part of currently available DNA data is made up of partial sequences, the majority of which are ESTs.

A DNA segment with repeat sequences is an indication of unlikely protein coding region, whereas a segment with apparent codon bias is one of an indicator being the protein coding region. Sequence similarity to other genes or gene products provides strong evidence for exons, and matches to template patterns may indicate the locations of functional sites on the DNA.

10.2. APPROACHES TO GENE IDENTIFICATION

Genomics conducts structural and functional studies of genomes (McKusick, 1997; Shapiro and Harris, 2000; Starkey and Elaswarapu, 2001). The former deals with the determination of DNA sequences and gene mapping, while the latter is concerned with the attachment of functional information to existing structural knowledge about DNA sequences. Genome database mining refers to the process of computational genome annotation by which uncharacterized DNA sequence is documented by the location along the DNA sequence involved in genome functionality. Two levels of computational genome annotation are structural annotation or gene identification and functional annotation or assignment of gene functionality. Structural annotation refers to the identification of ORFs and gene candidates in a DNA sequence using the computational gene discovery algorithm (Borodovsky and McIninch, 1993: Bunset and Guigo, 1996; Guigo et al., 1992; Huang et al., 1997). Functional annotation refers to the assignment of function to the predicted genes using sequence similarity searches against other genes of known function (Bork et al., 1998; Gelfand, 1995).

There are two approaches in the development of sequence analysis techniques for gene identification (Fickett, 1996). *Template methods* attempt to compose more or less concise descriptions of prototype objects and then identify genes by matching to such prototypes. The method identifies statistical patterns that distinguish coding from noncoding DNA sequences. A good example is the use of consensus sequences in identifying promoter elements or splice sites. Curation of data is a prerequisite to developing pattern recognition algorithms for identifying features of biological interest for which well-cleansed, nonredundant databases are needed. Information on nonredundant, gene-oriented clusters can be found at UniGene (http://www/ncbi.nlm.nih.gov/UniGene/) and TIGR Gene Indices (http://www.tigr.org/tdb/tdb.html). *Look-up methods*, on the other hand, attempt to identify a gene or gene component by finding a similar known object in available databases. The method relies on finding coding regions by their homology to known expressed sequences. An excellent example is the search for genes by trying to find a similarity between the query sequence and the contents of the sequence databases. Some representative human gene expression databases include Cellular Response Database (http://LH15.umbc.edu/crd), GeneCards (http://bioinformatics.weizmann.ac.il/cards/), Globin Gene Server (http://globin.csc.psu.edu), Human Developmental Anatomy (http://www.ana.ed.ac.uk/anatomy/database/), and Merck Gene Index (http://www.merck.com/mrl/merckgeneindex.2.html).

The basic information flow for an overall gene identification protocol follows:

1. *Masking Repetitive DNA.* A map of repeat locations shows where regulatory and protein-coding regions are unlikely to occur (Smit, 1996). It is therefore best to locate and remove interspersed and simple repeats from eukaryotic sequences as the first step in any gene identification analysis. Because such repeats rarely overlap promoters or the coding portions of exons, their locations can provide important negative information on the location of gene features. RepeatMasker (http://www.genome.washington.edu/analysistools/repeatmask.htm) provides annotation and masking of both interspersed and simple repeats.

2. *Database Searches.* Sequence similarity to other genes or gene products provides strong positive evidence for exons. Thus, searching for a known homologue is perhaps the most widely understood means of identifying new CDS. For protein-coding genes, translating the sequence in all six possible reading frames and using the result as a query against databases of amino acid sequences and functional motifs is usually the best first step for finding important matches. Once a homologue has been found, a program such as Procrustes http://www.hto.usc.edu/software/procrustes) may be used to make an optimal alignment between the known gene product and the new gene (Gelfand et al., 1996). It is shown (Green et al., 1993) that:

(a) Most ancient conserved regions, ACR (regions of protein sequences showing highly significant homologies across phyla), are already known and may be found in current databases.

(b) Roughly 20–50% of newly found genes contain an ACR that is represented in the databases.

(c) Rarely expressed genes are less likely to contain an ACR than moderately or highly expressed ones.

The EST databases probably contain fragments of a majority of all genes (Aaronson et al., 1996). Current estimates are that the public EST databases of human and mouse ESTs represent more than 80% of the genes expressed in these organisms. Thus they are an important resource for locating some part of most genes. Finding good matches to ESTs is a strong suggestion that the region of interest is expressed and can give some indication as to the expression profiles of the located genes. However, the problem of inferring function remains because an EST that represents a protein without known relatives will not be informative about gene function. An access to dbEST can be found at http://www.ncbi.nlm.nih.gov/dbEST/index.html.

3. Codon Bias Detection. Statistical regularity suggesting apparent "codon bias" over a region is one of the indicators of protein-coding regions. Information on codon usage can be found at the Codon usage database, http://www.kazusa.or.jp/codon/ or http://biochem.otago.ac.nz:800/Transterm/homepage.html. Most computational identification of protein-coding genes relies heavily on recognizing the somewhat diffuse regularities in protein-coding regions that are due to bias in codon usage. Some of the most informative coding measures are:

- Dicodon counts — that is, frequency counts for the occurrence of successive codon pairs
- Certain measure of periodicity — that is, the tendency of multiple occurrences of the same nucleotide to be found at distances of $3n$ bp
- A measure of homogeneity versus complexity — that is, counting long homopolymer runs
- Open reading frame occurrence (Fickett and Tung, 1992).

Many coding region detection programs are primarily the result of combining the numbers from one or more coding measures to form a single number called a discriminant. Typically, the discriminant is calculated in a sliding window (i.e., for successive subsequence of fixed length) and the result is plotted. To gain significant information from a coding measure discriminant, a DNA chain of more than 100 bases is required.

4. Detecting Functional Sites in the DNA. Because matches to template patterns may indicate the location of functional sites on the DNA, it would be more assuring if we are able to recognize the locations such as transcription factor binding sites and exon/intron junctions where the gene expression machinery interacts with the nucleic acids. One way to summarize the essential information content of these locations termed *signals* is to give the *consensus sequence*, consisting of the most common base at each position of an alignment of specific binding sites. Consensus sequence are very useful as mnemonic devices but are typically not very reliable for discriminating true sites from pseudosites. One technique for discriminating between true sites and pseudosites with a basis in physical chemistry is the position weight matrix (PWM). A score is assigned to each possible nucleotide at each possible position of the signal. For any particular sequence, considered as a possible occurrence of the signal, the appropriate scores are summed to give a score to a potential site. Under some circumstances this score may be approximately proportional to the energy of binding for a control ribonucleoprotein/protein.

Promoters. The promoter is an information-rich signal that regulates transcription, especially for eukaryotes. Computer recognition of promoters (Fickett and Hatzigeorgiou, 1997) is important partly for the advance it may provide in gene identification. Available programs include those depending primarily on simple oligonucleotide frequency counts and those depending on libraries describing transcription factor binding specificities together with some description of promoter structure. Information on eukaryotic polymerase II (mRNA synthesis) promoters can be found from EPD (Praz, et al., 2002) at http://www.epd.isb-sib.ch.

Intron Splice Sites. PWMs have been complied for splice sites in a number of different taxonomic groups (Senapathy et al., 1990), and these may be the best resource available for analysis in many organisms. Unfortunately, PWM analysis of the splice junction provides rather low specificity perhaps because of the existence of multiple splicing mechanisms and regulated alternative splicing. Many of the integrated gene identification services provide separate splice site predictions. ExInt database (http://intron.bic.nus.edu.sg/exint/extint.html) provides information on the exton–intron structure of eukaryotic genes. Various information resources related to introns can be found at EID (http://mcb.harvard.edu/gilbert/EID/) and Intron (http://nutmeg.bio.indiana.edu/intron/index.html) servers.

Translation Initiation Site. In eukaryotes, if the transcription start site is known, and there is no intron interrupting the 5′ UTR, Kozak's rule (Kozak, 1996) probably will locate the correct initiation codon in most cases. Splicing is normally absent in prokaryotes, yet because of the existence of multicitronic operons, promoter location is not the key information. Rather, the key is reliable localization of the ribosome binding site. The TATA sequence about 30 bp from the transcription start site may be used as a possible resource.

Termination Signals. The polyadenylation and translation termination signals also help to demarcate the extent of a gene (Fickett and Hatzigeorgiou, 1997).

Untranslated Regions. The 5′ and 3′ UTRs are portions of the DNA/RNA sequences flanking the final coding sequence (CDS). They are highly specific both to the gene and to the species from which the sequences are derived. Thus UTRs can be used to locate CDS—for example, eukaryotic polyadenylation signal (Wahle and Keller, 1996), mammalian pyrimidine-rich element (Baekelandt et al., 1997), mammalian AU-rich region (Senterre-Lesenfants et al., 1995), *E. coli* fbi (3′-CCGCGAAGTC-5′) element (French et al., 1997), and so on. Information on 5′ and 3′ UTRs of eukaryotic mRNAs can be found at UTRdb (http://bigrea.area.ba.cnr.it:8000/srs6/) (Pesolo et al., 2002).

Online bibliographies (indexed by years) on computational gene recognition are maintained at http://linkage.rockefeller.edu/wli/gene/right.html. Some computational gene discovery programs accessible through the WWW are listed in Table 10.1. Most programs are organism-specific and require a selection of organisms. Because no single prediction method is perfect, it is advisable to submit the query sequence to several different computational programs for the gene prediction.

10.3. GENE IDENTIFICATION WITH INTERNET RESOURCES

Computational analyses of DNA sequences/gene identification can be carried out by the use of Internet resources. Because no single tool can perform all the relevant sequence analysis leading to gene identification, it is recommended to submit the query sequence to the analysis of several software packages to make use of the best

TABLE 10.1. Web Servers for Gene Identification

Program	URL
Aat	http://genome.cs.mtu.edu/aat.html
BCM GeneFinder	http://dot.imgen.bcm.tmc.edu:9331/gene-finder/gfb.html
CGG GeneFinder	http://genomic.sanger.ac.uk/gf/gfb.html
GeneID	http://www1.imim.es/software/geneid/index.html
GeneMark at EBI	http://www2.ebi.ac.uk/genemark/
Genie	http://www.fruitfly.org/seq_tools/genie.html
GenLang	http://www.cbil.upenn.edu/genlang/genlang_home.html
GenScan	http://genes.mit.edu/GENSCAN.html
Grail/Grail 2	http://compbio.ornl.gov/gallery.html
Mzef	http://www.cshl.org/genefinder/
ORFgene	http://www.ncbi.nlm.nih.gov/gorf/gorf.html
Procrustes	http://www-hto.usc.edu/software/procrustes/wwwserv.html
WebGene	http://www.itba.mi.cnr.it/webgene/
WebdGeneMark	http://genemark.biology.gatech.edu/GeneMark/webgenemark.html

computational techniques. It must be emphasized that a gene predicted by computational methods must be viewed as a hypothesis subject to experimental verification.

10.3.1. Genomic Databases

The nucleotide sequences can be retrieved from one of the three IC (International Collaboration) nucleotide sequence repositories/databases: GenBank (http://www.ncbi.nlm.nih.gov/GenBabk/), EMBL Nucleotide Sequence Database (http://www.ebi.ac.uk/embl.html), and DNA Data Bank of Japan (http://www.ddbj.nig.ac.jp). The GenBank nucleotide sequence database can be accessed from the integrated database retrieval system of NCBI, Entrez (http://www.ncbi.nlm.nih.gov), by selecting the Nucleotide or Genome menu. The EMBL Nucleotide Sequence Database can be accessed from EBI outstation (http://www.ebi.ac.uk/) using SRS (Sequence Retrieval System) querying. The DNA Data Bank of Japan can be accessed from the DBGet link (http://www.genome.ad.jp/dbget/). To retrieve a nonredundant, gene-oriented cluster, conduct an organism/keyword search at UniGene of NCBI (http://www.ncbi.nlm.nih.gov/UniGene/).

The DNA sequence of an unknown gene often exhibits structural homology with a known gene. The sequence alignment is important for the recognition of patterns or motifs common to a set of functionally related DNA sequences and is of assistance in structure prediction. The multiple sequence alignment can be conducted at EBI using ClustalW tool (http://www.ebi.ac.uk/clustalW/) (Figure 10.1).

To perform ClustalW sequence alignment, upload sequences from the user's file in fasta format by clicking Upload (which opens the Browser box). Enter user's e-mail address and alignment title, choose the desired options, and click the Run ClustalW button. Record the URL address of posting "Your Job output results" (e.g., http://www.ebi.ac.uk/servicestmp/.html) from where the alignment, scores, and tree files can be retrieved. The ClustalW alignment is also available at DDBJ (http://spiral.gene.nig.ac.jp/homology/clustalw-e.shtml).

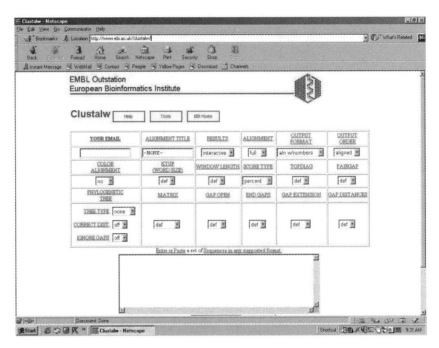

Figure 10.1. ClustalW server of EBI.

EST data are held in the dbEST database, which maintains its own format and identification number system and is accessible via the NCBI Web server, http://www.nbi.nlm.nih.gov/dbEST/. The sequence data, together with a summary of the dbEST annotation, are also distributed as a subsection of the primary DNA database. The publicly available EST analysis tools fall into three categories:

1. *Sequence similarity search tools:* Alignments of the query sequence with databases produce sequence similarity. The BLAST series of programs has variants that will translate DNA databases, translate the input sequence, or both. FASTA provides a similar suite of programs.

2. *Sequence assembly tools:* The iterative sequence alignment by incorporating the matched consensus sequence from the previous search into the subsequent round of search is called sequence assembly. The available tools are Staden assembler, TIGR assembler, and so on.

3. *Sequence clustering tools:* Sequence clustering tools are programs that take a large set of sequences and divide them into subsets or clusters, based on the extent of shared sequence identity in a minimum overlap region.

The genome information is available at Entrez and at TDB (http://www.tigr.org/tdb/tdb.html) of the Institute for Genomic Research (TIGR). The TDB provides DNA and protein sequence, gene expression, cellular role, protein family information, and taxonomic data of microbes, plants, and humans. The resources include microbial genome database, gene indices (analysis of public EST), and human chromosome genomic project with links worldwide. Other general as well as

TABLE 10.2. Some Genomic Databases

Database	URL	Sequence/Genome
AceDB	http://www.sanger.ac.uk/Software/Acedb/	*Cae.elegans*
AtDB	http://genome-www.stanford.edu/Arabidopsis	*Arabidopsis thaliana* genome
Celera	http://www.celera.com	Human genome data and tools
CropNet	http://synteny.nott.ac.uk	Crop plants genome mapping
EcoGene	http://bmb.med.miami.edu/EcoGene/EcoWeb	*Eschericia coli* K-12 sequences
EMGlib	http://pbil.univ-lyon1.fr/emglib/emglib.html	Bacterial and yeast genome
FlyBase	http://www.fruitfly.org	Drosophila sequence and genome
GDB	http://www.gdb.org	Human genes and genomic maps
HuGeMap	http://www.infobiogen.fr/services/Hugemap	Human genomic map
INE	http://rgp.dna.affrc.go.jp/gict/INE.html	Rice genomic maps and sequences
MitBase	http://www3.ebi.ac.uk/Research/Mitbase/	Mitochondrial genomes
NRSub	http://pbil.univ-lyon1.fr/nrsub/nrsub.html	*Bacillus subtilis* genome
RsGDB	http://utmmg.med.uth.tmc.edu/sphaeroides	*Rhodobacter sphaeroides* genome
Sanger	http://genomic.sanger.ac.uk/inf/infodb.shtml	Human, mammals, and eukaryotes
SDG	http://genome-www.stanford.edu/Saccharomyces	*Saccharomyces cerevisiae* genome
TIGR	http://www.tigr.org/tdb/mdb/mdb.html	Microbial genomes and links
Uwisc	http://www.genetics.wisc.edu/	*Escherichia coli* genome
ZmDB	http://zmdb.iastate.edu/	Maize genome

specialized genomic databases can be found online, and some of representative servers are given in Table 10.2.

The GeneQuiz of EBI at http://columbs.ebi.ac.uk:8765/ lists the completed genomes alphabetically according to species name with pointers to sequence databases at EBI. Similar information is also available at Genome Information Broker (http://gib.genes.ac.jp/gib_top.html). Various information and links to human genome project can be obtained at NCBI (http://www/ncbi.nlm.nih.gov/), Sanger Centre (http://genomic.sanger.ac.uk/), and GeneStudio (http://studio.nig.ac.jp/).

10.3.2. Gene Expression

Eukaryotic genes may contain noncoding repetitive elements. The removal of these repetitive elements facilitates database searches. RepeatMasker (http://www.washington.edu/uwgc.analysistools/repeatmask.htm) screens a query sequence against a library of repetitive elements and returns the masked DNA sequence. Select Submit data to RepeatMasker web server. Paste the query sequence, choose return format and method, pick organism, select analysis and output options, and then click the Submit Sequence button. The return lists repeat sequence (positions, matching repeats and their repeat classes), summary, and masked sequence with X's (if the option, "mask with X's" is selected) as shown in Figure 10.2.

To identify six frames ORF, launch ORF Finder at http://www.ncbi.nlm.nih.gov/gorf/gorf.html. Paste the query sequence on the query box (Sequence ID is followed by > in FASTA format), select genetic codes, and click OrfFind. The query returns with the bar graph showing ORFs and the list of six ORF translations (Figure 10.3). Press the SixFrames button, and then the View All button to display all six ORF (* indicates stop codon) of the query sequence. Alternatively, activate the frame (bar) to display the nucleotide and protein (translate) sequences as shown

```
>human lysozyme (sequence 511-1500 showing masked Alusx form 917-1224)
GCTACGTATGGAACAGACACTAGGAGAGAAGGAAGAAGAAGAAGGGGCTT
TGAGTGAATAGATGTTTTATTTCTTTGTGGGTTTGTATACTTACAATGGC
TAAAAACATCAGTTTGGTTCTTTATAACCAGAGATACCCGATAAAGGAAT
ACGGGCATGGCAGGGGAAAATTCCATTCTAAGTAAAACAGGACCTGTTGT
ACTGTTCTAGTGCTAGGAAGTTTGCTGGGTGCCTGAGATTCAATGGCACA
TGTAAGCTGACTGAAAGATACATTTGAGGACCTGGCAGAGCTCTCTCTCA
AGTCCTTGGTATGTGACTCCAGTTATTTCCCATTTTGAACTTGGTCTGAG
AGCCTAGAGTGATCAGTATTTTCTTGTCTTCAAGTCCCCTGCCGTGATGT
GGGATTTTTATTTTTAXXXXXXXXXXXXXXXXXXXXXXXXXXXXXXXXXXX
XXXXXXXXXXXXXXXXXXXXXXXXXXXXXXXXXXXXXXXXXXXXXXXXXXX
XXXXXXXXXXXXXXXXXXXXXXXXXXXXXXXXXXXXXXXXXXXXXXXXXXX
XXXXXXXXXXXXXXXXXXXXXXXXXXXXXXXXXXXXXXXXXXXXXXXXXXX
XXXXXXXXXXXXXXXXXXXXXXXXXXXXXXXXXXXXXXXXXXXXXXXXXXX
XXXXXXXXXXXXXXXXXXXXXXXXXXXXXXXXXXXXXXXXXXXXXXXXXXX
XXXXXXXXXXXXXXXXXXXXXXXXXXXGACATGGGATTTTTAACAGTGATGTT
TTTAAGAATATATTGAATTCCCTACACAAGAGCAGTAGGAACCTAGTTCC
CTTCAGTCACTCTTTGTATAGGATCCCAGAAACTCAGCATGAAATGTTTT
ATTATTTTTATCTACCTACTTGATTAACTATCTTTCATTTCTCCCACACA
ATTCAAGATGTGCCATGAGGAAAGTTATTTATAGTTTAGTACATAGTTGT
CGATGTAATAATCTCTGTAGTTTTCAGATTGAATTCAGACATTTCCCCTC
```

Figure 10.2. Removal of noncoding repetitive elements with RepeatMasker. The nucleotide sequence encoding human lysozyme is submitted to RepeatMasker and the sequence with masked(X) noncoding repetitive elements is returned (partial sequence is shown here).

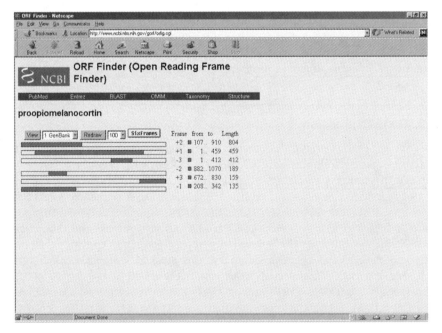

Figure 10.3. Identification of open reading frames with ORF Finder. The nucleotide sequence encoding human proopiomelanocortin mRNA (1071 bp) is submitted to ORF Finder. The return shows frame bars (from Frame +1 to Frame −3) with highlighted relative length of ORF and the list of ORF arranged according to their lengths.

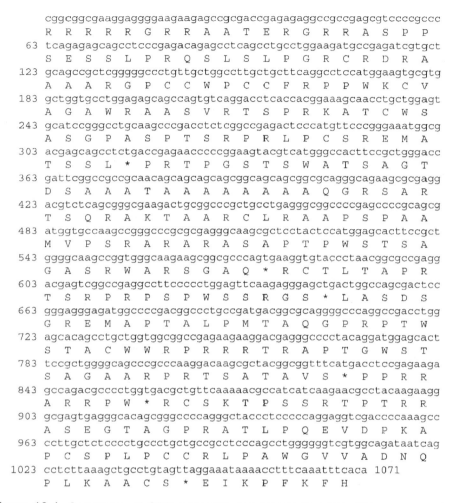

```
      cggcggcgaaggaggggaagaagagccgcgaccgagagaggccgccgagcgtccccgccc
      R   R   R   R   G   R   R   A   A   T   E   R   G   R   R   A   S   P   P
   63 tcagagagcagcctcccgagacagagcctcagcctgcctggaagatgccgagatcgtgct
      S   E   S   S   L   P   R   Q   S   L   S   L   P   G   R   C   R   D   R   A
  123 gcagccgctcggggggccctgttgctggccttgctgcttcaggcctccatggaagtgcgtg
      A   A   A   R   G   P   C   C   W   P   C   C   F   R   P   P   W   K   C   V
  183 gctggtgcctggagagcagccagtgtcaggacctcaccacggaaagcaacctgctggagt
      A   G   A   W   R   A   A   S   V   R   T   S   P   R   K   A   T   C   W   S
  243 gcatccgggcctgcaagcccgacctctcggccgagactcccatgttcccgggaaatggcg
      A   S   G   P   A   S   P   T   S   R   P   R   L   P   C   S   R   E   M   A
  303 acgagcagcctctgaccgagaacccccggaagtacgtcatgggccacttccgctgggacc
      T   S   S   L   *   P   R   T   P   G   S   T   S   W   A   T   S   A   G   T
  363 gattcggccgccgcaacagcagcagcagcggcagcagcggcgcagggcagaagcgcgagg
      D   S   A   A   A   T   A   A   A   A   A   A   A   Q   G   R   S   A   R
  423 acgtctcagcgggcgaagactgcggcccgctgcctgagggcggccccgagcccgcagcg
      T   S   Q   R   A   K   T   A   A   R   C   L   R   A   A   P   S   P   A   A
  483 atggtgccaagccgggcccgcgcgagggcaagcgctcctactccatggagcacttccgct
      M   V   P   S   R   A   R   A   R   A   S   P   T   P   W   S   T   S   A
  543 ggggcaagccggtgggcaagaagcggcgcccagtgaaggtgtaccctaacggcgccgagg
      G   A   S   R   W   A   R   S   G   A   Q   *   R   C   T   L   T   A   P   R
  603 acgagtcggccgaggccttcccccctggagttcaagagggagctgactggccagcgactcc
      T   S   R   P   R   P   S   P   W   S   S   R   G   S   *   L   A   S   D   S
  663 gggagggagatggccccgacggccctgccgatgacggcgcaggggcccaggccgacctgg
      G   R   E   M   A   P   T   A   L   P   M   T   A   Q   G   P   R   P   T   W
  723 agcacagcctgctggtggcggccgagaagaaggacgagggcccctacaggatggagcact
      S   T   A   C   W   W   R   P   R   R   R   T   R   A   P   T   G   W   S   T
  783 tccgctggggcagcccgcccaaggacaagcgctacggcggtttcatgacctccgagaaga
      S   A   G   A   A   R   P   R   T   S   A   T   A   V   S   *   P   P   R   R
  843 gccagacgcccctggtgacgctgttcaaaaacgccatcatcaagaacgcctacaagaagg
      A   R   R   P   W   *   R   C   S   K   T   P   S   S   R   T   P   T   R   R
  903 gcgagtgagggcacagcgggccccagggctaccctcccccaggaggtcgaccccaaagcc
      A   S   E   G   T   A   G   P   R   A   T   L   P   Q   E   V   D   P   K   A
  963 ccttgctctccctgccctgctgccgcctcccagcctggggggtcgtggcagataatcag
      P   C   S   P   L   P   C   C   R   L   P   A   W   G   V   V   A   D   N   Q
 1023 cctcttaaagctgcctgtagttaggaaataaaacctttcaaatttcaca 1071
      P   L   K   A   A   C   S   *   E   I   K   P   F   K   F   H
```

Figure 10.4. Sequences of ORF nucleotide and protein translate. The sequences for nucleotides and translated amino acids from Frame +3 ORF of proopiomelanocortin DNA (Figure 10.3) are displayed by clicking SixFrame button and then activating the desired frame bar. The initiation codon (sky) and stop codon (pink) suggests potential start and end positions of CDS/translate(s) for the Frame.

in Figure 10.4. The DNA sequence can also be translated into six frame protein sequences by using Translate tool at ExPASy (http://www.expasy.ch/tools/). Various outputs (nucleotide, amino acid, and nucleotide with amino acid sequences) can be requested (the stop codons are indicated by hyphones). The ORFGene tool of WebGene (http://125.itba.mi.cnr.it/~webgene/wwworfgene2.html) predicts ORF of a query nucleotide sequence by first predicting potential exons followed by homology search of translated amino acid fragments.

PEDANT (http://pedant.mips.biochem.mpg.de/) at Biomax Informatics provides information on proteins as the gene products which are either known or derived from the best Blast hits. Select the listed organism (under the categories of complete genomic sequences and unfinished genomic sequences) on the home page to open the searchable entries. These include lists of ORF, contigs, descriptions/ pointers to protein function (homologues, paralogues, prosite patterns, domains, and sequence blocks of ORFs) and protein structures (known 3D, SCOP domains, transmembrane, CATH classes, signal peptides of ORFs).

10.3.3. Detection of Functional Sites

The ability to predict promoter elements, splice sites and exons greatly facilitate gene identifications of query DNA sequences. The promoter sequences for chromosomal genes of plant (e.g., maize), chromosomal genes of fruit fly, chromosomal genes of vertebrates (e.g., human, cattle, chicken, frog, mouse, and rat), transposable elements, retroviruses, and viral genes can be retrieved from EPD at http://epd.isb-sib.ch/seq_download.html. On the EPD home page, select promoter subset from the list, specify the sequence segment relative to transcription start for extraction, choose format, and click Retrieve sequence to open query form. The database uses the SRS query system, and clicking Do Query (after selecting category/entering keywords) returns the query results. The promoter sequence is given on the SE line of the output (Embl format).

The nucleotide sequence analysis of Computational Genomics Group (CGG) at the Sanger Centre (http://genomic.sanger.ac.uk) provides comprehensive tools for finding/predicting functional sites and gene structure (Figure 10.5). These tools are also available at GeneFinder site of Baylor College of Medicine (BCM) (http://dot.-imgen.bcm.tmc.edu). To predict eukaryotic polymerase II promoter region, paste the query sequence on the nucleotide sequence analysis request page, enter the sequence name, select either TSSG or TSSW (recognition of human pol II promoter region) tool, and click the Perform search button. The request returns number and positions of the predicted promoter sites (e.g., DNA encoding for human lysozyme with 5648 bp):

```
>hmLyz
   Length of sequence-    5648
   Threshold for LDF-    4.00
       1 promoter(s)   were predicted
   Pos.:    1931 LDF-    7.48 TATA box predicted at    1899
   Transcription factor binding sites:
   for promoter at position -    1931
```

The TATA box prediction is also available with HCtata tool of WebGene (http://www.itba.mi.cnr.it/webgene/).

The exon–intron database can be downloaded from EID at http://www.mcb.-harvard.edu/gilbert/EID/, whereas online database search can be conducted at ExInt (http://intron.bic.nus.edu.sg/exint/exint.html). On the ExInt home page, click Search ExInt by keywords to open the query page. Choose Complete ExInt for the Database, choose Text word for Search field, and enter the keyword (protein name, e.g., hexokinase, lysozyme). Click the Submit button to receive search results. The output includes locus name, description (viz. GenBank), NCBI ID and pointer (nucleotide sequence), phase and position of introns, number, size, and length (in amino acids) of exons, nucleotide position for the introns, and protein sequence

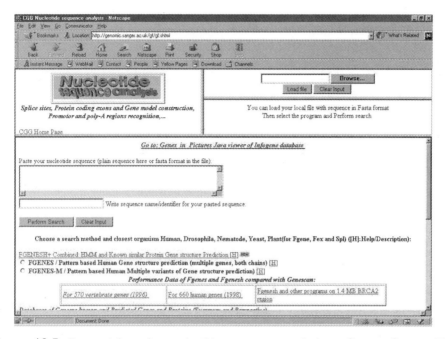

Figure 10.5. Request form for nucleotide sequence analysis at Sanger Center. The GeneFinder at Sanger Centre is a comprehensive gene identification site providing numerous tools for analyses of nucleotide sequences. These include FGENESH (hidden Markov model based on human/Drosophila/nematode/plant gene structure prediction), BESTORF (finding potential coding fragment in EST/mRNA), FGENE (finding gene by exon search and assembling), FEX (finding potential 5'-, internal, and 3'-coding exons), SPL (search for potential splice site), NSITE (recognition of regulatory motifs with statistics), POLYAH (recognition of 3'-end cleavage and polyadenylylation region), TSSG/ TSSW (recognition of human Pol II promoter and start of transcription), RNASPL (search for exon-exon junction positions), and HBR (recognition of human and *E. coli* sequences).

with link (intron which can be clicked to display nucleotide sequence of the intron. Numerous servers provide online prediction of exons such as GeneID http://www1.imim.es/software/geneid/index.html) and FEX tool of the Sanger Centre (Figure 10.6).

SpliceView tool of WebGene (http://125.itba.mi.cnr.it/~webgene/wwwsplice-view.html) applies consensus sequences to predict splicing signals. Select the organisms, enter DNA name, paste the query sequence, and click the Submit button. The online results consist of three windows: a graphical diagram of potential acceptor sites, a graphical presentation of potential donor sites, and tabular lists of potential donor and acceptor sites (a partial list is shown in Figure 10.7). The graphical presentations can be viewed at different threshold values by keying in the desired number, adjusting the side sliding bar, or simply choosing the desired scales (marginal, good, or excellent).

The SPL (search for potential splice sites) tool of the Sanger Centre predicts splice sites of an input query sequence. On the nucleotide sequence analysis page, paste the query sequence, enter the sequence name, select SPL tool, and click the

```
>hmLyz
 length of sequence -    5648
 # of potential exon:    4
  4140 -    4218 w= 12.74 ORF=  2 Num ORFs  1    4142 -    4216
  5072 -    5134 w=  6.90 ORF=  1 Num ORFs  3    5074 -    5133
  2025 -    2189 w=  5.81 ORF=  2 Num ORFs  2    2027 -    2188
  3873 -    3898 w=  4.05 ORF=  2 Last  exon     3875 -    3895
 Exon-       1 Amino acid sequence -       25aa
LLQDNIADAVACAKRVVRDPQGIRA
 Exon-       2 Amino acid sequence -       20aa
WHGEIVVKTEMSVSMFKVVE
 Exon-       3 Amino acid sequence -       54aa
MCLAKWESGYNTRATNYNAGDRSTDYGIFQINSRYWCNDGKTPGAVNACHLSCS
 Exon-       4 Amino acid sequence -       7aa
SSAKKNL
```

Figure 10.6. Prediction of exons with FEX at Sanger Centre. The nucleotide sequence of DNA encoding human lysozyme (5648 bp) is submitted to Sanger Centre for exon prediction with FEX. The output shows number of predicted exons, locations, and amino acid sequences of translates.

```
DONOR SITES:
POSITION            EXON INTRON      SCORE
     301            CTG GTCAGC        81.
     400            AAG GTGTGA        84.
     402            GGT GTGAGT        81.
     468            ACT GTAAGT        82.
     576            TTT GTGGGT        73.
     751            CAT GTAAGC        84.
     808            TTG GTATGT        82.
    1066            CAG GTGTGT        88.
    1190            GTG GTAGGA        80.
    1202            CAG GTGTGA        85.
      . .           ... ......        ..
ACCEPTOR SITES:
POSITION            INTRON   EXON     SCORE
      87            TTCTTCCAG AGCC     85.
     129            TTCCTGTAG ACTA     86.
     141            AATTCTTAG GACA     77.
     165            CTCTTCAAG CATT     79.
     328            ACCTAGCAG TCAA     77.
     384            ACGGTCCAG GGCA     85.
     494            TAATTCCAG AGAA     84.
     632            TATAACCAG AGAT     80.
     711            CTGTTCTAG TGCT     77.
     717            TAGTGCTAG GAAG     78.
      . .           .. ... .. ....     ..
```

Figure 10.7. Prediction of splice signals with SpliceView at WebGene. The nucleotide sequence of DNA encoding human lysozyme (5648 bp) is submitted to WebGene for prediction of splice signals with SpliceView tool. The potential donor sites and acceptor sites are listed (partial list) in the output.

Perform search button. The request returns the positions and scores (threshold) of the donor/acceptor sites (e.g., DNA encoding for human lysozyme with 5648 bp):

```
>hmLyz
  Length of sequence -     5648
Number of Donor sites:     5 Threshold: 0.76
     1     469     0.77
     2     752     0.76
     3     809     0.80
     4    1067     0.77
     5    4219     0.82
Number of Acceptor sites:    6 Threshold: 0.65
     1     946     0.66
     2    2024     0.74
     3    2187     0.66
     4    2728     0.68
     5    4139     0.84
     6    5071     0.78
```

To obtain information on untranslated regions (UTRs) at http://bio-www.ba.-cnr.it:8000/srs6/, select either UTRr (redundant) or UTRnr (nonredundant) to initiate the SRS query. Clicking Submit query (after selecting category/entering keywords) returns a list of hits (ID # starting with 5 for 5′-UTR and 3 for 3′-UTR). Choose the desired entry to receive an output (Embl format) with the sequence of UTR on the SQ line. The online prediction of Poly A region is available with POLYAH tool of the Sanger Centre and HCpolya tool of WebGene. On the nucleotide sequence analysis request page of the Sanger Centre, paste the query sequence, select POLYAH tool, and click the Perform search button. A list of potential polyadenylylation sites is returned (e.g., DNA encoding for human lysozyme with 5648 bp):

```
>hmLyz
  Length of sequence-     5648
      6 potential polyA sites were predicted
Pos.:    2581 LDF-    1.50
Pos.:    3248 LDF-    4.11
Pos.:    3325 LDF-    3.93
Pos.:    4093 LDF-    7.40
Pos.:    4789 LDF-    2.55
Pos.:    4804 LDF-    2.92
```

10.3.4. Integrated Gene Identification

The gene structure of an unknown nucleotide sequence can be identified by alignment against target protein sequences. Procrustes (http://www-hto.usc.edu/software/procrustes/wwwserv.html) performs spliced alignment of the predicted protein translations of a query nucleotide sequence against the target protein sequences (Gelfand et al., 1996). On the query page, enter the name of the query DNA and paste the query nucleotide sequence, then enter the names of the target protein and paste the amino acid sequence of the target protein. Click the Run Procrustes button. The user may request to receive the results by e-mail or online. The online results are kept in http://www.hto.usc.edu/tmp/fileid.html for approximately 10 hours after completing the task. The report includes pairwise presentations of spliced alignment score, positions and length of exons, translation of predicted gene based on the target protein, and spliced alignment with the target protein as shown in Figure 10.8.

Target Protein *pig lyz*
Spliced Alignment score=532 (65.5%)

Overprediction	Expected: 96%
Less than...	probability
5%	84%
10%	89%
20%	92%

	From	To	Length	
Start	333	468	136	---
---	2025	2189	165	---
---	4140	4218	79	---
---	5072	5135	64	Stop

Translation of Predicted Gene (based on *pig lyz*)
```
  1  MKALIVLGLV LLSVTVQGKV FERCELARTL KRLGMDGYRG ISLANWMCLA KWESGYNTRA
 61  TNYNAGDRST DYGIFQINSR YWCNDGKTPG AVNACHLSCS ALLQDNIADA VACAKRVVRD
121  PQGIRAWVAW RNRCQNRDVR QYVQGCGV
```

Spliced Alingment with *pig lyz*

```
====> Exon #1 (333..468) Score = 78.3%
target    1 KVYDRCEFARILKKSGMDGYRGVSLAN
            **||*** ** **| *******|****
source   19 KVFERCELARTLKRLGMDGYRGISLAN


====> Exon #2 (2025..2189) Score = 78.6%
target   28 WVCLAKWESDFNTKAINRNVG--STDYGIFQINSRYWCNDGKTPKAVNACHISCK
            *|******* |**|* * * * ********************* ******|**
source   46 WMCLAKWESGYNTRATNYNAGDRSTDYGIFQINSRYWCNDGKTPGAVNACHLSCS


====> Exon #3 (4140..4218) Score = 72.7%
target   81 VLLDDDLSQDIECAKRVVRDPQGIKA
            **|*|||| | ************|*
source  101 ALLQDNIADAVACAKRVVRDPQGIRA


====>Exon #4 (5072..5135) Score = 79.5%
target  107 WVAWRTHCQNKDVSQYIRGC
            ***** |***|** **||**
source  127 WVAWRNRCQNRDVRQYVQGC
```

Figure 10.8. Gene identification by Procrustes. The nucleotide sequence encoding human lysozyme is used as a query sequence to identify its gene structure against known protein sequence (i.e., pig lysozyme protein). The output includes sequence aignment of the source (predicted translate) versus target protein (pig lysozyme).

```
ID    POMC
OS    Homo sapiens
CC    Search in direct strand
CC    Results of gene prediction by GenView computer system.
FH    Key              Location/Qualifiers
FT    CDS              1..906
FT                     /note=potential exon predicted by GenView system
CC    Translated peptide
CC    AAAKEGKKSRDRERPPSVPALREQPPETEPQPAWKMPRSCCSRSGALLLALLLQASMEVR
CC    GWCLESSQCQDLTTESNLLECIRACKPDLSAETPMFPGNGDEQPLTENPRKYVMGHFRWD
CC    RFGRRNSSSSGSSGAGQKREDVSAGEDCGPLPEGGPEPRSDGAKPGPREGKRSYSMEHFR
CC    WGKPVGKKRRPVKVYPNGAEDESAEAFPLEFKRELTGQRLREGDGPDGPADDGAGAQADL
CC    EHSLLVAAEKKDEGPYRMEHFRWGSPPKDKRYGGFMTSEKSQTPLVTLFKNAIIKNAYKK
CC    G
SQ    1071 BP; 217 A; 355 C; 361 G; 138 T; 0 other;
      agcggcggcg aaggagggga agaagagccg cgaccgagag aggccgccga gcgtccccgc     60
      cctcagagag cagcctcccg agacagagcc tcagcctgcc tggaagatgc cgagatcgtg    120
      ctgcagccgc tcggggcccc tgttgctggc cttgctgctt caggcctcca tggaagtgcg    180
      tggctggtgc ctggagagca gccagtgtca ggacctcacc acggaaagca acctgctgga    240
      gtgcatccgg gcctgcaagc ccgacctctc ggccgagact cccatgttcc cgggaaatgg    300
      cgacgagcag cctctgaccg agaacccccg gaagtacgtc atgggccact tccgctggga    360
      ccgattcggc cgccgcaaca gcagcagcag cggcagcagc ggcgcagggc agaagcgcga    420
      ggacgtctca gcgggcgaag actgcggccc gctgcctgag ggcggccccg agccccgcag    480
      cgatggtgcc aagccgggcc cgcgcgaggg caagcgctcc tactccatgg agcacttccg    540
      ctggggcaag ccggtgggca agaagcggcg cccagtgaag gtgtacccta acggcgccga    600
      ggacgagtcg gccgaggcct tccccctgga gttcaagagg gagctgactg gccagcgact    660
      ccgggaggga gatggccccg acggccctgc cgatgacggc gcaggggccc aggccgacct    720
      ggagcacagc ctgctggtgg cggccgagaa gaaggacgag ggcccctaca ggatggagca    780
      cttccgctgg ggcagcccgc ccaaggacaa gcgctacggc ggtttcatga cctccgagaa    840
      gagccagacg cccctggtga cgctgttcaa aaacgccatc atcaagaacg cctacaagaa    900
      gggcgagtga gggcacagcg ggccccaggg ctaccctccc ccaggaggtc gaccccaaag    960
      ccccttgctc tcccctgccc tgctgccgcc tcccagcctg gggggtcgtg gcagataatc   1020
      agcctcttaa agctgcctgt agttaggaaa taaaaccttt caaatttcac a             1071
//
```

Figure 10.9. Output of GeneView tool of WebGene. The DNA encoding proopiomelanocortin mRNA (1071 bp) is submitted to gene prediction by GeneView at WebGene. The output adapts GenBank format.

GeneBuilder tool of WebGene (http://125.itba.mi.cnr.it/~webgene/genebuilder.-html) also derives the predicted gene from a query nucleotide sequence by comparing with an input protein sequence. Select (radio selections) prediction features (exon analysis, potential coding region identification, repeated element mapping, EST mapping, splice site prediction, potential coding regions, TATA box prediction, and poly-A site prediction) and click the Start analysis button. The GeneView tool of WebGene (http://125.itba.mit.cnr.it/~webgene/wwwgene.html) takes the query sequence and returns the predicted CDS in the GenBank format as shown in Figure 10.9.

The generalized hidden Markov model trained on human genes of Genie (Henderson et al., 1997) at http://www.fruitfly.org/seq_tools/genie.html accepts the input sequence in fasta format and returns the predicted exons by e-mail. The discriminant analysis at MZEF (http://argon.cshl.org/genefinder/human.htm) identifies protein coding regions in a human genome by returning the predicted internal coding exons (Zhang, 1997). The GeneFinder server (http://argon.cshl.org/genefinder/Pombe/pombe.htm) also predicts internal exons, introns, and ORF of a query sequence from *Schizosaccharomyces pombe*.

GeneID (Guigó, 1998) at http://www1.imim.es/software/geneid/index.html predict splice sites, codons, and exons along the query sequence using PWM. Exons are built and scored as the sum of the scores of the sites, plus the log-likelihood ratio of a Markov model for coding DNA. The gene structure is then assembled from the assembled exons. On the Web server page of GeneID at http://www1.imim.es/ geneid.html (Figure 10.10), paste the query sequence in fasta format (starting with > Name of sequence), choose prediction options and output options (acceptor sites, donor sites, start codons, stop codons, first exons, internal exons, terminal exons, and single gene). Click the Submit button to display predicted gene structures in a text listing (positions, sequences, and scores) and a graphical presentation.

The discriminant analysis (Mclachlan, 1992) based on pattern recognition of GeneFinder (Solovyev et al., 1994) is accessible at CGG of the Sanger Centre (http://genomics.sanger.ac.uk/gf/gfb.html) and Baylor College of Medicine (http:// dot.imgen.bcm.tmc.edu:9331/gene-finder/gfb.html). The GeneFinder offers various programs for predicting functional sites (e.g., SPL, TSSG/TSSW, FEX, POLYAH and NSITE etc.) and the gene structure such as FGENESH (hidden Markow model based gene structure prediction for multiple genes) and FGENE (finding gene by exon search and assembling). On the analysis request page of CGG at the Sanger Centre or BCM, paste the query sequence, enter the sequence name, choose an organism (human, *Drosophila*, *C.elegns*, yeast or plant), select the program (FGENESH or FGENE), and click the Perform Search button. The results with the predicted gene structure and statistics are returned (Figure 10.11).

The heuristic approach (Stagle, 1971) for the gene prediction of GeneMark can be accessed online at WebGeneMark (http://genemark.biology.gatech.edu/Gene-Mark/webgenemark.html) and at EBI (http://www2.ebi.ac.uk/genmark/). On

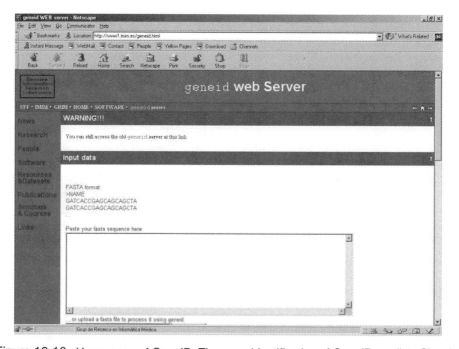

Figure 10.10. Home page of GeneID. The gene identification of GeneID predicts Signals (acceptor sites, donor sites, start codons, and stop codons) and exons (first exons, internal exons, terminal exons, all exons) and single genes.

```
Fgenesh
FGENESH 1.0 Prediction of potential genes in Human genomic DNA
Seq name: hmLyz
Length of sequence:  5648   GC content: 37 Zone: 1
Number of predicted genes 1 in +chain 1 in -chain 0
Number of predicted exons 4 in +chain 4 in -chain 0
Positions of predicted genes and exons:

 G Str Feature     Start      End    Score       ORF             Len

  1 +   1 CDSf     333  -    468     14.74     333  -     467      135
  1 +   2 CDSi    2025  -   2189      2.33    2027  -    2188      162
  1 +   3 CDSi    4140  -   4218      5.70    4142  -    4216       75
  1 +   4 CDSl    5072  -   5138     -0.42    5073  -    5138       66
  1 +     PolA    5194            -1.88

Predicted protein(s):
>FGENESH    1    4 exon (s)     333   -   5138    148 aa, chain +
MKALIVLGLVLLSVTVQGKVFERCELARTLKRLGMDGYRGISLANWMCLAKWESGYNTRA
TNYNAGDRSTDYGIFQINSRYWCNDGKTPGAVNACHLSCSALLQDNIADAVACAKRVVRD
PQGIRAWVAWRNRCQNRDVRQYVQGCGV



Fgene
>hmLyz
 length of sequence -    5648
 number of predicted exons -   4
 positions of predicted exons:

   333  -     468 w=   1.63  ORF:   333  -     467
  2025  -    2189 w=   4.83  ORF:  2027  -    2188
  4140  -    4218 w=  12.32  ORF:  4142  -    4216
  5072  -    5138 w=   4.50  ORF:  5073  -    5138

 Length of Coding region-     447bp          Amino acid sequence -     148aa
MKALIVLGLVLLSVTVQGKVFERCELARTLKRLGMDGYRGISLANWMCLAKWESGYNTRA
TNYNAGDRSTDYGIFQINSRYWCNDGKTPGAVNACHLSCSALLQDNIADAVACAKRVVRD
PQGIRAWVAWRNRCQNRDVRQYVQGCGV*
```

Figure 10.11. Outputs from FGENESH and FGENE of GeneFinder. The nucleotide sequence of DNA encoding human lysozyme (5648 bp) is submitted to Sanger Centre for gene identification with FGenesh (hidden Markov model) and Fgene (pattern recognition) tools. The identical translate is predicted by the two programs.

the WebGeneMark query page, enter the sequence name, paste the query sequence, select running options and output options, and then click the Start GeneMark button. The following result (e.g., DNA encoding human lysozyme) is returned:

```
Sequence length: 5648
GC Content: 36.97%
Window length: 96
Window step: 12
Threshold value: 0.500
- - -
Matrix: H. sapiens, 0.00 < GC < 0.46 - Order 4
Matrix author: JDM
Matrix order: 4
```

```
Sequence InsR : 4044 bp : 65.68% C+G : Isochore 4 (57 - 100 C+G%)

Predicted genes/exons:
Gn.Ex Type S .Begin ...End .Len Fr Ph I/Ac Do/T CodRg P.... Tscr..
----- ---- - ------ ------ ---- -- -- ---- ---- ----- ----- ------
1.01 Init +   2424   2610  187  2  1   85   84   314 0.991  28.07
1.02 Term +   3397   3542  146  2  2   48   42   185 0.984   8.92
1.03 PlyA +   3596   3601    6                                1.05

Predicted peptide sequence(s):
>InsR|GENSCAN_predicted_peptide_1|110_aa
MALWMRLLPLLALLALWGPDPAAAFVNQHLCGSHLVEALYLVCGERGFFYTPKTRREAED
LQVGQVELGGGPGAGSLQPLALEGSLQKRGIVEQCCTSICSLYQLENYCN

Predicted coding sequence(s):
>InsR|GENSCAN_predicted_CDS_1|333_bp
atggccctgtggatgcgcctcctgcccctgctggcgctgctggccctctggggacctgac
ccagccgcagcctttgtgaaccaacacctgtgcggctcacacctggtggaagctctctac
ctagtgtgcggggaacgaggcttcttctacacacccaagacccgccgggaggcagaggac
ctgcaggtggggcaggtggagctgggcggggggccctggtgcaggcagcctgcagcccttg
gccctggagggggtccctgcagaagcgtggcattgtggaacaatgctgtaccagcatctgc
tccctctaccagctggagaactactgcaactag
```

Explanation
Gn.Ex : gene number, exon number (for reference)
Type :   Init = Initial exon (ATG to 5' splice site)
         Intr = Internal exon (3' splice site to 5' splice site)
         Term = Terminal exon (3' splice site to stop codon)
         Sngl = Single-exon gene (ATG to stop)
         Prom = Promoter (TATA box / initation site)
         PlyA = poly-A signal (consensus: AATAAA)
S   :   DNA strand (+ = input strand; - = opposite strand)
Begin : beginning of exon or signal (numbered on input strand)
End :   end point of exon or signal (numbered on input strand)
Len :   length of exon or signal (bp)
Fr :   reading frame (a forward strand codon ending at x has frame x mod 3)
Ph :   net phase of exon (exon length modulo 3)
I/Ac :   initiation signal or 3' splice site score (tenth bit units)
Do/T :   5' splice site or termination signal score (tenth bit units)
CodRg : coding region score (tenth bit units)
P   :   probability of exon (sum over all parses containing exon)
Tscr :   exon score (depends on length, I/Ac, Do/T and CodRg scores)

Figure 10.12. Prediction of gene structure by GenScan.

List of Open reading frames predicted as CDSs, shown with alternate starts
(regions from start to stop codon w/ coding function >0.50)

Left end	Right end	DNA Strand	Coding Frame	Avg Prob	Start Prob
3	167	complement	fr 2	0.65	0.89
3	110	complement	fr 2	0.56	0.04

```
[grail2exons -> Exons]
      St Fr Start     End ORFstart ORFend    Score    Quality
   1-  f 1   107      910      1      910    92.000   excellent
[grail2exons -> Shadow Exons]
      St Fr Start     End ORFstart ORFend    Score    Quality
   2-  r 0   447      548     307     552    70.000       good
   3-  r 2   181      423      54     644    60.000       good
[grail2exons -> Exon Translations]
   4-  MPRSCCSRSGALLLALLLQASMEVRGWCLESSQCQDLTTESNLLECIRAC
       KPDLSAETPMFPGNGDEQPLTENPRKYVMGHFRWDRFGRRNSSSSGSSGA
       GQKREDVSAGEDCGPLPEGGPEPRSDGAKPGPREGKRSYSMEHFRWGKPV
       GKKRRPVKVYPNGAEDESAEAFPLEFKRELTGQRLREGDGPDGPADDGAG
       AQADLEHSLLVAAEKKDEGPYRMEHFRWGSPPKDKRYGGFMTSEKSQTPL
       VTLFKNAIIKNAYKKGE
[grail2exons -> Forward Clusters]
   Frm Start    End   Score
    1    107    910     92
    1    107    906     77
    1    107    333     44
    .     . .    . .     ..
    1    773    906     55
    1    773    910     52
    1    827    906     50
   Frm Start    End   Score
    2    181    423     60
   Frm Start    End   Score
    2    181    644     46
    0    447    548     70
    0    447    552     51
```

Figure 10.13. Prediction of exons by Grail. Grail provides eight features for gene analysis. They are Grail1Exons, Grail1aExons, Grail2Exons, PolyA site, CpG Islands, Repetitive DNA, Simple repeats, and Frame shift error. To predict exons Grail1Exons and Grail1aExons utilize human and mouse models while Grail2Exons employs human, mouse, Arabidopsis, and Drosophila as organism models. The analysis result is derived from the query nucleotide sequence encoding proopiomelanocortin mRNA (1071 bp). The prepended number to the lines is used for GenQuest search if the searcheable feature is available.

```
List of Regions of interest
(regions from stop to stop codon w/ a signal in between)
LEnd    REnd    Strand        Frame
-----   -----   ----------    -----
    3    197    complement    fr 2
 1946   2203    direct        fr 2
 3938   4291    direct        fr 2
 4971   5159    complement    fr 2
```

On the EBI GeneMark page, enter e-mail address, choose options (e.g., exons, ORFs, and regions in nucleotide_transcripts or protein_translations), paste or upload the query sequence, and click the Run GeneMark button. The query returns with lists of regions of interests (from start to stop codons), protein-coding exons (between acceptor and donor sites), and pointers to requested output files which can be viewed/saved.

GenScan (Burge and Karlin, 1997) combines information about gene feature statistics with a probabilistic model to predict gene structure. The Web site is accessible at http://genes.mit.edu/GENSCAN.html. Paste the query sequence, select organism and print option and then click run GENSCAN button. The results (online or by e-mail) include clickable view of gene structure in pdf or ps images, a table listing predicted gene types (initial, internal and terminal exons, single-exon gene, promoter and poly A signal), their length, beginning and end points, reading frames and scores of the predicted features. The table is followed by the amino acid sequence of the predicted translate and the nucleotide sequence of the predicted CDS (Figure 10.12).

Grail (Uberbacher et al., 1993) combines a variety of signals and content information using a neural network to perform gene analysis. The gene recognition of Grail can be accessed at http://compbio.ornl.gov/Grail-1.3/Toolbar.html.

The gene recognition of Grail (Uberbacher et al., 1993) can be accessed at http://compbio.ornl.gov/Grail-1.3/Toolbar.html. Click the New Grail Analysis menu on the tool bar to select an organism (human, mouse, *Arabidopsis*, or *E. coli*). Click Generate Request Form to open the query form. Select analysis options, paste the query sequence, and click the Submit Request button. After the completion of analysis, click Display Results Window to view/save the results with a listing of exons, prediction scores, and exon translations (Figure 10.13).

10.4. WORKSHOPS

1. A genetic map depicts the linear arrangement of genes or genetic marker sites along a chromosome. Two types of genetic maps — genetic linkage maps (genetic maps) and physical maps — are identified. Genetic linkage maps represent distances between markers based on recombination frequencies, while physical maps represent distances between markers based on numbers of nucleotides. Retrieve genetic and physical maps with loci for alcohol dehydrogenase from one of completed microbial genomes.

2. Retrieve DNA encoding human alcohol dehydrogenase isozymes in fasta format and perform multiple alignment with ClustalW.

3. Identify the repeat elements (locations and types) of the following nucleotide sequence and mask these regions with X.

```
>Human liver glucokinase (ATP:D-hexose 6-phosphotransferase),
complete cds
AAGCCCTGGGCTGCCAGCCTCAGGCAGCTCTCCATCCAAGCAGCCGTTGCTGCCACAGGC
GGGCCTTACGCTCCAAGGCTACAGCATGTGCTAGGCCTCAGCAGGCAGGAGCATCTCTGC
CTCCCAAAGCATCTACCTCTTAGCCCCTCGGAGAGATGGCGATGGATGTCACAAGGAGCC
AGGCCCAGACAGCCTTGACTCTGCCAGACTCTCCTCTGAACTCGGGCCTCACATGGCCAA
CTGCTACTTGGAACAAATCGCCCCTTGGCTGGCAGATGTGTTAACATGCCCAGACCAAGA
TCCCAACTCCCACAACCCAACTCCCAGGTAGAGCAGATCCTGGCAGAGTTCCAGCTGCAG
GAGGAGGACCTGAAGAAGGTGATGAGACGGATGCAGAAGGAGATGGACCGCGGCCTGAGG
CTGGAGACCCATGAAGAGGCCAGTGTGAAGATGCTGCCCACCTACGTGCGCTCCACCCCA
GAAGGCTCAGAAGTCGGGGACTTCCTCTCCCTGGACCTGGGTGGCACTAACTTCAGGGTG
ATGCTGGTGAAGGTGGGAGAAGGTGAGGAGGGGCAGTGGAGCGTGAAGACCAAACACCAG
ACGTACTCCATCCCCGAGGACGCCATGACCGGCACTGCTGAGATGCTCTTCGACTACATC
TCTGAGTGCATCTCCGACTTCCTGGACAAGCATCAGATGAAACACAAGAAGCTGCCCCTG
```

```
GGCTTCACCTTCTCCTTTCCTGTGAGGCACGAAGACATCGATAAGGGCATCCTTCTCAAC
TGGACCAAGGGCTTCAAGGCCTCAGGAGCAGAAGGGAACAATGTCGTGGGGCTTCTGCGA
GACGCTATCAAACGGAGAGGGGACTTTGAAATGGATGTGGTGGCAATGGTGAATGACACG
GTGGCCACGATGATCTCCTGCTACTACGAAGACCATCAGTGCGAGGTCGGCATGATCGTG
GGCACGGGCTGCAATGCCTGCTACATGGAGGAGATGCAGAATGTGGAGCTGGTGGAGGGG
GACGAGGGCCGCATGTGCGTCAATACCGAGTGGGGCGCCTTCGGGGACTCCGGCGAGCTG
GACGAGTTCCTGCTGGAGTATGACCGCCTGGTGGACGAGAGCTCTGCAAACCCCGGTCAG
CAGCTGTATGAGAAGCTCATAGGTGGCAAGTACATGGGCGAGCTGGTGCGGCTTGTGCTG
CTCAGGCTCGTGGACGAAAACCTGCTCTTCCACGGGGAGGCCTCCGAGCAGCTGCGCACA
CGCGGAGCCTTCGAGACGCGCTTCGTGTCGCAGGTGGAGAGCGACACGGGCGACCGCAAG
CAGATCTACAACATCCTGAGCACGCTGGGGCTGCGACCCTCGACCACCGACTGCGACATC
GTGCGCCGCGCCTGCGAGAGCGTGTCTACGCGCGCTGCGCACATGTGCTCGGCGGGGCTG
GCGGGCGTCATCAACCGCATGCGCGAGAGCCGCAGCGAGGACGTAATGCGCATCACTGTG
GGCGTGGATGGCTCCGTGTACAAGCTGCACCCCAGCTTCAAGGAGCGGTTCCATGCCAGC
GTGCGCAGGCTGACGCCCAGCTGCGAGATCACCTTCATCGAGTCGGAGGAGGGCAGTGGC
CGGGGCGCGGCCCTGGTCTCGGCGGTGGCCTGTAAGAAGGCCTGTATGCTGGGCCAGTGA
GAGCAGTGGCCGCAAGCGCAGGGAGGATGCCACAGCCCCACAGCACCCAGGCTCCATGGG
GAAGTGCTCCCCACACGTGCTCGCAGCCTGGCGGGGCAGGAGGCCTGGCCTTGTCAGGAC
CCAGGCCGCCTGCCATACCGCTGGGGAACAGAGCGGGCCTCTTCCCTCAGTTTTTCGGTG
GGACAGCCCCAGGGCCCTAACGGGGGTGCGGCAGGAGCAGGAACAGAGACTCTGGAAGCC
CCCCACCTTTCTCGCTGGAATCAATTTCCCAGAAGGGAGTTGCTCACTCAGGACTTTGAT
GCATTTCCACACTGTCAGAGCTGTTGGCCTCGCCTGGGCCCAGGCTCTGGGAAGGGGTGC
CCTCTGGATCCTGCTGTGGCCTCACTTCCCTGGGAACTCATCCTGTGTGGGGAGGCAGCT
CCAACAGCTTGACCAGACCTAGACCTGGGCCAAAAGGGCAGGCCAGGGGCTGCTCATCAC
CCAGTCCTGGCCATTTTCTTGCCTGAGGCTCAAGAGGCCCAGGGAGCAATGGGAGGGGGC
TCCATGGAGGAGGTGTCCCAAGCTTTGAATACCCCCCAGAGACCTTTTCTCTCCCATACC
ATCACTGAGTGGCTTGTGATTCTGGGATGGACCCTCGCAGCAGGTGCAAGAGACAGAGCC
CCCAAGCCTCTGCCCCAAGGGGCCCACAAAGGGGAGAAGGGCCAGCCCTACATCTTCAGC
TCCCATAGCGCTGGCTCAGGAAGAAACCCCAAGCAGCATTCAGCACACCCCAAGGGACAA
CCCCATCATATGACATGCCACCCTCTCCATGCCCAACCTAAGATTGTGTGGGTTTTTTAA
TTAAAAATGTTAAAAGTTTTAAAAAAAAAA
```

4. Given a human DNA with the following nucleotide sequence.

```
AGCGGCGGCGAAGGAGGGGAAGAAGAGCCGCGACCGAGAGAGGCCGCCGAGCGTCCCCGCCCTCAGAGAG
CAGCCTCCCGAGACAGAGCCTCAGCCTGCCTGGAAGATGCCGAGATCGTGCTGCAGCCGCTCGGGGGCCC
TGTTGCTGGCCTTGCTGCTTCAGGCCTCCATGGAAGTGCGTGGCTGGTGCCTGGAGAGCAGCCAGTGTCA
GGACCTCACCACGGAAAGCAACCTGCTGGAGTGCATCCGGGCCTGCAAGCCCGACCTCTCGGCCGAGACT
CCCATGTTCCCGGGAAATGGCGACGAGCAGCCTCTGACCGAGAACCCCCGGAAGTACGTCATGGGCCACT
TCCGCTGGGACCGATTCGGCCGCCGCAACAGCAGCAGCAGCGGCAGCAGCGGCGCAGGGCAGAAGCGCGA
GGACGTCTCAGCGGGCGAAGACTGCGGCCCCGCTGCCTGAGGGCGGCCCCGAGCCCCGCAGCGATGGTGCC
AAGCCGGGCCCGCGCGAGGGCAAGCGCTCCTACTCCATGGAGCACTTCCGCTGGGGCAAGCCGGTGGGCA
AGAAGCGGCGCCCAGTGAAGGTGTACCCTAACGGCGCCGAGGACGAGTCGGCCGAGGCCTTCCCCCTGGA
GTTCAAGAGGGAGCTGACTGGCCAGCGACTCCGGGAGGGAGATGGCCCCCGACGGCCCTGCCGATGACGGC
GCAGGGCCCAGGCCGACCTGGAGCACAGCCTGCTGGTGGCGGCCGAGAAGAAGGACGAGGGCCCCTACA
GGATGGAGCACTTCCGCTGGGGCAGCCCGCCCAAGGACAAGCGCTACGGCGGTTTCATGACCTCCGAGAA
GAGCCAGACGCCCCTGGTGACGCTGTTCAAAAACGCCATCATCAAGAACGCCTACAAGAAGGGCGAGTGA
GGGCACAGCGGGCCCCAGGGCTACCCTCCCCCAGGAGGTCGACCCCAAAGCCCCTTGCTCTCCCCTGCCC
TGCTGCCGCCTCCCAGCCTGGGGGGTCGTGGCAGATAATCAGCCTCTTAAAGCTGCCTGTAGTTAGGAAA
TAAAACCTTTCAAATTTCACA
```

Identify the functionally active frame of translation by comparing the results of (a) six frame translations with Translate tool of ExPASy or ORF prediction with ORF Finder of NIH and (b) ORFGene of WebGene.

5. Submit the above nucleotide sequence to appropriate server for predicting the following functional sites/signals: (a) promoter, (b) TATA signal, and (c) poly-A region.

6. Predict the splice sites (acceptor and donor sites) of the above nucleotide sequence with WebGene or GeneID and SPL of Sanger Centre. Compare the results of their splice site predictions.

7. The human lactate dehydrogenase gene with the following nucleotide sequence has been isolated.

```
>Human testis-specific lactate dehydrogenase (LDHC4, LDHX)
CGCCTCAACTGTCGTTGGTGTATTTTTCTGGTGTCACTTCTGTGCCTTCCTTCAAAGGTTCTCCAAATGT
CAACTGTCAAGGAGCAGCTAATTGAGAAGCTAATTGAGGATGATGAAAACTCCCAGTGTAAAATTACTAT
TGTTGGAACTGGTGCCGTAGGCATGGCTTGTGCTATTAGTATCTTACTGAAGGATTTGGCTGATGAACTT
GCCCTTGTTGATGTTGCATTGGACAAACTGAAGGGAGAAATGATGGATCTTCAGCATGGCAGTCTTTTCT
TTAGTACTTCAAAGGTTACTTCTGGAAAAGATTACAGTGTATCTGCAAACTCCAGAATAGTTATTGTCAC
AGCAGGTGCAAGGCAGCAGGAGGGAGAAACTCGCCTTGCCCTGGTCCAACGTAATGTGGCTATAATGAAA
ATAATCATTCCTGCCATAGTCCATTATAGTCCTGATTGTAAAATTCTTGTTGTTTCAAATCCAGTGGATA
TTTTGACATATATAGTCTGGAAGATAAGTGGCTTACCTGTAACTCGTGTAATTGGAAGTGGTTGTAATCT
AGACTCTGCCCGTTTCCGTTACCTAATTGGAGAAAAGTTGGGTGTCCACCCCACAAGCTGCCATGGTTGG
ATTATTGGAGAACATGGTGATTCTAGTGTGCCCTTATGGAGTGGGGTGAATGTTGCTGGTGTTGCTCTGA
AGACTCTGGACCCTAAATTAGGAACGGATTCAGATAAGGAACACTGGAAAAATATCCATAAACAAGTTAT
TCAAAGTGCCTATGAAATTATCAAGCTGAAGGGGTATACCTCTTGGGCTATTGGACTGTCTGTGATGGAT
CTGGTAGGATCCATTTTGAAAAATCTTAGGAGAGTGCACCCAGTTTCCACCATGGTTAAGGGATTATATG
GAATAAAAGAAGAACTCTTTCTCAGTATCCCTTGTGTCTTGGGGCGGAATGGTGTCTCAGATGTTGTGAA
AATTAACTTGAATTCTGAGGAGGAGGCCCTTTTCAAGAAGAGTGCAGAAACACTTTGGAATATTCAAAAG
GATCTAATATTTTAAATTAAAGCCTTCTAATGTTCCACTGTTTGGAGAACAGAAGATAGCAGGCTGTGTA
TTTTAAATTTTGAAAGTATTTTCATTGATCTTAAAAAATAAAAACAAATTGGAGACCTG
```

Predict its most likely protein translation by submitting the DNA sequence along with the following amino acid sequences of human lactate dehydrogenase isozymes to Procustes analyses.

```
>gi|1070432|pir||DEHULH L-lactate dehydrogenase (EC 1.1.1.27) chain H -
human
MATLKEKLIAPVAEEEATVPNNKITVVGVGQVGMACAISILGKSLADELALVDVLEDKLKGEMMDLQHGS
LFLQTPKIVADKDYSVTANSKIVVVTAGVRQQEGESRLNLVQRNVNVFKFIIPQIVKYSPDCIIIVVSNP
VDILTYVTWKLSGLPKHRVIGSGCNLDSARFRYLMAEKLGIHPSSCHGWILGEHGDSSVAVWSGVNVAGV
SLQELNPEMGTDNDSENWKEVHKMVVESAYEVIKLKGYTNWAIGLSVADLIESMLKNLSRIHPVSTMVKG
MYGIENEVFLSLPCILNARGLTSVINQKLKDDEVAQLKKSADTLWDIQKDLKDL
>gi|65922|pir||DEHULM L-lactate dehydrogenase (EC 1.1.1.27) chain M - human
MATLKDQLIYNLLKEEQTPQNKITVVGVGAVGMACAISILMKDLADELALVDVIEDKLKGEMMDLQHGSL
FLRTPKIVSGKDYNVTANSKLVIITAGARQQEGESRLNLVQRNVNIFKFIIPNVVKYSPNCKLLIVSNPV
DILTYVAWKISGFPKNRVIGSGCNLDSARFRYLMGERLGVHPLSCHGWVLGEHGDSSVPVWSGMNVAGVS
LKTLHPDLGTDKDKEQWKEVHKQVVESAYEVIKLKGYTSWAIGLSVADLAESIMKNLRRVHPVSTMIKGL
YGIKDDVFLSVPCILGQNGISDLVKVTLTSEEEARLKKSADTLWGIQKELQF
>gi|65927|pir||DEHULC L-lactate dehydrogenase (EC 1.1.1.27) chain X - human
MSTVKEQLIEKLIEDDENSQCKITIVGTGAVGMACAISILLKDLADELALVDVALDKLKGEMMDLQHGSL
FFSTSKVTSGKDYSVSANSRIVIVTAGARQQEGETRLALVQRNVAIMKSIIPAIVHYSPDCKILVVSNPV
DILTYIVWKISGLPVTRVIGSGCNLDSARFRYLIGEKLGVHPTSCHGWIIGEHGDSSVPLWSGVNVAGVA
LKTLDPKLGTDSDKEHWKNIHKQVIQSAYEIIKLKGYTSWAIGLSVMDLVGSILKNLRRVHPVSTMVKGL
YGIKEELFLSIPCVLGRNGVSDVVKINLNSEEEALFKKSAETLWNIQKDLIF
```

8. Compare exon translation of a human ribonuclease DNA with the following sequence predicted by FEX tool of GeneFinder, Genie and Grail.

```
>Homo sapiens ribonuclease, RNase A (pancreatic)
GAATTCCGGGTTTGAAAAGGAGTTCTAGGGAAGAAGAGAGTTAGTTAGCACATCAATGGGAGCAGGGCTC
TTACCCCACGTGGTGTTACATATATATTATTTTCATACATGGTTTCTGGCTCATAAGTTCCTTAGCCCTT
GCTATAGTCTTTTGTGTTCGGTCTTAAGGGCAGGACTGTACTCTTCCCTCACCTTTCTAATTGTGCATCT
TAAGACCTTCCCCAGAGAGGGTGGTGCCCTGTAGTTGTGGGAAGGAATGCTGGCATCATGAAGCTTCCAT
AAAAACCCGAGAAACGAGCTTCTGGATAGCTGGACACATGGAGGTCCTGGAGGGTGGAGCCCAGGGAGGC
ATGGAAGCTCCACAGCCCTTCCCCCATACCTTACCCTATTTCCTCTGTATCCTTTGTAATATCCTTTATG
ATAAACCAGCAAATGTGTGTAAATGTTTCCCTAAGGTCTGTGGCCACTCCAGCAAATTAATTGAACCTAA
AGAGGGGGTCGTGGGAACCCCAACTTGAAGCCAGTCAGTCAGAAGTTCTGGATGTCCAGACTTCAGACTG
GTGTCTGAAAGGGTGGAGGCAGTCTTGGGGACCGAGCCCCCAATCTATGGGATCTGACACTATCTCCAGT
AGTGTTGGAATTGAGTCACCAGCGTGTCCACTGGTTAGTGTGTGAGAAACTCCCTACCATTGGTCACAGA
AGTCTTCTTCTGTGTTGATAGTTGTAGTGTGACAGCAGAGGAAAAACAAAGTCAGAAAGAGTTTTCCCGA
ACACACCCAATTTCTCCATTTTACTATCCATTTCCACAAACACTGACTACAATAGAAGTATAAAAATTAC
TCCACTGCATCATTCAGCTTTCCATCTCTCTCAGACACCAAGCTGCAGATCCAGGTCACTTTGTAGGTCA
CCACCTAGAGGGGAGGAAGACCTCGCTTTGGAGAGTGGGAATAAAACGCTCGTGGAAAAGGGTACACGCT
TTTCTGGGAAAGTGAGGCCACCATGGCTCTGGAGAAGTCTCTTGTCCGGCTCCTTCTGCTTGTCCTGATA
CTGCTGGTGCTGGGCTGGGTCCAGCCTTCCCTGGGCAAGGAATCCCGGGCCAAGAAATTCCAGCGGCAGC
ATATGGACTCAGACAGTTCCCCCAGCAGCAGCTCCACCTACTGTAACCAAATGATGAGGCGCCGGAATAT
GACACAGGGGCGGTGCAAACCAGTGAACACCTTTGTGCACGAGCCCCTGGTAGATGTCCAGAATGTCTGT
TTCCAGGAAAAGGTCACCTGCAAGAACGGGCAGGGCAACTGCTACAAGAGCAACTCCAGCATGCACATCA
CAGACTGCCGCCTGACAAACGGCTCCAGGTACCCCAACTGTGCATACCGGACCAGCCCGAAGGAGAGACA
CATCATTGTGGCCTGTGAAGGGAGCCCATATGTGCCAGTCCACTTTGATGCTTCTGTGGAGGACTCTACC
TAAGGTCAGAGCAGCGAGATACCCCACCTCCCTCAACCTCATCCTCTCCACAGCTGCCTCTTCCCTCTTC
CTTCCCTGCTGTGAAAGAAGTAACTACAGTTAGGGCTCCTATTCAACACACACATGCTTCCCTTTCCTGA
GCCGGAATTC
```

9. Retrieve nucleotide sequence encoding mRNA of human (Homo sapiens) soluble aldehdyde dehdrogenase 1 (ALDH1: gi11429705) and perform gene identification with GeneFinder.

10. Retrive nucleotide sequence encoding mRNA of human mitochondrial aldehyde dehydrogenase 2 (ALDH 2: gi11436532) and carry out gene identification with GeneMark.

REFERENCES

Aaronson, J., Eckman, B., Blevins, R. A., Borkowski, J. A., Myerson, J., Imran, S., and Elliston, K. O. (1996) *Genome Res.* **6**:829–845.

Baekelandt, I. N., Goritchenko, L., and Benowitz, L. I. (1997) *Nucleic Acids Res.* **25**:1281–1288.

Bork, P., Dandekar, T., Diaz-Lazoz, Y., Eisenhaber, F., Huynen, M., and Yuen, Y., (1998) *J. Mol. Biol.* **283**:707–725.

Borodovsky, M. Y., and McIninch, J. D. (1993) *Comput. Chem.* **17**:123–133.

Bunge, C., and Karlin, S. (1997) *J. Mol. Biol.* **268**:78–94.

Bunset, M., and Guigo, R. (1996) *Genomics* **34**:353–367.

Cowell, I. G., and Austin, C. A., Eds. (1997) *cDNA Library Protocols.* Humana Press, Totowa, NJ.

Fickett, J. W. (1996a) *Trends Genet.* **12**:3165–320.

Fickett, J. W. (1996b) *Comput. Chem.* **20**:103–118.

Fickett, J. W., and Hatzigeorgiou, A. G. (1997) *Genome Res.* **7**:861–878.

Fickett, J. W., and Tung, C.-S. (1992) *Nucleic Acids Res.* **20**:6441–6450.

French, T., Gultyaev, A. P., and Gerdes, K. (1997) *J. Mol. Biol.* **273**:38–51.

Gelfand, M. S. (1995) *J. Comput. Biol.* **2**:87–115.

Gelfand, M. S., Mironov, A. A., and Pevzner, P. A. (1996) *Proc. Natl. Acad. Sci. U.S.A.* **93:**9061–9066.

Green, P., Lipman, D., Hillier, L. Waterston, R., States, D., and Claverie, J. M. (1993) *Science* **229:**1711–1716.

Guigó, R. (1998) *J. Comput. Biol.* **5:**681–702.

Guig, R., Knudsen, S., Drake, N., and Smith, T. (1992) *J. Mol. Biol.* **226:**141–157.

Henderson, J., Salzberg, S., and Fasman, K. H. (1997) *J. Comput. Biol.* **4:**127–142.

Huang, X., Adams, M. D., Zhou, H., and Kerlavage, A. (1997) *Genomics* **46:**37–45.

Kozak, M. (1996) *Mamm. Genome* **7:**563–574.

McKusick, V. A. (1997) *Genomics* **45:**244–249.

Mclachlan, G. J. (1992) *Discriminant Analysis and Statistical Pattern Recognition*, John Wiley & Sons, New York.

Pesalo, G., Liuni, S., Grillo, G., Licciulli, F., Mignone, F., Gissi, C., and Saccone, C. (2002) *Nucleic Acids Res.* **30:**335–340.

Praz, V., Périer, R., Bonnard, C., and Bucher, P. (2002) *Nucleic Acids Res.* **30:**322–324.

Rival, E., Dauchet, M., Delehaye, J. P., and Delgrange, O. (1996) *Biochimie* **78:**315–322.

Senapathy, P., Shapiro, M. B., and Harris, N. L. (1990) *Meth. Enzymol.* **183:**252–278.

Senterre-Lesenfants, S., Alag, A. S., and Sobel, M. E. (1995) *J. Cell Biochem.* **58:**445–454.

Shapiro, L., and Harris, T. (2000) *Current Opin. Biotechnol.* **11:**31–35.

Smit, A. F. A. (1996) *Curr. Opin. Genet. Dev.* **6:**743–749.

Solovyev, V. V., Salamov, A. A., and Lawrence, C. B. (1994) *Nucleic Acids Res.* **22:**5156–5163.

Stagle, J. R. (1971) *Artificial Intelligence: The Heuristic Programming Approach.* McGraw Hill, New York.

Starkey, M. P., and Elaswarapu, R., Eds. (2001) *Genomic Protocols.* Humana Press, Totowa, NJ.

Uberbacher, E. C., Einstein, J. R., Guan, X., and Mural, R. J. (1993) in *The Second International Conference on Bioinformatics, Supercomputing and Complex Genome Analysis*, edited by Lim, H. A., Fickett, J. W., Cantor, C. R., and Robbins, R. J. World Scientific, Singapore, pp. 465–476.

Wahle, E., and Keller, W. (1996) *Trends Biochem. Sci.* **21:**247–150.

Zhang, M. Q. (1997) *Proc. Natl. Acad. Sci. U.S.A.* **94:**565–568.

11

PROTEOMICS: PROTEIN SEQUENCE ANALYSIS

Proteomics is concerned with the analysis of the complete protein complements of genomes. Thus proteomics includes not only the identification and quantification of proteins, but also the determination of their localization, modifications, interactions, activities, and functions. This chapter focuses on protein sequences as the sources of biochemical information. Protein sequence databases are surveyed. Similarity search and sequence alignments using the Internet resources are described.

11.1. PROTEIN SEQUENCE: INFORMATION AND FEATURES

Proteome refers to protein complement expressed by a genome. Thus proteomics concerns with the analysis of complete complements of proteins. It is the study of proteins that are encoded by the genes of a cell or an organism. Such study includes determination of protein expression, identification and quantification of proteins as well as characterization of protein structures, functions and interactions. The functional classification of proteins in genomes (i.e., proteomes) can be accessed from the Proteome Analysis Database at http://ebi.ac.uk/proteome/ (Apweiler et al., 2001).

The unique characteristic of each protein is the distinctive sequence of amino acid residues in its polypeptide chain. The amino acid sequence is the link between the genetic message in DNA and the three-dimensional structure that performs a protein's biological function. The sequence comparison among analogous proteins yields insights into the evolutionary relationships of proteins and protein function.

The comparison of sequences between normal and mutant proteins provides invaluable information toward identification of critical residues involved in protein function as well as detection and treatment of diseases. The knowledge of amino acid sequences is essential to the protein's three-dimensional structure, mechanism of actions, and design.

The general strategy for determining the amino acid sequence of a protein (Findley and Geisow, 1989; Needleman, 1975; Walsh et al., 1981) involves several steps:

1. Separation and purification of individual polypeptide chain from a protein containing subunits.
2. Cleavage of intramolecular disulfide linkage(s).
3. Determination of amino acid composition of purified polypeptide chain.
4. Analyses of N- and C-terminal residues.
5. Specific cleavage of the polypeptide chain reproducibly into manageable fragments and purification of resulting fragments.
6. Sequence analyses of the peptide fragments.
7. Repeat steps 5 and 6 using a different cleavage procedure to generate a different and overlapping set of peptide fragments and their sequence analyses.
8. Reconstruction of the overall amino acid sequence of the protein from the sequences in overlapping fragments.
9. Localization of the disulfide linkage(s).

Alternatively, protein sequence information can be derived from translating the nucleotide sequences of DNA that encode the proteins.

11.1.1. Protein Identity Based on Composition and Properties

A number of useful computational tools have been developed for predicting the identity of unknown proteins based on the physical and chemical properties of amino acids and vice versa. Many of these tools are available through the Expert Protein Analysis System (ExPASy) at http://www.expasy.ch (Appel et al., 1994) and other servers.

Rather than using an amino acid sequence to search SWISS-PROT, AACompIdent of ExPASy Proteomic tools (http://www.expasy.ch/tools/) uses the amino acid composition of an unknown protein to identify known proteins of the same composition. The program requires the desired amino acid composition, the pI and molecular weight of the protein (if known), the appropriate taxonomic class, and any special keywords. The user must select from one of six amino acid constellations that influence how the analysis is performed. For each sequence in the database, the algorithm computes a score based on the difference in compositions between the sequence and the query composition. The results, returned by e-mail, are organized as three ranked lists. Because the computed scores are a measure of difference, a score of zero implies that there is exact correspondence between the query composition and that sequence entry. AACompSim (ExPASy Proteomic tools) performs a similar type of analysis, but rather than using an experimentally derived amino acid composition as the basis of searches, the sequence of a Swiss-Prot protein is used instead. A theoretical pI and molecular weight are computed prior to computing the difference score using Compute pI/MW.

11.1.2. Physicochemical Properties Based on Sequence

Compute pI/MW (ExPASy Proteomic tools) is a tool that calculates the isoelectric point (pI) and molecular weight of an input sequence. Molecular weights are calculated by the addition of the average isotopic mass of each amino acid in the sequence plus that of one water molecule. The sequence can be furnished by the user in fasta format, or by a Swiss-Prot identifier. If a sequence is furnished, the tool automatically computes the pI and molecular weight for the entire length of the sequence. If a Swiss-Prot identifier is given, the user may specify a range of amino acids so that the computation is done on a fragment rather than on the entire protein. The absorption coefficient and pI value are also calculated at aBi (http://www.up.univ-mrs.fr/~wabim/d_abim/compo-p.html). If "courbe de titrage" is checked, the titration curve based on the query sequence is returned.

PeptideMass (ExPASy Proteomic tools), which is designed for use in peptide mapping experiments, determines the cleavage products of a protein after exposure to a specific protease or chemical reagent. The enzymes and reagents available for cleavage via PeptideMass are trypsin, chymotrypsin, Lys C, cyanogen bromide, Arg C, Asp N, and Glu C.

The AAindex database http://www.genome.ad.jp/dbget/aaindex.html (Kawashima and Kanehisa, 2000) collects physicochemical properties of amino acids such as molecular weight, bulkiness, polarity, hydrophobicity, average area buried/exposed, solvent accessibility, and secondary structure parameters (propensities for amino acid residues to form α-helix, β-strand, reverse turn and coil structures of proteins). The physicochemical properties can be plotted along the sequence using ProtScale at ExPASy (http://www.expasy.ch/cgi-bin/protscale.pl) .

TGREASE calculates the hydrophobicity — that is, the propensity of the amino acid to bury itself in the core of the protein and away from surrounding water. This tendency, coupled with steric and other considerations, influence how a protein ultimately folds. As such, the program finds application in the determination of putative transmembrane sequences as well as in the prediction of buried regions of globular proteins. TGREASE (ftp://ftp.virginia.edu/pub/fasta/) can be downloaded and run on DOS-based computers. The method relies on a hydropathy scale (Kyte and Doolittle, 1982). Amino acids with higher positive scores are more hydrophobic; those with more negative scores are more hydrophilic. A moving average, or hydropathic index, is then calculated across the protein. The window length is adjustable, with a span of 7 to 11 residues recommended to minimize noise and maximize information content. The results are then plotted as hydropathic index versus residue number. TMpred of EMBNet at http://www.ch.embnet.org/software/TMPRED_form.html and TMAP (Milpetz et al., 1995) at http://www.embl-heidelberg.de/tmap/tmap_info.html predicts possible transmembrane regions and their topology from the amino acid sequence.

11.1.3. Sequence Comparison

When we compare the sequences of proteins, we analyze the similarities and differences at the level of individual amino acids with the aim of inferring structural, functional, and evolutionary relationships among the sequences under study. The most common comparative method is sequence alignment, which provides an explicit mapping between the residues of two or more sequences.

Sequence identity refers to the occurrence of exactly the same nucleotide/amino acid in the same position in two sequences. *Sequence similarity* refers to the sequences aligned by applying possible substitutions scored according to the probability with which they occur. *Sequence homology* reflects the evolutionary relationship between sequences. Two sequences are said to be homologous if they are derived from a common ancestral sequence. Thus, similarity is an observable quantity that refers to the presence of identical and similar sites in the two sequences while homology refers to a conclusion drawn from similarity data that two sequences share a common ancestor. The changes that occur during divergence from the common ancestor can be characterized as substitutions, insertions, and deletions. Residues that have been aligned, but are not identical, would represent substitutions. Regions in which the residues of one sequence correspond to nothing in the other would be interpreted as either an insertion into one sequence or a deletion from the other (INDEL). *Analogy* refers to either protein structures that share similar folds but have no demonstrable sequence similarity or to proteins that share catalytic residues with almost exactly equivalent spatial geometry without sequence or structural similarity. It is useful to distinguish between homologous proteins as *orthologues* that perform the same function in different species, and as *paralogues* that perform different but related functions within one organism. Sequence comparison of orthologous proteins leads to the study of molecular systematics, and therefore the construction of phylogenetic trees has revealed relationships among different species. On the other hand, the study of paralogous proteins has provided insights into the underlying mechanisms of evolution because paralogous proteins arose from single genes via successive duplication events. In general, the three-dimensional structures are more likely to be conserved than are the corresponding amino acid sequences between distantly related proteins.

Apart from strict identity in pattern matching, one approach of tolerance in the sequence comparison can be introduced by considering amino acid residues as member of groups with shared biochemical properties as shown in Table 11.1.

The presence of the regular repeated structure regions in proteins permits the comparison of their structural characteristics that display common features. A search for structural similarity of proteins can be processed at various levels of complexity depending upon the field of application. Two approaches are considered in the similarity comparison of protein sequences:

1. Sequence comparison to detect common features. This may involve mapping profiles of conformational preferences, accessibility, hydrophilicity/hydrophobicity, and so on. The profiles are compared. Another useful topography is a dot–plot similarity matrix that compares sequences of two proteins according to the predefined criterion of similarity, and the similarities are identified as a succession of consecutive dots parallel to the diagonal.

2. Sequence alignment that defines the best one-to-one correspondence between amino acid sequences of two or more proteins. This scores correspondence

TABLE 11.1. Classification of Amino Acids
According to their Biochemical Properties

Property	Residue
Small	Ala, Gly
Hydroxyl/phenol	Ser, Thr, Tyr
Thiol/thioether	Cys, Met
Acidic/amide	Asp, Asn, Gln, Glu
Basic	Arg, His, Lys
Polar	Ala, Cys, Gly, Pro, Ser, Thr
Aromatic	Phe, Trp, Tyr
Alkyl hydrophobic	Ile, Leu, Met, Val

on the entire sequences with appropriate accommodations for insertions and
deletions.

11.2. DATABASE SEARCH AND SEQUENCE ALIGNMENT

11.2.1. Primary Database

It is much easier and quicker to produce sequence information than to determine
3D structures of proteins in atomic detail. As a consequence, there is a protein
sequence/structure deficit. In order to benefit from the wealth of sequence information, we must establish, maintain, and disseminate sequence databases; provide
user-friendly software to access the information, and design analytical tools to
visualize and interpret the structural/functional clues associated with these data.

There are different classes of protein sequence databases. Primary and secondary
databases are used to address different aspects of sequence analysis. Composite
databases amalgamate a variety of different primary sources to facilitate sequence
searching efficiently. The primary structure (amino acid sequence) of a protein is
stored in primary databases as linear alphabets that represent the constituent
residues. The secondary structure of a protein corresponding to region of local
regularity (e.g., α-helices, β-strands, and turns), which in sequence alignments are
often apparent as conserved motifs, is stored in secondary databases as patterns. The
tertiary structure of a protein derived from the packing of its secondary structural
elements which may form folds and domains is stored in structure databases as sets
of atomic coordinates. Some of the most important protein sequence databases are
PIR (Protein Information Resource), SWISS-PROT (at EBI and ExPASy), MIPS
(Munich Information Center for Protein Sequences), JIPID (Japanese International
Protein Sequence Database), and TrEMBL (at EBI) .

The PIR Protein Sequence Database (Barker et al., 2001; Wu et al., 2002)
developed at the National Biomedical Research Foundation (NBRF) has been
maintained by PIR-International Protein Sequence Database (PSD), which is the
largest publicly distributed and freely available protein sequence database. The
consortium includes PIR at the NBRF, MIPS, and JIPID. PIR-International
provides online access at http://pir.georgetown.edu to numerous sequence and
auxiliary databases. These include PSD (annotated and classified protein sequences),
PATCHX (sequences not yet in PSD), ARCHIVE (sequences as originally reported

in a publication), NRL_3D (sequences from 3D structure database PDB), PIR-ALN (sequence alignment of superfamilies, families, and homology domains), RESID (post-translational modifications with PSD feature information), iProClass (non-redundant sequences organized according to superfamilies and motifs), and ProtFam (sequence alignments of superfamilies). The PIR server offers three types of database searches:

1. Interactive text-based search that allows Boolean queries of text field
2. Standard sequence similarity search including Peptide Match, Pattern Match, BLAST, FASTA, Pairwise Alignment, and Multiple Alignment
3. Advanced search that combines sequence similarity and annotation searches or evaluate gene family relationships

Each PIR entry consists of Entry (entry ID), Title, Alternate_names, Organism, Date, Accession (accession number), Reference, Function (description of protein function), Comment (e.g., enzyme specificity and reaction, etc.), Classification (super-family), Keywords (e.g., dimer, alcohol metabolism, metalloprotein, etc.), Feature (lists of sequence positions for disulfide bonds, active site and binding site amino acid residues, etc.), Summary (number of amino acids and the molecular weight), and Sequence (in PIR format, Chapter 4). In addition, links to PDB, KEGG, BRENDA, WIT, alignments, and iProClass are provided.

The Munich Information Center for Protein Sequences (MIPS) (Mewes et al., 2000) collects and processes sequence data for the PIR-International Protein Sequence Database project. Access to the database is provided through its Web server, http://www.mips.biochem.mpg.de/. The implementation of PrIAn (Protein Input and Annotation) data input has greatly increased database entries of PIR-International.

SWISS-PROT (Bairoch and Apweiler, 2000) is a protein sequence database that, from its inception in 1986, was produced collaboratively by the Department of Medical Biochemistry at the University of Geneva and the EMBL. The database is now maintained collaboratively by Swiss Institute of Bioinformatics (SIB) and EBI/EMBL. SWISS-PROT provides high-level annotations, including descriptions of the function of the protein and of the structure of its domains, its post-translational modifications, its variants, and so on. The database can be accessed from http://expasy.hcuge.ch/sprot/sprot-top.html or numerous mirror sites. In 1966, Translated EMBL (TrEMBL) was created as a computer-annotated supplement to SWISS-PROT (Bleasby et al., 1994).

Each SWISS-PROT entry consists of general information about the entry (e.g., entry name and date, accession number), Name and origin of the protein (e.g., protein name, EC number and biological origin), References, Comments (e.g., catalytic activity, cofactor, subuit structure, subcellular location and family class, etc.), Cross-reference (EMBL, PIR, PDB, Pfam, ProSite, ProDom, ProtoMap, etc.), Keywords, Features (e.g., active site, binding site, modification, secondary structures, etc.), and Sequence information (amino acid sequence in Swiss-Prot format, Chapter 4).

To facilitate sequence efficiency, a variety of different primary sources are amalgamated to composite databases such as NRDB (PDB, Swiss-Prot, PIR, and GenPept), OWL (GenBank, PIR, Swiss-Prot, and NRL-3D), and MIPSX (PIR-MIPS, Swiss-Prot, NRL-3D, Kabat, and PseqIP). NRDB is built at the NCBI. The database is comprehensive and up-to-date; however, it is not nonredundant but rather nonidentical. OWL (http://www.bioinf.man.ac.uk/dbbrowser/OWL/) is a nonredundant protein sequence database (Bleasby et al., 1994). The database is a composite of four major primary sources that are assigned a priority with regard to their level of annotations and sequence validation with SWISS-PROT being the highest priority. MIPSX is a merged database produced at MIPS.

11.2.2. Secondary Database

There are many secondary (or pattern) databases that contain the biologically significant information of sequence analyses from the primary sources. Because there are several different primary databases and a variety of ways to analyze protein sequences, the type of information stored in each of the secondary databases is different. However, the creation of most secondary databases is based on a common principle that multiple alignments may identify homologous sequences within which conserved regions or motifs may reflect some vital information crucial to the structure or function of these homologous proteins. Motifs have been exploited in various ways to build diagnostic patterns for particular protein families. If the structure and function of the family are known, searches of pattern databases may offer a fast track to the inference of biological function of the unknown. Notwithstanding, the pattern databases should be used to augment primary database searches.

The secondary database, PROSITE (Hofmann et al., 1999), which uses SWISS-PROT, is maintained collaboratively at the Swiss Institute of Bioinformatics (http://expasy.hcuge.ch/sprot/prosite.html) and Swiss Institute for Experimental Cancer Research (http://www.isrec.isb-sib.ch/). It is rationalized that protein families could be simply and effectively characterized by the single most conserved motif observable in a multiple alignment of known homologous proteins. Such motifs usually associate with key biological functions such as enzyme active sites, ligand binding sites, structural signatures, and so on. Entries are deposited in PROSITE in two files: the data file and the documentation file. In the data file, each entry contains an identifier (ID) and an accession number (AC). A title or description of the family is contained in the DE line, and the pattern itself resides on PA lines. The following NR lines provide technical details about the derivation and diagnostic performance of the pattern. Large numbers of false-positives and false-negatives indicate poorly performing patterns. The comment (CC) lines furnish information on the taxonomic range of the family, the maximum number of observed repeat, functional site annotations, and so on. The following DR lines list the accession number and SWISS-PROT identification codes of all the true matches to the pattern (denoted by T) and any possible matches (denoted by P). The last DO line points to the associated documentation file. The documentation file begins with the accession number and cross-reference identifier of its data file. This is followed by a free format description of the family including details of the pattern, the biological role of selected motif(s) if known, and appropriate bibliographic references.

Most protein families are characterized by several conserved motifs. The PRINTS fingerprint database was developed to use multiple conserved motifs to build diagnostic signatures of family membership (Attwood et al., 1998). If a query sequence fails to match all the motifs in a given fingerprint, the pattern of matches formed by the remaining motifs allows the user to make a reasonable diagnosis. The PRINTS can be accessed by keyword and sequence searches at http://www.bioinf.

man.ac.uk/dbbrowser/PRINTS/. There are three sections in a PRINTS entry. In the first section, each fingerprint is given an identifying code, an accession number, number of motifs in the fingerprint, a list of cross-linked databases and bibliographic information. In the second section, the diagnostic performance is given in the summary and the table of results. The last section lists seed motifs and final motifs that result from the iterative database scanning. Each motif is identified by an ID code followed by the motif length. The aligned motifs themselves are then provided, together with the corresponding source database ID code, the location in the parent sequence of each fragment (ST), and the interval between the fragment and its preceding neighbor (INT).

In the BLOCKS database (Henikoff et al., 1998), the motifs or blocks are created by automatically detecting the most highly conserved regions of each protein family. Initially three conserved amino acids are identified. The resulting blocks are then calibrated against SWISS-PROT to obtain a measure of the likelihood of a chance match. The database is accessible by keyword and sequence searches via the BLOCKS Web server at the Fred Hitchinson Cancer Research Center (http://www.blocks.fhcrc.org/). The Block includes the SWISS-PROT IDs of the constituent sequences, the start position of the fragment, the sequence of the fragment, and a score or weight indicating a measure of the closeness of the relationship of that sequence to others in the block (100 being the most distant).

Another resource derived from BLOCKS and PRINTS is IDENTIFY, which, instead of encoding the exact information at each position in an alignment, tolerates alternative residues according to a set of prescribed groupings of related biochemical properties. IDENTIFY and its search software, eMOTIF (Huang and Brutlag, 2001), are accessible for use from the protein function Web server of the Biochemistry Department at Stanford University, http://dna.Stanford.edu/identify/. PANAL (http://mgd.ahc.umn.edu/panal/run_panal.html) of Computational Biology Centers at the University of Minnesota is a combined resource for ProSite, BLOCKS, PRINTS, and Pfam (Bateman et al., 2000).

The application of the secondary databases to study protein structure and function will be discussed in the following chapter (Chapter 12).

11.2.3. Sequence Alignment and Similarity Search

Pairwise comparison of two sequences is a fundamental process in sequence analysis. It defines the concepts of sequence identity, similarity, and homology as applied to two proteins, DNA or RNA sequences. Database interrogation can take the form of text queries or sequence similarity searches. Typically, the user employs a query sequence to conduct sequence similarity search so that the relationships between the query sequence (probe) and another sequence (target) can be quantified and their similarity assessed.

A sequence consists of letters selected from an alphabet. The complexity of the alphabet is defined as the number of different letters it contains. The complexity is 4 for nucleic acids and 20 for proteins. Sometimes additional characters are used to indicate ambiguities in the identity of a particular base or residue — for example, B for aspartate or asparagine and Z for glutamate or glutamine (in nucleic acids: R for A or G, and Y for C, T, or U).

Similarity Scoring. A comprehensive alignment must account fully for the positions of all residues in the sequences under comparison. To achieve this, gaps may be inserted in a manner as to maximize the number of identical matches. However the result of such process is biologically meaningless. Thus scoring penalties are introduced to minimize the number of gaps that are opened and extension penalties are also incurred when a gap has to be extended. The total alignment score is a function of the identity between aligned residues and the incurred gap penalties. By doing this, only residue identities are considered, resulting in a unitary matrix that weights identical residue matches with a score of 1 and other elements being 0. In order to improve diagnostic performance, the scoring potential of the biologically significant signals is enhanced so that they can contribute to the matching process without amplifying noise. A need to balance between high-scoring matches that have only mathematical significance and lower-scoring matches that are biologically meaningful is the essential requisite of sequence analysis. Scoring matrices that weight matches between nonidentical residues based on evolutionary substitution rates have been devised to address this need. Such tools may increase the sensitivity of the alignment, especially in a situation where sequence identity is low. One of the most popular scoring matrices is Dayhoff mutation data (MD) matrix (Dayhoff et al., 1978).

The MD score is based on the concept of the *point accepted mutation* (PAM). Thus, one PAM is a unit of evolutionary divergence in which 1% of the amino acids have been changed. If these changes were purely random, the frequencies of each possible substitution would be determined simply by the overall frequencies of the different amino acid (background frequencies). However, in related proteins, the observed substitution frequencies (target frequencies) are biased toward those that do not seriously disrupt the protein's function. In other words, these are mutations that have been accepted during evolution. Thus substitution scores in the mutation probability matrix are proportional to the natural log of the ratio of target frequencies to background frequencies. Mutation probability matrices corresponding to longer evolutionary distance can be derived by multiplication of this matrix of probability values by itself the appropriate number of times (n times for a distance of n PAM). Table 11.2 shows the values connecting the overall proportion of

TABLE 11.2. PAM Scale for Amino Acid Residues

Observed % Difference	Evolutionary Distance in PAM
1	1
10	11
20	23
25	30
30	38
40	56
50	80
60	112
70	159
75	195
80	246
85	328

observed mismatches in amino acid residues with the corresponding evolutionary distance in PAM (PAM scale).

The 250 PAM (250 accepted point mutations per 100 residues) matrix gives similarity scores equivalent to 80% difference between two protein sequences. Because the sequence analysis is aimed at identifying relationships in the twilight zone (20% identity), the MD for 250 PAM has become the default matrix in many analysis packages. The appropriate similarity score for a particular pairwise comparison at any given evolutionary distance is the logarithm of the odds that a particular pair of amino acid residues has arisen by mutation. These odds can be derived via dividing the values in the mutation probability matrix by the overall frequencies of the amino acid residues. In the matrix, residues likely undergoing mutations, neutral (random mutations), and unlikely undergoing mutations have values of greater than 0, equal to 0, and less than 0, respectively.

There are two approaches of sequence alignments: A global alignment compares similarity across the full stretch of sequences, while a local alignment searches for regions of similarity in parts of the sequences.

Global Alignment. In this approach (Needleman and Wunsch, 1970), a 2D matrix is constructed by comparing two sequences that are placed along the x- and the y-axes, respectively (Table 11.3). Cells representing identities are scored 1, and those with mismatches are scored 0 to populate the 2D array with 0's and 1's.

An operation of successive summation of cells begins at the last cell (S,T) in the matrix progressing to the next row or column by adding the maximum number from two constituent subpaths to the cell. For the cell (R, R) in the example, the original score is 1. The maximum value of the two subpaths is 5, thus the new value for the cell (R, R) becomes $5 + 1 = 6$. This process is repeated for every cell in the matrix (Table 11.4).

An alignment is generated by working through the completed matrix, starting at the highest-scoring element at the N-terminal and following the path of high

TABLE 11.3. Unitary Scoring Matrix

	C	H	E	M	I	C	A	L	R	E	A	G	E	N	T
C	1	0	0	0	0	1	0	0	0	0	0	0	0	0	0
H	0	1	0	0	0	0	0	0	0	0	0	0	0	0	0
E	0	0	1	0	0	0	0	0	0	1	0	0	1	0	0
M	0	0	0	1	0	0	0	0	0	0	0	0	0	0	0
I	0	0	0	0	1	0	0	0	0	0	0	0	0	0	0
S	0	0	0	0	0	0	0	0	0	0	0	0	0	0	0
T	0	0	0	0	0	0	0	0	0	0	0	0	0	0	1
R	0	0	0	0	0	0	0	0	1	0	0	0	0	0	0
Y	0	0	0	0	0	0	0	0	0	0	0	0	0	0	0
A	0	0	0	0	0	0	1	0	0	0	1	0	0	0	0
G	0	0	0	0	0	0	0	0	0	0	0	1	0	0	0
E	0	0	1	0	0	0	0	0	0	1	0	0	1	0	0
N	0	0	0	0	0	0	0	0	0	0	0	0	0	1	0
T	0	0	0	0	0	0	0	0	0	0	0	0	0	0	1
S	0	0	0	0	0	0	0	0	0	0	0	0	0	0	0

TABLE 11.4. Completed Maximum-Match Matrix

	C	H	E	M	I	C	A	L	R	E	A	G	E	N	T
C	11	9	8	7	6	7	6	6	6	5	4	3	2	1	0
H	9	10	8	7	6	6	6	6	6	5	4	3	2	1	0
E	8	8	9	7	6	6	6	6	5	6	4	3	2	1	0
M	7	7	7	8	6	6	6	6	5	5	4	3	2	1	0
I	6	6	6	6	7	6	6	6	5	5	4	3	2	1	0
S	6	6	6	6	6	6	6	6	5	5	4	3	2	1	0
T	6	6	6	6	6	5	6	6	5	5	4	3	2	1	1
R	5	5	5	5	5	5	5	5	6	5	4	3	2	1	0
Y	5	5	5	5	5	5	5	5	5	5	4	3	2	1	0
A	4	4	4	4	4	4	5	4	4	4	5	3	2	1	0
G	3	3	3	3	3	3	3	3	3	3	3	4	2	1	0
E	2		3	2	2	2	2	2	3	2	2	3	1	0	
N	1	1	1	1	1	1	1	1	1	1	1	1	1	2	0
T	0	0	0	9	0	0	0	0	0	0	0	0	0	0	1
S	0	0	0	0	0	0	0	0	0	0	0	0	0	0	0

scores through to the C-terminal. For example:

```
CHEMICALREAGENT-
|||||   |||||
CHEMIST-RYAGENTS
```

Local Alignment. Two sequences that are distantly related to each other may exhibit small regions of local similarity, though no satisfactory overall alignment can be found. The Smith and Waterman algorithm (Smith and Waterman, 1981) is designed to find these common regions of similarity and has been used as the basis for many subsequent algorithms.

The FASTA (Lipman and Pearson, 1985) and BLAST (Altschul et al., 1990; Altschul et al., 1997) programs are local similarity search methods that concentrate on finding short identical matches, which may contribute to a total match. FASTA uses *ktups* (1 or 2 amino acids for proteins and 4 or 6 nucleotides for nucleic acids) while BLAST uses words (3 amino acids for proteins and 11 nucleotides for nucleic acids) of short sequences in the database searching. In a typical output from a FASTA search, the name and version of the program are given at the top of the file with the appropriate citation. The query sequence together with the name and version of the search database are indicated. A range of parameters used by the algorithm and the run-time of the program are printed. Following these statistics come the results of the database search with a list of a user-defined number of matches with the query sequence and information for the source database (identifier, ID code, accession number, and title of the matched sequences). The length in amino acid residues of each of the retrieved matches is indicated in brackets. Various initial and optimized score and *E*(expected) value are given. After the search summary, the output presents the complete pairwise alignments of a user-defined number of hits with the query sequence. Within the alignments, identities are indicated by the ':' character while similarity is indicated by the '.' character.

The BLAST (Basic Local Alignment Search Tool) (Altschul et al., 1990) searches two sequences for subsequences of the same length that form an ungapped alignment. The algorithm searches and calculates all segment pairs between the query and the database sequences above a scoring threshold. The resulting high-scoring pairs (HSPs) form the basis of the ungapped alignments that characterize BLAST output. In the gapped BLAST (Altschul et al., 1997), the algorithm seeks only one ungapped alignment that makes up a significant match. Dynamic programming is used to extend a central pair of aligned residues in both directions to yield the final gapped alignment to speed up the initial database search. In the typical BLAST output, the name and version of the program together with the appropriate citation are given at the top of the file. The query sequence and the name of the search database are then indicated. Next come the results of the database search with a list of a user-defined number of matches with the query sequence and information for the source database (identifier, ID code, accession number, and title of the matched sequences). The highest score of the set of matching segment pairs for the given sequence with the number of HSPs denoted by the parameter N is given. This is followed by a p(probability) value. After the search summary, the output presents the pairwise alignments of the HSPs for each of the user-defined number of hits with the query sequence. For each aligned HSP, the beginning and end locations within the sequence are marked and identities between them are indicated by the corresponding amino acid symbols.

Multiple Sequence Alignment. The goal of multiple sequence alignment is to generate a concise summary of sequence data in order to make an informed decision on the relatedness of sequences to a gene family. Alignments should be regarded as models that can be used to test hypothesis. There are two main approaches on the construction of alignments. The first method is guided by the comparison of similar strings of amino acid residues, taking into account their physicochemical properties, mutability data, and so on. The second method results from comparison at the level of secondary or tertiary structure where alignment positions are determined solely on the basis of structural equivalence. Alignments generated from these rather different approaches often show significant disparities. Each is a model that reflects a different biochemical view of sequence data. The multiple sequence analysis draws together related sequences from various species and expresses the degree of similarity in a relatively concise format. The ClustalW algorithm (Higgins et al., 1996) is one of the most widely used multiple sequence alignment program (where W stands for weighting). This program takes an input set of sequences (the simplest format is fasta format) and calculates a series of pairwise alignments, comparing each sequence to every other sequence, one at a time. Based on these comparisons, a distance matrix is calculated, reflecting the relatedness of each pair of sequences. The distance matrix, in turn, serves in the calculation of a phylogenetic tree that can be weighted to favor closely related alignment of the most closely related sequences, then realigning with the addition of the next sequence, and so on. The Clustal X (Jeanmougin et al., 1995) is the portable windows interface Clustal program for multiple sequence alignment.

11.3. PROTEOMIC ANALYSIS USING INTERNET RESOURCES: SEQUENCE AND ALIGNMENT

11.3.1. Sequence Databases

The Protein Information Resources (PIR) (Wu et al., 2002) of NBRF in collaboration with MIPS and JIPID produces the annotated protein sequence database in the PIR-MIPS International Protein Sequence Database (PSD). The PSD is a comprehensive annotated and nonredundant protein sequence database. Its annotation includes concurrent cross-references to other sequence, structure, genomic and citatation databases, as well as functional descriptions and structural features. The PIR-International database is accessible at the PIR site, http://pir. georgetown.edu, and at the MIPS site, http://www.mips.biochem.mpg.de.

To open the search protein sequence database page at PIR (Figure 11.1), select Text/Entry Search under Search Databases. Enter the search keyword string by using Boolean expression or by using one of the five options of textboxes marked Search string and then click the Submit button. After an initial return of hits, you may modify the number of entries on the list, and view of the output. Pick the desired entry and the annotated sequence to receive the return in the default PIR format. An output in the fasta format can be requested. Links to the structure (PDB), alignment, and protein classification (at PIR and MIPS) are provided on the output page.

The PIR site also provides facilities for sequence similarity search (BLAST or FASTA), alignment (ClustalW), and analysis (ProClass, ProtFam, RESID, SSEARCH) of proteins. To perform the similarity search, select Blast or Fasta to

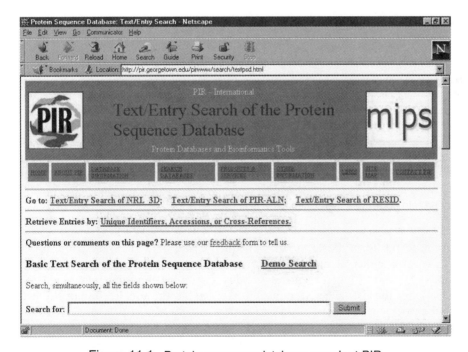

Figure 11.1. Protein sequence database search at PIR.

DB	ID	Title	Score	e-value	#aa	%idn	Ov.lap
PIR	A42343	alcohol dehydrogenase - Baltic cod	703	0.0	375	100	375
PIR	JC4967	alcohol dehydrogenase class III - gil...	450	e-126	376	62	375
PIR	S68061	alcohol dehydrogenase class III - Ind...	447	e-125	373	64	370
PIR	A33419	alcohol dehydrogenase class III -horse	440	e-123	373	62	371
PIR	DEHUC2	alcohol dehydrogenase 5 [validated]	438	e-122	374	63	372
PIR	DERTA	alcohol dehydrogenase - rat	437	e-122	373	63	372
PIR	A56643	alcohol dehydrogenase 2 - mouse	431	e-120	374	62	372
PIR	S51187	alcohol dehydrogenase class III - Atl...	430	e-120	376	61	373
PIR	A49662	alcohol dehydrogenase class III - com...	412	e-114	378	58	373
PIR	T04164	formaldehyde dehydrogenase (glutathione)	401	e-111	381	56	378
PIR	S51357	alcohol dehydrogenase Fdh - fruit fly...	400	e-111	379	58	376
PIR	S71244	alcohol dehydrogenase class III - Ara...	394	e-109	379	56	375
PIR	DEMSAA	alcohol dehydrogenase A - mouse	391	e-108	375	57	375
PIR	A35837	alcohol dehydrogenase - Japanese quail	390	e-108	375	57	376
PIR	DECHA1	alcohol dehydrogenase - chicken	387	e-107	375	56	376
PIR	A26468	alcohol dehydrogenase I - rat	386	e-107	376	56	376
PIR	A49107	alcohol dehydrogenase I - deer mouse	385	e-106	375	55	381
PIR	DEHOAL	alcohol dehydrogenase E - horse	385	e-106	375	55	375
PIR	S62640	alcohol dehydrogenase - Indian cobra	385	e-106	375	56	374
PIR	T03289	formaldehyde dehydrogenase (glutathione)	384	e-106	381	55	377

```
DEHOAL alcohol dehydrogenase (EC 1.1.1.1) E - horse
       Length = 375
Score = 385 bits (1036), Expect = e-106
Identities = 196/374 (52%), Positives = 260/374 (69%), Gaps = 2/374 (0%)

Query 1    ATVGKVIKCKAAVAWEANKPLVIEEIEVDVPHANEIRIKIIATGVCHTDLYHLFEGKHKD 60
           +T GKVIKCKAAV WE KP IEE+EV  P A+E+RIK++ATG+C +D  H+   G
Sbjct 2    STAGKVIKCKAAVLWEEKKPFSIEEVEVAPPKAHEVRIKMVATGICRSD-DHVVSGTLVT 60

Query 61   GFPVVLGHEGAGIVESVGPGVTEFQPGEKVIPLFISQCGECRFCQSPKTNQCVKGWANES 120
              PV+ GHE AGIVES+G GVT  +PG+KVIPLF  QCG+CR C+ P+ N C+K    +
Sbjct 61   PLPVIAGHEAAGIVESIGEGVTTVRPGDKVIPLFTPQCGKCRVCKHPEGNFCLKNDLSMP 120

Query 121  PDVMSPKETRFTCKGRKVLQFLGTSTFSQYTVVNQIAVAKIDPSAPLDTVCLLGCGVSTG 180
           M    +RFTC+G+   FLGTSTFSQYTVV++I+VAKID ++PL+ VCL+GCG STG
Sbjct 121  RGTMQDGTSRFTCRGKPIHHFLGTSTFSQYTVVDEISVAKIDAASPLEKVCLIGCGFSTG 180

Query 181  FGAAVNTAKVEPGSTCXXXXXXXXXXXXXXXXCHSAGAKRIIAVDLNPDKFEKAKVFGATD 240
           +G+AV  AKV  GSTC           C +AGA RII VD+N DKF KAK  GAT+
Sbjct 181  YGSAVKVAKVTQGSTCAVFGLGGVGLSVIMGCKAAGAARIIGVDINKDKFAKAKEVGATE 240

Query 241  FVNPNDHSEPISQVLSKMTNGGVDFSLECVGNVGVMRNALESCLKGWGVSVLVGW-TDLH 299
           VNP D+ +PI +VL++M+NGGVDFS E +G + M AL C +GVSV+VG   D
Sbjct 241  CVNPQDYKKPIQEVLTEMSNGGVDFSFEVIGRLDTMVTALSCCQEAYGVSVIVGVPPDSQ 300

Query 300  DVATRPIQLIAGRTWKGSMFGGFKGKDGVPKMVKAYLDKKVKLDEFITHRMPLESVNDAI 359
           +++  P+ L++GRTWKG++FGGFK KD VPK+V  ++ KK LD  ITH +P E +N+
Sbjct 301  NLSMNPMLLLSGRTWKGAIFGGFKSKDSVPKLVADFMAKKFALDPLITHVLPFEKINEGF 360

Query 360  DLMKHGKCIRTVLS 373
           DL++ G+ IRT+L+
Sbjct 361  DLLRSGESIRTILT 374
```

Figure 11.2. PIR BLAST Search results for Baltic cod alcohol dehydrogenase. A list of hits (only top 20 hits are shown; the EC numbers are deleted from the entries) is followed by pairwise alignments of similar sequences (only the sequence alignment between the query protein and horse liver E isozyme is shown).

open the search form. Paste the query sequence and click the Submit button. The search returns a list of hits (PIR Search results as shown in Figure 11.2) with option(s) for modifying the entries (modify query by choosing refine, add to or remove from the list). The Modify option(s) enable the user to design the list of desired entries.

SWISS-PROT (Hofmann et al., 1999) is a curated protein sequence database maintained by the Swiss Institute of Bioinformatics and is a collaborative partner of EMBL. The database consists of SWISS-PROT and TrEMBL, which consists of entries in SWISS-PROT-like format derived from the translation of all CDS in the

EMBL Nucleotide Sequence Database. SWISS-PROT consists of core sequence data with minimal redundancy, citation and extensive annotations including protein function, post-translational modifications, domain sites, protein structural information, diseases associated with protein deficiencies and variants. SWISS-PROT and TrEMBL are available at EBI site, http://www.ebi.ac.uk/swissprot/, and ExPASy site, http://www.expasy.ch/sprot/. From the SWISS-PROT and TrEMBL page of ExPASy site, click "Full text search" (under Access to SWISS-PROT and TrEMBL) to open the search page (Figure 11.3). Enter the keyword string (use Boolean expression if required), check SWISS-PROT box, and click the Submit button. Select the desired entry from the returned list to view the annotated sequence data in Swiss-Prot format. An output in the fasta format can be requested. Links to BLAST, feature table, some ExPASy proteomic tools (e.g., Compute pI/Mw, ProtParam, ProfileScan, ProtScale, PeptideMass, ScanProsite), and structure (SWISS-MODEL) are provided on the page.

The amino acid sequences can be searched and retrieved from the integrated retrieval sites such as Entrez (Schuler et al., 1996), SRS of EBI (http://srs.ebi.ac.uk/), and DDBJ (http://srs.ddbj.nig.ac.jp/index-e.html). From the Entrez home page (http://www.ncbi.nlm.nih.gov/Entrez), select Protein to open the protein search page. Follow the same procedure described for the Nucleotide sequence (Chapter 9) to retrieve amino acid sequences of proteins in two formats; GenPept and fasta. The GenPept format is similar to the GenBank format with annotated information, reference(s), and features. The amino acid sequences of the EBI are derived from the SWISS-PROT database. The retrieval system of the DDBJ consists of PIR, SWISS-PROT, and DAD, which returns sequences in the GenPept format.

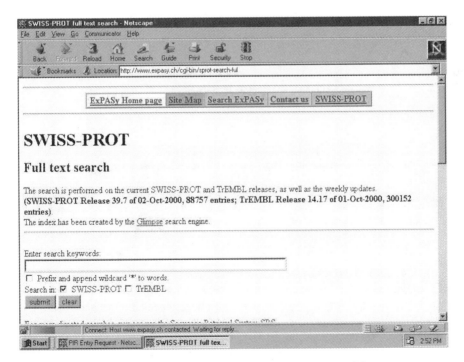

Figure 11.3. SWISS-PROT Search page of ExPASy.

11.3.2. Sequence Information Tools

The comprehensive molecular biology server, ExPASy (Expert Protein Analysis System), provides Proteomic tools (http://www.expasy.ch/tools) that can be accessed directly or from the ExPASy home page by selecting Identification and Characterization under Tools and Software Packages. The Protein identification and characterization tools of the Proteomics tools (Figure 11.4) provide facilities to:

- Identify a protein by its amino acid composition (AACompIdent)
- Compare the amino acid composition of an entry with the database (AACompSim)
- Predict potential post-translational modifications (FindMod)
- Calculate the mass of peptides for peptide mapping (PeptideMass)
- Evaluate physicochemical parameters from a query sequence (ProtParam)
- Scan for PROSITE (ScanProsite)
- Calculate/plot amino acid scales (ProtScale)

The identification of a protein from its amino acid composition and physicochemical properties (e.g., pI and molecular weight) is a valuable step prior to

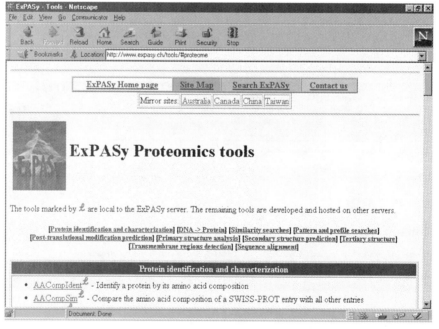

Figure 11.4. ExPASy Proteomic tools. ExPASy server provides various tools for proteomic analysis which can be accessed from ExPASy Proteomic tools. These tools (locals or hyperlinks) include Protein identification and characterization, Translation from DNA sequences to protein sequences. Similarity searches, Pattern and profile searches, Post-translational modification prediction, Primary structure analysis, Secondary structure prediction, Tertiary structure inference, Transmembrane region detection, and Sequence alignment.

TABLE 11.5. Search Result of AACompIdent for Lipoamide Dehydrogenase from *Schizosaccharomyces pombe*

The closest SWISS-PROT entries (in terms of AA composition) for the keyword FLAVOPROTEIN and the species SCHIZOSACCHAROMYCES POMBE:

Rank	Score	Protein	(pI	Mw)	Description
1	2	DLDH_SCHPO	8.29	50793	DIHYDROLIPOAMIDE DEHYDROGENASE.
2	16	GSHR_SCHPO	6.90	49999	GLUTATHIONE REDUCTASE (EC 1.6.4.2) (GR)
3	25	YDGE_SCHPO	6.20	62103	PUTATIVE FLAVOPROTEIN C26F1.14C.
4	31	DCP1_SCHPO	6.11	63294	PUTATIVE PYRUVATE DECARBOXYLASE C13A11
5	34	ODPA_SCHPO	6.72	41421	PYRUVATE DEHYDROGENASE E1 COMPONENT
6	40	ILVB_SCHPO	7.85	57643	ACETOLACTATE SYNTHASE.
7	43	PYRD_SCHPO	9.44	44526	DIHYDROOROTATE DEHYDROGENASE.
8	46	DCP2_SCHPO	5.71	64787	PROBABLE PYRUVATE DECARBOXYLASE C1F8.07C
9	47	ETFD_SCHPO	8.78	65780	PROBABLE ELECTRON TRANSFER FLAVOPROTEIN
10	53	ETFA_SCHPO	5.48	32424	PROBABLE ELECTRON TRANSFER FLAVOPROTEIN
11	54	TRXB_SCHPO	5.19	34618	THIOREDOXIN REDUCTASE (EC 1.6.4.5).
12	56	MTHS_SCHPO	5.91	72140	METHYLENETETRAHYDROFOLATE REDUCTASE 2
13	63	ODPB_SCHPO	5.29	35852	PYRUVATE DEHYDROGENASE E1 COMPONENT BETA
14	67	OYEA_SCHPO	8.77	43813	PUTATIVE NADPH DEHYDROGENASE C5H10.04
15	67	GPDM_SCHPO	6.24	68229	GLYCEROL-3-PHOSPHATE DEHYDROGENASE
16	77	OYEB_SCHPO	5.83	44983	PUTATIVE NADPH DEHYDROGENASE C5H10.10
17	79	MTHR_SCHPO	5.17	69012	PROBABLE METHYLENETETRAHYDROFOLATE
18	80	NCPR_SCHPO	5.41	76775	NADPH-CYTOCHROME P450 REDUCTASE (EC 1.6
19	91	MT10_SCHPO	5.69	111353	PROBABLE SULFITE REDUCTASE [NADPH]
20	92	COQ6_SCHPO	9.29	52160	PUTATIVE UBIQUINONE BIOSYNTHESIS

initiating sequence determination. On the AACompIdent page, select a Constellation to be used (e.g., Constellation 0 for all amino acids) to open the request form. Enter pI value, molecular weight, biological source of the protein (OS or OC), the ExPASy recognizable keyword (e.g., dehydrogenase, kinase), your e-mail address, and amino acid composition (in molar %). Click the Run AACompIdent button. The search result is returned by e-mail as shown in Table 11.5.

PROPSEARCH (http://www.embl-heidelberg.de/prs.html) is designed to find the putative protein family if alignment methods fail (Hobohm and Sander, 1995). The program uses the amino acid composition of a protein taken from the query sequence to discern members of the same protein family by weighing 144 different physical properties including molecular weight, the content of bulky residues, average hydrophobicity, average charge, and so on, in performing the analysis. This collection of physical properties is called the query vector, and it is compared against the same type of vector precomputed for every sequence in the target databases (SWISS-PROT and PIR). The search results are returned by e-mail.

An amino acid scale refers to a numerical value assigned to each amino acid such as polarity, hydrophobicity, accessible/buried area of the residue, and propensity to form secondary structures (e.g., amino acid parameter of AAIndex). For example, selecting (clicking) ProtScale from the list of ExPASy Proteomics tools opens the query page with a list of predefined amino acid scales. Paste the query

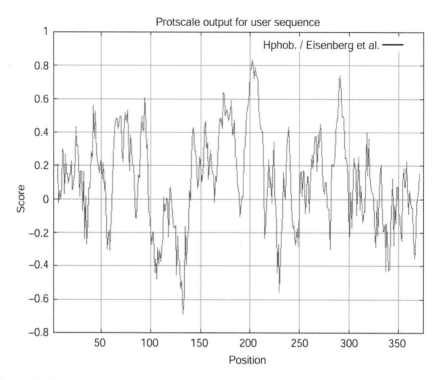

Figure 11.5. ProtScale (ExPASy Proteomic Tools) output of hydrophobicity parameter. The hydrophobicity profile of cod alcohol dehydrogenase is generated with ProtScale using Eisenberg hydrophobic scale. The individual values for the 20 amino acids used in the profile are also listed in the report.

sequence into the sequence box, select the desired scale and window size, set your choice (yes/no) for normalizing the scale from 0 to 1, and click the Submit button. The query returns with a list of scales used for amino acids and the profile plot (score versus residue position as shown in Figure 11.5), which can be saved as image file (protscal.gif or protscal.ps). In addition, numerous links are made to various servers for similarity searches, pattern and profile searches, post-translational modification predictions, structural analyses, and sequence alignments.

11.3.3. Sequence Alignment

One of the most common analyses of protein sequences is the similarity/homology search. It allows a mapping of information from known sequences to novel ones, especially the functional sites of the homologous proteins. Dynamic programming that is a general optimization technique has been applied to sequence alignment and related computational problems. A dynamic programming algorithm finds the best solution by breaking the original problem into smaller sequentially dependent subproblems that are solved first. Each intermediate solution is stored in a table along with a score and the sequence of solutions that yields the highest score is chosen. The search for similarity/homology is well supported by the Internet

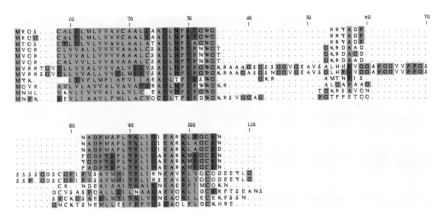

Figure 11.6. Amino acid sequence alignment at EBI. Amino acid sequences of type C lysozyme are submitted to ClustalW tool of EBI, and the alignment is retrieved from the notified URL.

accessible tools. A query sequence may be entered to conduct homology search using the BLAST server at ExPASy or NCBI. Launch Basic BLAST at NCBI (http://www.ncbi.nlm.hih.gov/BLAST/) as previously shown in Figure 9.1. Paste or upload the query sequence (fasta format is preceded by >seqname). Choose the program (blastn for nucleic acids and blastp for proteins) and database (nr for non-redundant). Set to either Perform ungapped alignment or Perform CDD Search (Protein only). You may choose the formatting options now or later. Enter the address to receive reply by e-mail. After clicking the Search button, you receive a confirmation message, "Your request has been successfully submitted and put into the Blast Queue. Query = seqname" with the request ID which is used to retrieve the search results. The output includes alignment scores plot, list of sequences producing significant alignments (accession ID, name, score, and E value), and pairwise alignments (if this Alignment view is chosen).

Similarity search can be performed with the FASTA program, though BLAST is easier to use. Both FASTA and BLAST homology searches are also available at EBI, PIR and DDBJ. The small P values indicate homologous sequences.

The multiple sequence alignment of homologous proteins can be accomplished with ClustalW, which is one of the most commonly used programs. A number of ClustalW servers are available. These include EBI of EMBL (http://www2.ebi.ac.uk/clustalw/), PIR (http://pir.georgetown.edu/), IBC of Washington University in St. Louis (http://www.ibc.wustl.edu/msa/clustal.html), and PredictProtein of Columbia University (http://cubic.bioc.columbia.edu/predictprotein/). BCM server (http://dot.imgen.bcm.tme.edu.9331/multi-align/multi-align.html) and DDBJ (http://www.ddbj.nig.ac.jp/). The returned alignment via e-mail can be saved as alignfile.ps (e.g., Figure 11.6 for lysozyme C from EBI).

To perform multiple sequence alignment online (at DDBJ), select (click) Search and Analysis to access DDBJ Homology Search System (http://spiral.genes.nig.ac.jp/homology/tope.html). Select ClustalW to open the query form (Figure 11.7). Select the desired options, upload the file containing multiple sequences, pick WWW, and click Send. The returned output includes pairwise alignment score, multiple sequence alignment and tree file.

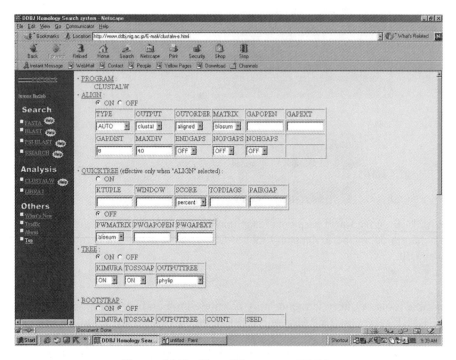

Figure 11.7. ClustalW server at DDBJ.

11.4. WORKSHOPS

1. A purified mammalian protein with a molecular weight of $40,000 \pm 1000$ and a pI of 8.0 (± 0.5 is composed (an amino acid composition in mol %) of Ala, 7.5; Arg, 3.2; Asn, 2.1; Asp, 4.5; Cys, 3.7; Gln, 2.1; Glu, 5.6; Gly, 10.2; His, 1.9; Ile, 6.4; Leu, 6.7; Lys, 8.0; Met, 2.4; Phe, 4.8; Pro, 5.3; Ser, 7.0; Thr, 6.4; Trp, 0.5; Tyr, 1.1; and Val, 10.4. Submit these data to the Internet server that searches its database to identify the protein.

2. Analyze peptide fragments (PeptideMass of ExPASy) produced by treating a protein with the following amino acid sequence with chymotrypsin, proteinase K, and trypsin,. Deduce the specificities of these enzymes.

```
>gi|35497|emb|CAA78820.1| protein kinase C gamma [Homo sapiens]
QLEIRAPTADEIHVTVGEARNLIPMDPNGLSDPYVKLKLIPDPRNLTKQKTRTVKATLNPVWNETFVFNL
KPGDVERRLSVEVWDWDRTSRNDFMGAMSFGVSELLKAPVDGWYKLLNQEEGEYYNVPVADADNCSLLQK
FEACNYPLELYERVRMGPSSSPIPSPSPSPTDPKRCFFGASPGRLHISDFSFLMVLGKGSFGKVMLAERR
GSDELYAIKILKKDVIVQDDDVDCTLVEKRVLALGGRGPGGRPHFLTQLHSTFQTPDRLYFVMEYVTGGD
LMYHIQQLGKFKEPHAAFYAAEIAIGLFFLHNQGIIYRDLKLDNVMLDAEGHIKITDFGMCKENVFPGTT
TRTFCGTPDYIAPEIIAYQPYGKSVDWWSFGVLLYEMLAGQPPFDGEDEEELFQAIMEQTVTYPKSLSRE
AVAICKGFLTKHPGKRLGSGPDGEPTIRAHGFFRWIDWERLERLEIPPPFRPRPCGRSGENFDKFFTRAA
PALTPPDRLVLASIDQADFQGFTYVNPDFVHPDARSPTSPVPVPVM
```

3. Given the following amino acid sequence, estimate (ProtParam of ExPASy) its amino acid composition, numbers of charged residues, extinction coefficient, estimated half-life, and instability index of the protein. Elaborate briefly how extinction coefficient, half-life and instability index are estimated.

```
>gi|68532|pir||SYBYDC aspartate--tRNA ligase (EC 6.1.1.12), cytosolic -
yeast
MSQDENIVKAVEESAEPAQVILGEDGKPLSKKALKKLQKEQEKQRKKEERALQLEAEREAREKKAAAEDT
AKDNYGKLPLIQSRDSDRTGQKRVKFVDLDEAKDSDKEVLFRARVHNTRQQGATLAFLTLRQQASLIQGL
VKANKEGTISKNMVKWAGSLNLESIVLVRGIVKKVDEPIKSATVQNLEIHITKIYTISETPEALPILLED
ASRSEAEAEAAGLPVVNLDTRLDYRVIDLRTVTNQAIFRIQAGVCELFREYLATKKFTEVHTPKLLGAPS
EGGSSVFEVTYFKGKAYLAQSPQFNKQQLIVADFERVYEIGPVFRAENSNTHRHMTEFTGLDMEMAFEEH
YHEVLDTLSELFVFIFSELPKRFAHEIELVRKQYPVEEFKLPKDGKMVRLTYKEGIEMLRAAGKEIGDFE
DLSTENEKFLGKLVRDKYDTDFYILDKFPLEIRPFYTMPDPANPKYSNSYDFFMRGEEILSGAQRIHDHA
LLQERMKAHGLSPEDPGLKDYCDGFSYGCPPHAGGGIGLERVVMFYLDLKNIRRASLFPRDPKRLRP
```

4. ProtScale tool of ExPASy computes amino acid scale (physicochemical properties/parameters) and presents the result in a profile plot. Perform ProtScale computations to compare the hydrophobicity/polarity profiles with %buried residues/%accessible residues profiles for human serine protease with the following amino acid sequence.

```
>gi|2318115|gb|AAB66483.1| serine protease [Homo sapiens]
MKKLMVVLSLIAAAWAEEQNKLVHGGPCDKTSHPYQAALYTSGHLLCGGVLIHPLWVLTAAHCKKPNLQV
FLGKHNLRQRESSQEQSSVVRAVIHPDYDAASHDQDIMLLRLARPAKLSELIQPLPLERDCSANTTSCHI
LGWGKTADGDFPDTIQCAYIHLVSREECEHAYPGQITQNMLCAGDEKYGKDSCQGDSGGPLVCGDHLRGL
VSWGNIPCGSKEKPGVYTNVCRYTNWIQKTIQAK
```

5. Scan the relative mutability of the protein with the following amino acid sequence:

```
>gi|67414|pir||LZPY lysozyme (EC 3.2.1.17) c - pigeon
KDIPRCELVKILRRHGFEGFVGKTVANWVCLVKHESGYRTTAFNNNGPNSRDYGIFQINSKYWCNDGKTR
GSKNACNINCSKLRDDNIADDIQCAKKIAREARGLTPWVAWKKYCQGKDLSSYVRGC
```

Is there any correlation between the relative mutability and polarity, average flexibility and/or average buried area of the amino acid residues?

6. Search the Web site to predict transmembrane topology of rhodopsin with the following sequence and compare the membrane-spanning regions with the hydrophobicity profiles of ProScale

```
>gi|10720173|sp|Q9YH05OPSD_DIPAN RHODOPSIN
MNGTEGPFFYVPMVNTTGIVRSPYEYPQYYLVNPAAYAALGAYMFLLILVGFPINFLTLYVTIEHKKLRT
PLNYILLNLAVADLFMVLGGFTTTMYTSMHGYFVLGRLGCNIEGFFATLGGEIALWSLVVLAIERWVVVC
KPISNFRFGENHAIMGLAFTWTMAMACAAPPLVGWSRYIPEGMQCSCGIDYYTRAEGFNNESFVIYMFIC
HFTIPLTVVFFCYGRLLCAVKEAAAAQQESETTQRAEKEVTRMVIMMVIAFLVCWLPYASVAWYIFTHQG
SEFGPVFMTIPAFFAKSSSIYNPMIYICLNKQFRHCMITTLCCGKNPFEEEEGASTASKTEASSVSSSSV
SPA
```

7. Perform a similarity search of PIR database for a protein with the following amino acid sequence:

```
>gi|67428|pir||LZGSG lysozyme (EC 3.2.1.17) g - goose
RTDCYGNVNRIDTTGASCKTAKPEGLSYCGVSASKKIAERDLQAMDRYKTIIKKVGEKLCVEPAVIAGII
SRESHAGKVLKNGWGDRGNGFGLMQVDKRSHKPQGTWNGEVHITQGTTILINFIKTIQKKFPSWTKDQQL
KGGISAYNAGAGNVRSYARMDIGTTHDDYANDVVARAQYYKQHGY
```

8. Perform a pairwise sequence alignment for the following analogous proteins:

```
>gi|539057|pir||A61024 alcohol dehydrogenase (EC 1.1.1.1) - wheat
MATAGKVIECKAAVAWEAGKPLSIEEVEVAPPHAMEVRVKILYTALCHTDVYFWEAKGQTPVFPRILGHE
AGGIVESVGEGVTELVPGDHVLPVFTGECKDCAHCKSEESNLCDLLRINVDRGVMIGDGQSRFTINGKPI
FHFVGTSTFSEYTVIHVGCLAKINPEAPLDKVCVLSCGISTGLGATLNVAKPKKGSTVAIFGLGAVGLAA
MEGARMAGASRIIGVDLNPAKYEQAKKFGCTDFVNPKDHTKPVQEVLVEMTNGGVDRAVECTGHIDAMIA
AFECVHDGWGVAVLVGVPHKEAVFKTYPMNFLNERTLKGTFFGNYKPRTDLPEVVEMYMRKELELEKFIT
HSVPFSQINTAFDLMLKGEGLRCIMRMDQ

>gi|82347|pir||S01893 alcohol dehydrogenase (EC 1.1.1.1) 1 - barley
MATAGKVIKCKAAVAWEAGKPLTMEEVEVAPPQAMEVRVKILFTSLCHTDVYFWEAKGQIPMFPRIFGHE
AGGIVESVGEGVTDVAPGDHVLPVFTGECKECPHCKSAESNMCDLLRINTDRGVMIGDGKSRFSIGGKPI
YHFVGTSTFSEYTVMHVGCVAKINPEAPLDKVCVLSCGISTGLGASINVAKPPKGSTVAIFGLGAVGLAA
AEGARIAGASRIIGVDLNAVRFEEARKFGCTEFVNPKDHTKPVQQVLADMTNGGVDRSVECTGNVNAMIQ
AFECVHDGWGVAVLVGVPHKDAEFKTHPMNFLNERTLKGTFFGNFKPRTDLPNVVEMYMKKELEVEKFIT
HSVPFSEINTAFDLMAKGEGIRCIIRMDN
```

9. Retrieve amino acid sequences (in fasta format) of human alcohol dehydrogenase isozymes and perform multiple alignment with ClustalW to evaluate their homology. Identify the amino acid substitutions among the seven isozymes.

10. Perform a BLAST similarity search for a human enzyme with the following amino acid sequence:

```
MALEKSLVRLLLLVLILLVLGWVQPSLGKESRAKKFQRQHMDSDSSPSSSSTYCNQMMRRRNMTQGRCKP
VNTFVHEPLVDVQNVCFQEKVTCKNGQGNCYKSNSSMHITDCRLTNGSRYPNCAYRTSPKERHIIVACEG
SPYVPVHFDASVEDST
```

Identify its isozymes and subject these isozymes to multiple alignment with ClustalW to evaluate their homology.

REFERENCES

Altschul, S. F. Gish, W., Miller, W., Myers, E. W., and Lipman, D. J. (1990) *J. Mol. Biol.* **215**:403–410.

Altschul, S. F., Madden, T. L., Schaffer, A. A., Zhang, J., Zhang, A., Miller, W., and Lipmen, D. J. (1997) *Nucleic Acids Res.* **25**:2289–3402.

Appel, R. D., Bairoch, A., and Hochstrasser, D. F. (1994) *Trends Biochem. Sci.* **19**:258–260.

Apweiler, R. et al. (2001) *Nucleic Acids Res.* **29**:37–40.

Attwood, T. K., Beck, M. E., Flower, D. R., Scordis, P., and Selley, J. (1998) *Nucleic Acids Res.* **26**: 304–308.

Bairoch, A., and Apweiler, R. (2000) *Nucleic Acids Res.* **28**:45–48.

Barker, W. C., Garavelli, J. S., Hou, Z., Huang, H., Ledley, R. S., McGarvey, P. B., Mews, H.-W., Orcutt, B. C., Pfeiffer, F., Tsugita, A., Vinayaka, C. R., Xiao, C., Yeh, L.-S. L., and Wu, C. (2001). *Nucleic Acids Res.* **29**:29–32.

Bateman, A., Birney, E., Durbin, R., Eddy, S. R., Howe, K. L., and Sonnhammer, E. L. L. (2000) *Nucleic Acids Res.* **28**:263–266.

Bleasby, A. J., Akrigo, D.. and Attwood, T. K. (1994) *Nucleic Acids Res.* **22**:3574–3577.

Dayhoff, M. O., Schwartz, R. M., and Orcutt, B. C. (1978) In *Atlas of Protein Sequence and Structure*, Vol. 5, Suppl. 3, M. O. Dayhoff (Ed.) NBRF, Washington, D.C., p. 345.

Findley, J. B. C., and Geisow, M. J., Eds. (1989) *Protein Sequencing: A Practical Approach.* IRL Press/Oxford University Press, Oxford.

Henikoff, S., Pietrokowski, S., and Henikoff, J. G. (1998) *Nucleic Acids Res.* **26**:309–312.

Higgins, D. G., Thompson, J. D., and Gibson, T. J. (1996). *Meth. Enzymol.* **226**:383–402.

Hobohm, U., and Sander, C. (1995) *J. Mol. Biol.* **251**:390–399.

Hofmann, K., Bucher, P., Falquet, L., and Bairoch, A. (1999) *Nucleic Acids Res.* **27**:215–219.

Huang, J. Y., and Brutlag, D. L. (2001) *Nucleic Acids Res.* **29**:202–204.

Jeanmougin, F., Thompson, J. D., Gouy, M., Higgins, D. G., and Gibson, T. J. (1995) *Trends Biochem. Sci.* **23**:403–405.

Kawashima, S., and Kanehisa, M. (2000) *Nucleic Acids Res.* **28**:374.

Kyte, J., and Doolittle, R. F. (1982) *J. Mol. Biol.* **157**:105–132.

Lipman, D. J., and Pearson, W. R. (1985) *Science* **227**:1435–1441.

Mewes, H. W., Frishman, D., Gruber, C., Geier, B., Haase, D., Kaps, A., Lemcke, K., Mannhaupt, G., Pfeiffer, F., Schüller, C., Stocker, S., and Well, B. (2000) *Nucleic Acids Res.* **28**:37–40.

Milpetz, F., Argos, P., and Persson, B. (1995) *Trends Biochem. Sci.* **20**:204–205.

Needleman, S. B. Ed. (1975) *Protein Sequence Determination.* Springer-Verlag, New York.

Needleman, S. B., and Wunsch, C.D. (1970) *J. Mol. Biol.* **48**:443–453.

Schuler, G. D., Epstein, J. A., Ohkawa, H., and Kans, J. A. (1996) *Meth. Enzymol.* **266**:141–162.

Smith, T. F., and Waterman, M. S. (1981) *J. Mol. Biol.* **147**:195–197.

Walsh, K. A., Ericsson, L. H., Parmelee, D. C., and Titani, K. (1981) *Annu. Rev. Biochem.* **50**:261–284.

Wu, C. H., Huang, H., Arminski, L., Castro-Alvear, J., Chen, Y., Hu, Z.-Z., Ledley, R. S., Lewis, K. C., Mewes, H.-W., Orcutt, B. C., Suzek, B. E., Tsugita, A., Vinayaka, C. R., Yeh, L.-S., Zhang, J., and Barker, W. C. (2002) *Nucleic Acids Res.* **30**:27–30.

Wyatt, A. L. (1995) *Success with Internet.* Boyd and Fraser, Danvers, MA. *Enzymol.* **266**:383–402.

12

PROTEOMICS: PREDICTION OF PROTEIN STRUCTURES

Computation proteome annotation is the process of proteome database mining, which includes structure/fold prediction and functionality assignment. Methodologies of secondary structure prediction and problems of protein folding are discussed. Approaches to identify functional sites are presented. Protein structure databases are surveyed. Secondary structure predictions and pattern/fold recognition of proteins using the Internet resources are described.

12.1. PREDICTION OF PROTEIN SECONDARY STRUCTURES FROM SEQUENCES

The enormous structural diversity of proteins begins with different amino acid sequences (primary structure) of polypeptide chains that fold into complex 3D structures. The final folded arrangement of the polypeptide chain is referred to as its conformation (secondary and tertiary structures). It appears that the information for folding to the native conformation is present in the amino acid sequences (Anfinsen, 1973); however, a special class of proteins known as *chaperons* is required to facilitate *in vivo* folding of a protein to form its native conformation (Martin and Hartl, 1997).

Large-scale sequencing projects produce data of genes; hence they produce protein sequences at an amazing pace. Although experimental determination of protein three-dimensional structure has become more efficient, the gap between the number of known sequences and the number of known structures is rapidly increasing. Protein structure prediction aims at reducing this sequence-structure gap. Methods for abstracting protein structures from their sequences can be aimed at secondary structures or 3D structures (folding problems).

There are four approaches to secondary structure prediction (Fasman, 1989):

1. Empirical statistical methods that use parameters derived from known 3D structures (Chou and Fasman, 1974).
2. Methods based on physicochemical properties of amino acid residues (Lim, 1974) such as volume, exposure, hydrophobicity/hydrophilicity, charge, hydrogen bonding potential, and so on.
3. Methods based on prediction algorithms that use known structures of homologous proteins to assign secondary structures (Garnier et al., 1978; Zvelebil et al., 1987).
4. Molecular mechanical methods that use force field parameters to model and assign secondary structures (Pittsyn and Finkelstein, 1983).

The requirements for hydrogen bond preservation in the folded structure result in the cooperative formation of regular hydrogen bonded secondary structure regions in proteins. The secondary structure specifies regular polypeptide chain-folding patterns of helices, sheets, coils, and turns which are combined/folded into tertiary structure. The main chain of an α helix forms the backbone with the side chains of the amino acids projecting outward form the helix. The backbone is stabilized by the formation of hydrogen bonds between the CO group of each amino acid and the NH group of the residue four positions forward ($n + 4$), creating a tight, rodlike structure. Some residues form α helices better than others: Alainine, glutamine, leucine, and methionine are commonly found in α helices, whereas proline, glycine, tyrosine, and serine are not. Proline is commonly thought of as a helix breaker because its pyrrole ring structure disrupts the formation of hydrogen bonds. α Helices vary considerably in length (4–40 residues) in proteins with the average length of about 10 residues corresponding to 3 turns. The most common location for an α helix in a protein is along the outside of the protein with one side of the helix facing the solvent and the other side toward the hydrophobic interior of the protein. The β-sheet structure is built up from a combination of several regions of polypeptide chain usually 5 to 10 residues long in an almost fully extended conformation and stabilized by hydrogen bonding between adjacent strands. The overall structure formed through the interaction of these individual β strands is known as a β-pleated sheet, which is generally right-twisted.

The regular secondary structures, α helices and β sheets, are connected by coil or loop regions of various lengths and irregular shape. A combination of secondary structure elements forms the stable hydrophobic core of the protein molecule. The loop regions are at the surface of the molecule and frequently participate in forming binding/catalytic sites. Loop regions joining two adjacent secondary structures that change directions are known as turns or reverse turns (or β bends or hairpins for connecting two adjacent antiparallel β strands). Most reverse turns involve four successive amino acid residues (often with Gly or Pro at the second or third position). Almost all proteins of larger than 60 amino acid residues contain one or more loops of 6 to 16 residues referred to as Ω loops (necked-in shape), which may contain reverse turns located on the surface of protein molecules.

The propensities of the 20 amino acid residues over three structural states (α, β, turn) predict segments of a secondary structure when a cluster of consecutive preferences along the sequence show a mean preference greater than the threshold value in the empirical statistical method of Chou and Fasman (Chou and Fasman, 1974). An example of the physicochemical approach (Lim, 1974) is the use of polar and nonpolar properties of residue along the sequence. In an α helix for example, successive side chains occur at rotation of nearly 100° to form helical wheel (Schiffer and Edmundson, 1967). Polar residues are usually on one side of the wheel to face the external aqueous environment, and hydrophobic residues are usually on the other side to pack against the protein's internal core. This pattern of hydrophilic/hydrophobic residues can be searched for to predict a helix. The information approach uses known structures and collects information on the significant pairwise dependence of an amino acid in a given position with number of residues defined by the window (e.g., 7–12 residues) on either side of it in a particular secondary structural setting. The method of Garnier, Osguthorpe, and Robson (GOR) (Garnier et al., 1978) considers the effect that residues of a given position within the region eight residues N-terminal to eight residues C-terminal have on the structure of that position. Thus a profile that quantifies the contribution of the residue type toward the probability of one of four states — α helix (H), β sheet (E), turn (T), and coil (C) — exits for each residue type. Four probabilities are calculated for each residue in the sequence by summing information from the 17 local residues, $i \pm 8$. The statistical information derived from proteins of known structure is stored in four (17×20) matrices for H, E, T, and C. For any residue, the predicted state is the one with the largest probability value with reference to the known structures. The refinement of this approach includes the information from the alignment of homologous sequences (Garnier et al., 1996).

An important approach that has been applied frequently in structure predictions is the *neural network*. Basically a neural network (Bharath, 1994; van Rooij et al., 1996), a realm of artificial intelligence (AI), gives computational processes the ability to "learn" in an attempt to approximate human learning, whereas most computer programs execute their instructions blindly in a sequential manner. The neural networks have been applied extensively to solve problems that require analysis of patterns and trends, such as secondary structure prediction of proteins. Every neural network has an *input layer* and an *output layer*. In the case of secondary structure prediction, the input layer would be information from the sequence itself, and the output layer would be the probabilities of whether a particular residue could form a particular structure. Between the input and output layers would be one or more hidden layers in which the actual *learning* takes place. This is accomplished by providing a training data set for the network. Here, an appropriate training set would be all sequences for which three-dimensional structures have been deduced. The network can process this information to look for what are possibly relationships between an amino acid sequence and the structures they can form in a particular context. Kneller et al. (1990) discussed an application of neural networks to protein secondary structure prediction. Rost and Sander (1993) combined neural networks with multiple sequence alignments to improve the success rate of the prediction in the prediction program (PHD at PredictProtein server). Molecular mechanical method for predicting protein structures is the topic of chapter 15.

A number of servers offer various methods to predict secondary structures of proteins. Secondary structure prediction of ExPASy Proteomic tools (http://

TABLE 12.1. Web Servers for Protein Secondary Structure Prediction

Server	URL
BCM	http://www.hgsc.bcm.tmc.edu/search_launcher/
BMERC	http://bmerc-www.bu.edu/psa/index.html
Jpred	http://jura.ebi.ac.uk:8888/index.html
nnPredict	http://www.cmpharm.ucsf.edu/~nomi/nnpredict.html
NPS@	http://npsa-pbil.ibcp.fr
Predator	http://www.embl-heidelberg.de/cgi/predator_serv.pl
PredictProtein	http://cubic.bioc.columbia.edu/predictprotein/
PSIpred	http://insulin.brunel.ac.uk/psipred/
SSThread	http://www.ddbj.nig.ac.jp/E-mail/ssthread/www_service.html

www.expasy.ch/tools) lists various prediction methods and pointers to their servers. Some of prediction servers and their URL are listed in Table 12.1. Results are returned either online (Jpred, BCM, nnPredict, and NPS@) or by e-mail (BMERC, Predator, PredictProtein, and PSIpred).

12.2. PROTEIN FOLDING PROBLEMS AND FUNCTIONAL SITES

12.2.1. Fold Recognition and Structure Alignment

It is generally accepted that there are a finite number of different protein folds, and it is very likely for a newly determined structure to adopt a familiar fold rather than a novel one. This limitation to protein folds presumably derives from the relatively few secondary structure elements in a given domain and the fact that the arrangement of these elements is greatly constrained by folding environments, protein functions and probably evolution. Because proteins sharing at least 30% of their amino acid sequence will adopt the same fold and will often exhibit similar functions, protein homology has been an important quality in the identification of related proteins with a common fold during database searches. Thus a sequence can be compared with the known 3D structure of homologous protein using one of two basic strategies (Johnson et al., 1996). In the first approach, one-dimensional sequences are compared with one-dimensional templates or profiles derived from known 3D structures (Chapter 12). The template provides a set of scores for matching a position within the structure with those in the sequence. Thus threading methods (Jones et. al., 1992) fit a probe sequence onto a backbone of a known structure and evaluate the compatibility by using pseudo-energy potential and residue environmental information (Shi et. al., 2001). The second strategy involves an evaluation of the total energy of a sequence of the molecule which is folded similarly to the known 3D structure (Chapter 15).

Each amino acid has distinct attributes such as size, hydrophobicity/hydrophilicity, hydrogen bonding capacity, and conformational preferences that allow it to contribute to a protein fold. Attempts are being made to interpret the protein fold in terms of amino acid descriptors. Structural alignments are more accurate than sequence alignments, and the local physicochemical environment of every residue within each 3D structure is directly obtainable (e.g., AAindex at http://www.

genome.ad.jp/aaindex/ or ProtScale of ExPASy Proteomic tools at http://www.-expasy.ch/tools/). In sequence alignments, a gap penalty is typically assigned as either a constant value or one that consists of two components: a constant component and a length-dependent component. However, gaps do not occur with equal frequency everywhere within the protein fold. They are rarely found within elements of regular secondary structure, but rather most often within loop regions that are exposed to solvent at the surface of proteins. Likewise, a hydrogen-bonded polar amino acid buried in the protein core is highly conserved, whereas these same residues at the surface of a protein that is exposed to solvent are not.

When representative structures for most of the common folds become available, the structure alignment will certainly be the most powerful approach to protein fold recognition. Thus, computer graphics with its interactive capabilities emerged as an attractive tool for displaying molecular shapes or superimposing active molecules to detect their common substructures. For two related structures, the main question is finding the best geometric transformation in order to superimpose their frameworks together as closely as possible and evaluate their degree of similarity. If a protein with a known sequence but unknown structure can be matched to one of the common folds, it is possible to model the 3D structure of the protein and learn something about its function.

12.2.2. Protein Folding Classes

A study of the amino acid sequences in homologous proteins proves very informative not only in establishing the evolutionary relationships of different species but also in deciphering the complicated question of how amino acid sequence is translated into a specific 3D structure. Programs to predict secondary structures can now distinguish helix, sheet, and turn residues with over 60% accuracy. Thus, a hierarchical classification of proteins (Richardson, 1981) can be employed to assess the quality of our ability to predict protein structures from amino acid sequences. Most protein classifications refer to structure domains. The domain is a region of protein that has relatively little interaction with the rest of the protein, and therefore its structure is essentially independent. Typically, domains are collinear in sequence, but occasionally one domain will have another inserted into it, or two homologous domains will intertwine by swapping some topologically equivalent parts of their chains.

A classification system, such as SCOP (Lesk and Chothia, 1984), categorizes structure domains based on secondary structural elements within a protein into α structure (made up primarily from α helices), β structure (made up primarily from β strands), α/β structure (comprised of primarily β strands alternating with α helices), and $\alpha + \beta$ structure (comprised of a mixture of isolated α helices and β strands). In this classification (Brenner et al., 1996; Lesk, 1991), only the core of the domain is considered. Therefore, it is possible for an all-α structure to have very small amount of β strand outside the α-helical core. Similarly, an all-β protein may have a small presence of α or 3_{10} helix. The SCOP (http://scop.mrc-lmb.cam.ac.uk/scop/) can be summarized as follows:

[1] α-Helical Proteins. Most proteins contain some helix. In a significant number of proteins, the secondary structure is limited to α helices. For example:

- There is a class of small proteins (e.g., haemerythrin, Figure 12.1a) that have the form of a four-helix bundle.
- Globins (myoglobin, hemoglobin) are rich in helices and all but one (3_{10}) are α helices.
- The cytochrome Cs are another family of haem proteins with exclusively helical structures.
- Very large proteins (e.g., citrate synthase) can also have a purely helical secondary structure.
- Membrane attached receptor proteins contains seven transmembrane helices (e.g., bacteriorhodopsin, Figure 12.1b).

For individual pairs of interacting helices, certain regularities in the patterns of tertiary-structural interactions between pairs of helices have been observed.

[2] β-Sheet Proteins. Domains in which the secondary structure is almost exclusively β-sheet tend to contain two sheets packed face to face. There are two major classes: those in which the strands are almost parallel and those in which the strands are almost perpendicular. The strands of sheet can vary in direction (parallel and antiparallel) and in connectivity, giving great topological variety.

(2.1) Parallel β-Sheet Proteins. The most common arrangement in this class contains four strands for each sheet to form the natural twist such as prealbumin. However, the directions of the strands are not all equivalent and the connectivities of the strands are different, such as transthyretin (Figure 12.1c).

(2.2) Orthogonal β-Sheet Proteins. An alternative way of packing two β sheets together is with the strands in the two sheets almost perpendicular. Each domain of the serine proteases (e.g., trypsin, Figure 12.1d) shows this arrangement.

(2.3) Other β-Sheet Proteins. Influenza neuraminidase contains a very unusual β-sheet propellor-like structure. Ascorbate oxidase (Figure 12.1e) is a large β-sheet protein that contains parallel β-sheet domains, and interleukin-1β may be described as containing both a barrel and a propeller.

[3] α + β Proteins. Many proteins such as lysozyme (Figure 12.1f) contain both α helices and β sheets, but do not have the special structures created by alternating $\beta-\alpha-\beta$ patterns. In the sulfhydryl proteases, such as papain and actinidin (Figure 12.1g), the strands of sheets and the helices tend to be segregated in different regions of space.

[4] α/β Proteins. The supersecondary structure consisting of a $\beta-\alpha-\beta$ unit with the hydrogen-bonded parallel β strands forms the basis of many enzymes, especially those that bind nucleotides (e.g., adenylate kinase, Figure 12.1h) or related molecules. The strands form a parallel β sheet. In some cases, there is a linear $\beta-\alpha-\beta-\alpha-\beta-\cdots$ arrangement, but in other cases the β sheet closes on itself with the last strand hydrogen-bonded to the first.

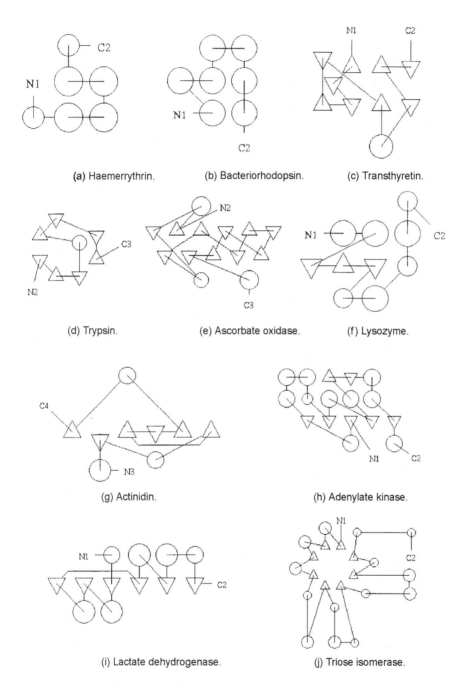

(a) Haemerrythrin. (b) Bacteriorhodopsin. (c) Transthyretin.

(d) Trypsin. (e) Ascorbate oxidase. (f) Lysozyme.

(g) Actinidin. (h) Adenylate kinase.

(i) Lactate dehydrogenase. (j) Triose isomerase.

Figure 12.1. Protein topology cartoons of representative protein domains. The 2D schematic drawings of some representative domains are illustrated in TOPS cartoons (originally retrieved from TOPS server) in which the domain start at N_i and ends at C_{i+1}. α- (and 3_{10}) Helices and β strands are represented by circles and triangles, respectively, according to the symbolisms described in Chapter 4.

(4.1) Linear or Open β–α–β Proteins. Many proteins that bind nucleotides contain a domain made up of six β–α units with a special topology. The long loop between β_C and β_D tends to create a natural pocket for the nucleotide ligand. The N–H groups in the last turn of the α_A helix are well-positioned to form hydrogen bonds to phosphate oxygen atoms of the ligand. The NAD-binding domains of horse liver alcohol dehydrogenase and lactate dehydrogenase (Figure 12.1i) are typical. Other dehydrogenases have very similar domains. Flavodoxin and adenylate kinase contains a variation on the theme. They have five strands instead of six. Dihydrofolate reductase has eight strands.

(4.2) Closed β–α–β Barrel Structures. Chicken triose phosphate isomerase (Figure 12.1j) is typical of a large number of structures that contain eight β–α units in which the strands form a sheet wrapped around into a closed structure, cylindrical in topology. The helices are on the outside of the sheet.

[5] Irregular Structures. There are structures that contain very few of their residues in helices and sheets. These tend to be stabilized by disulfide bridges or by metal ligands. For instance: (a) In the case of wheat germ agglutinin, there are numerous disulfide bridges. (b) In the case of ferridoxin, there are ion–sulfur clusters. (c) The kringle structure occurring in many proteins contains disulfide bridges as well as several short stretches of two-stranded β sheet.

[6] Multidomain Structures. Proteins may have multiple domains, but the different domains of these proteins have never been seen independently of each other, therefore accurate determination of their boundaries is not possible and perhaps not meaningful.

The CATH protein domain database (http://www.biochem.ucl.ac.uk/bsm/cath) is a hierarchical classification of protein domain structures into evolutionary families and structural groupings depending on sequence and structure similarity (Pearl et al., 2000). The protein domains are classified according to four major levels.

Level 1: *Class.* Three major classes — namely, mainly α, mainly β, and both α and β — are recognized by considering composition or residues in α helix, β-strand and the secondary structure packing, plus class 4 for few secondary structures,

Level 2: *Architecture.* Description of the gross arrangement of secondary structures in 3D space independent of their connectivity.

Level 3: *Topology* or *Fold.* Protein domains with significant structural similarity but no sequence or functional similarity. They have similar number and arrangement of secondary structures and similar connectivity.

Level 4: *Homologous Superfamily* or *Evolutionary Superfamily.* Proteins either having significant sequence similarity (35%) or high structural similarity and some sequence identity (20%).

Only protein structures solved to resolution better than 3.0 Å from the Protein Data Bank are considered. The 3D templates are generated with CORA (Conserved Residue Attributes) for recognition of structural relatives in each fold group (Orengo, 1999). The CATH Architectural descriptions that denote the arrangements of secondary structures are given in Table 12.2.

TABLE 12.2. Top Two Levels of CATH Classification

I.D. No.	Architecture Description	Example
Class 1, Mainly Alpha:		
10	Non-bundle	Aldehyde dehydrogenase, domain 2
20	Bundle	ATP Synthase, domain 2
25	Horseshoe	70-kD Soluble lytic transglycosylase, domain 1
40	Alpha solenoid	Peridinine chlorophyll protein, chain M
50	Alpha/alpha barrel	Glycosyl hydrolase, domain 2
Class 2, Mainly Beta:		
10	Ribbon	Trypsin inhibitor
20	Single sheet	Rubrerythrin, domain 2
30	Roll	Phosphomannose isomerase, domain 1
40	Barrel	Endoglucanase V
50	Clam	Bacteriochlorophyll-a protein
60	Sandwich	Galactose oxidase, domain 1
70	Distorted sandwich	Topoisomerase I, domain 3
80	Trefoil	Acidic fibroblase growth factor, subunit A
90	Orthogonal prism	Agglutinin, subunit A
100	Aligned prism	Vitelline membrane outer layer protein I, subunit A
102	3 Layer sandwich	Rieske iron–sulfur protein
110	4 Propeller	Hemopexim
120	4 Propeller	Neuraminidase
130	7 Propeller	Methylamine dehydrogenase, chain H
140	8 Propeller	Methanol dehydrogenase
150	2 Solenoid	Alkaline protease, subunit P, domain 1
160	3 Solenoid	UDP-NAc-glucosamine acyltransferase, domain 1
170	Complex	Phosphoenol pyruvate carboxykinase, domain 1
Class 3, Alpha and beta:		
10	Roll	Elastase, domain 1
15	Superroll	Bactericidal permeability incr. Protein, domain 1
20	Barrel	Aconitase, domain 4
30	2 Layer sandwich	α-Amylase, domain 2
40	3 Layer (aba) sandwich	Lysozyme
50	3 Layer (bba) sandwich	Restriction endonuclease, domain 2
60	4 Layer sandwich	Deoxyribonuclease I, subunit A
65	Alpha–beta prism	UDP-NAG-Cbov, transferase, chain A, domain 1
70	Box	Proliferating cell nuclear antigen
75	5 Stranded propeller	L-Arg/Gly aminotransferase, chain A
80	Horseshoe	Ribonuclease inhibitor
85		Sulfite reductase hemoprotein, domain 2
90	Complex	Glutamine synthetase, domain 1
Class 4, Few secondary structures:		
10	Irregular	Glucose oxidase, domain 2

TABLE 12.3. Web Servers for Protein Structure Analysis

Web Site	Service	URL
3D_PSSM	Protein folding recognition	http://www.bmm.icnet.uk/~3dpssm/
ASTRAL	Protein structure analysis	http://astral.stanford.edu/
CATH	Domain classes, 3D fold templates	http://www.biochem.ucl.ac.uk/bsm/cath/
Doe-Mbi	Fold recognition	http://fold.doe-mbi.ucla.edu/Login/
Enzyme DB	Enzyme structures, active sites	http://www.biochem.ucl.ac.uk/bsm/enzyme/index.html
LIBRA I	Folding, accessibility	http://www.ddbj.nig.ac.jp/E-mail/libra/LIBRA_I.html
ModBase	Comparative protein structure model	http://guitar.rockefeller.edu/modbase/
PDBSum	Summary and analysis of PDB	http://www.biochem.ucl.ac.uk/bsm/pdbsum/
Pfam	Protein domain families	http://www.sanger.ac.uk/Software/Pfam/
PredictProtein	PHDs and globularity	http://cubic.bioc.columbia.edu/predictprotein/
ProDom	Protein domain analysis	http://www.toulouse.inra.fr/prodom.html
Relibase	Receptor ligand interactions	http://relibase.ebi.ac.uk/
SCOP	Protein structure classification	http://scop.mrc-lmb.cam.ac.uk/scop/

The Protein Data Bank (PDB) is the primary structure database serving as the international repository for the processing and distribution of 3D structures of biomacromolecules (Bernstein et al., 1977). The database is operated by the Research Collaboratory for Structural Bioinformatics (RCSB) and is accessible from the primary RCSB site at http://www.rcsb.org/pdb/ (Berman et al., 2000). Most of the structure fold/motif/domain databases (Conte et al., 2000) and analysis servers (Brenner et al., 2000; Hofmann et al., 2000; Kelley et al., 2000; Shi et al., 2001) utilize 3D-structure information from PDB and sequence information from primary sequence databases. Some of these databases/analysis servers and their URL are listed in Table 12.3.

12.2.3. Functional Sites

An important aspect in the sequence analysis of proteins is the detection of functional sites such as the active site, ligand binding site, signal sequence/cleavage site, and post-translational modification sites. Generally, a prior knowledge concerning the chemical nature of these sites is available. For example, the active triad (Ser-His-Asp/Glu) of serine proteases, the Arg residue of $NADP^+$ binding site, and the Asn/Gln residue of glycosylation site are known, and detection of these functional sites can be accomplished by the alignment of a query sequence against sequences with known candidate sites. When studying the specificity of molecular functional sites, it has been common practice to create consensus sequences from alignments and then to choose the most common amino acid residue(s) representative at the given position(s).

The known functional sites are described in the FEATURE lines of the annotated sequence files such as PIR, GenPept, and Swiss-Prot. The PROSITE database (http://www.expasy.ch/sprot/prosite.html) consists of biologically significant patterns/profiles/signatures that may aid in identifying the query sequence to the known family of proteins (Hofmann et al., 2000). The database that is composed of two ascii (text) files, PROSITE.dat and PROSITE.doc, can be downloaded using FTP from ExPASy (ftp://www.expasy.ch/databases/prosite/) or Swiss Institute for

Experimental Cancer Research (ISREC) at ftp://ftp.isrec.isbsib.ch/sib-isrec/profiles or EBI (ftp://ftp.ebi.ac.uk/pub/databases/profiles/). The residue types that are identified as ProSites are generally enzyme active sites, prosthetic group attachment sites, ligand/metal binding sites, post-translational modification sites, and protein family signature residues. The similarity search for the highly conserved sequences/regions of each protein family is available from the BLOCKS database (Henikoff and Henikoff, 1994), which serves to identify the functional/signature regions of the query sequence. The database is accessible at the BLOCKS Web server (http://www.blocks.fhcrc.org/) of the Fred Hitchinson Cancer Research Center. Pfam is a protein families database containing functional annotation (Bateman et. al., 2000). The database is available at http://www.sanger.ac.uk/Pfam.

The active site topology of enzymes can be displayed/saved by accessing the Enzyme Structure Database (http://www.biochem.ucl.ac.uk/bsm/enzymes/index.html), and the binding site of receptors can be analyzed at Relibase (http://relibase.ebi.ac.uk/reli-cgi/rll?/reli-cgi/query/form_home.pl) as discussed earlier (Chapter 6). The center for Biological Sequence Analysis (CBS) at http://www.cbs.dtu.dk/ offers methods for predicting post-translational modifications including signal peptide cleavage (Nielsen et al., 1999), O-glycosylation (Gupta et al., 1999), and phosphorylation (Kreegipuu et al., 1999) sites. Prediction of mitochondrial targeting sequences is available at http://www.mips.biochem.mg.de/cgi-bin/proj/medgen/mitofilter. TargetP (http://www.cbs.dtu.dk/services/TargetP/) assigns the subcellular location of proteins based on the predictions of the N-terminal chloroplast transit peptide, mitochondrial targeting peptide, or secretory signal peptide.

Proteins with intracellular half-life ($t_{1/2} = 0.693/k$, where k is the first-order rate constant for the proteolytic degradation of a cellular protein) of less than two hours are found to contain region rich in proline, glutamic acid, serine, and threoinine, termed PEST (Rodgers et al., 1986). The PEST region is generally flanked by clusters of positively charged amino acids. The PEST sequence search is available at http://www.icnet.uk/LRITu/projects/pest/.

12.3. PROTEOMIC ANALYSIS USING INTERNET RESOURCES: STRUCTURE AND FUNCTION

12.3.1. Structure Database

Protein Data Bank (PDB) at http://www.rcsb.org/pdb/) is the worldwide archive of structural data of biomacromolecules. PDB was established at Brookhaven National Laboratories (BNL) in 1971 (Bernstein et al., 1977). In October 1998, the management of the PDB became the responsibility of the Research Collaboratory for Structural Bioinformatics (RCSB) (Berman et al., 2000; Westbrook et al., 2002). Two methods, SearchLite (simple keyword search) and SearchFields (advanced search), are available to search and retrieve atomic coordinates (pdb files). If you know the PDB identification code (4-character identifier of the form [0–9][A–Z or 0–9] [A–Z or 0–9] [A–Z or 0–9], e.g., 1LYZ for hen's egg-white lysozyme and 153L for goose lysozyme), enter the identification code into the PDB ID box and hit the Explore button. The Structure Explore page for this entry is returned. If the PDB ID is not known, clicking SearchLite opens the query page. Enter the keyword (name of ligand/biomacromolecule or author) and click the Search button. In the

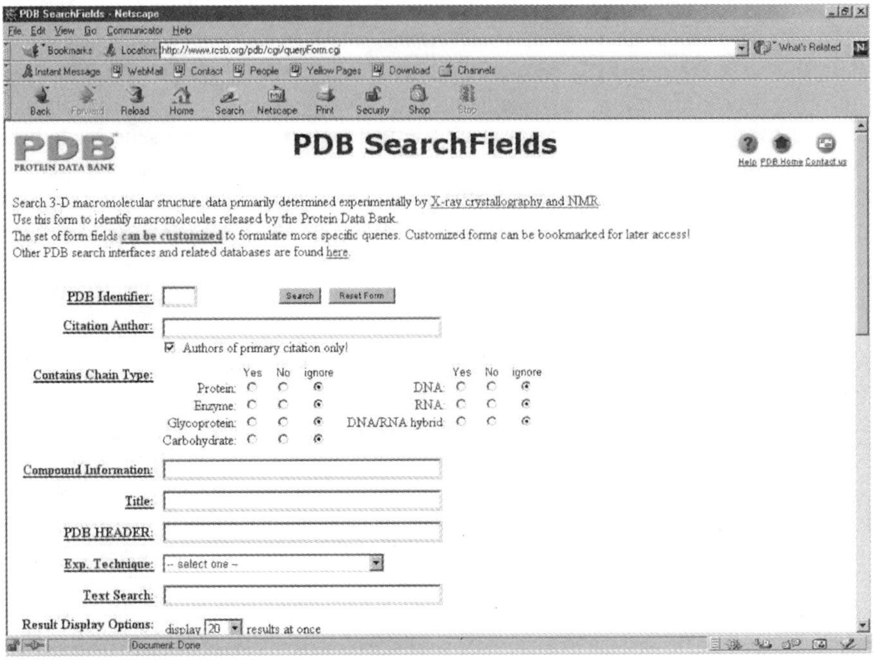

Figure 12.2. Advanced search for PDB file. From the PDB home page, select Advanced search field to open the PDB Search Fields form as shown. The advanced search enables the user to customize search fields including PDB identifier, author, chain types (protein, enzyme, DNA, and RNA, etc.), compound information, PDB header, experimental technique, text search, and result display options.

advanced search, clicking SearchField opens the query form (Figure 12.2). Construct combination of your query options including PDB ID, citation author, chain type (for protein, enzyme, carbohydrate, DNA or RNA), compound information, PDB header, and experimental technique used. Clicking the Search button returns you to the search page with a list of hits from which you select the desired entry to access the Summary information of the selected molecule.

From the Summary information, the user can choose one of many options including View Structure, Download/Display File, Structural Neighbors (links to CATH, CE, PSSP, SCOP and VAST), Geometry, or Sequence Details. Select Download/Display File, then choose PDB text and PDB noncompression format to retrieve the pdb file in text format. To view the structure online, select View Structure followed by choosing one of the 3D display options. The display can be saved as image.jpg/image.gif. The Structure Neighbors provide links to CATH (fold classification by domain), CE (representative structure comparison, structure alignments, structure superposition tool), FSSP (fold tree, domain dictionary, sequence neighbors, structure superposition), SCOP (Class, fold, superfamily, and family classification), and VAST (representative structure comparison, structure alignments, structure superposition tool). The Geometry option provides tables of dihedral angles, bond angles, and bond lengths.

Molecular modeling database (MMDB) of Entrez is a subset of 3D structures from PDB recorded in asn.1 format (Wang et. al., 2002). The database that provides the link between 3D structures and sequences can be accessed at http://www.ncbi.nlm.nih.gov/Entrez/structure.html. Enter author's last name or text words/keywords (e.g., zinc finger, DNA protein complex, or topoisomerase) or PDB ID (if it is known) and click the Search or Go button. A list of hits is returned. Select the desired entries and then click the PDB ID to receive the MMDB Structure summary (Figure 12.3). Choose options (radios) to view or save the structure file as structure.cgi (Cn3D), structure.kin (KineMage), or structure.pdb (RasMol). Sequence similarities or structure similarities can be viewed/saved by clicking the respective chain designation of Sequence neighbors or Structure neighbors. This returns the sequence display summary or VAST structure neighbors, respectively.

The structural information about nucleic acids can be obtained from PDB or Nucleic Acid Database (NDB) at http://ndbserver.rutgers.edu/NDB/ndb.html. On the NDB home page, choose Search→Nucleic acid database search→Quick search to open the search form for the database (Figure 12.4). Activate Classification list box (listing of the database by classes: DNA, DNA/RNA, Peptide nucleic acid, Peptide nucleic acid/DNA, Protein/DNA, Protein/RNA, Ribosome, Ribozyme, RNA, and tRNA), highlight the desired class, and click the Execute Selection button to open a list of database for the class. Highlight an entry and click the Display Selection button. An information page providing NDB ID, Compound name, Sequence, Citation, Crystal information, Coordinates, and Views is returned. Click

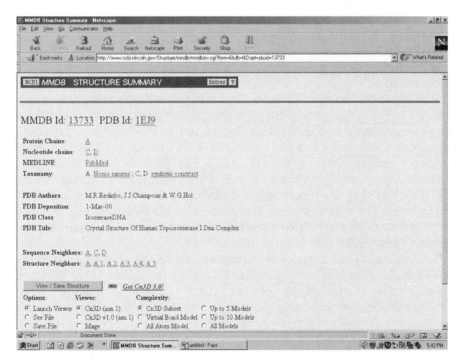

Figure 12.3. MMDB Structure summary. A search for the DNA protein complex from MMDB returns a list of hits from which human topoisomerase I-DNA complex (1EJ9) is selected. The structure summary offers options for viewing and saving the structure via Cn3D, (Kine)Mage, or RasMol (PDB) as well as Sequence neighbors or Structure neighbors.

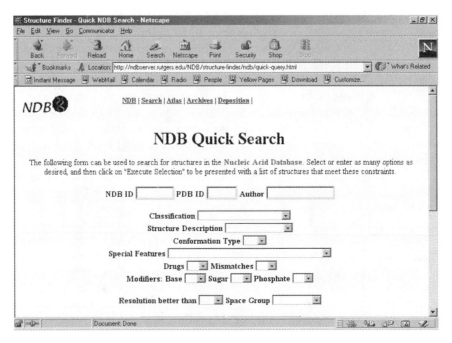

Figure 12.4. Quick search for nucleic acid database at NBD. Search for the database can be performed by making choice(s) from the list(s) of pop-up box(es).

the link, coordinates for the asymmetric unit to open the coordinate file and save it as nuclacid.pdb.

WPDB (Shindyalov and Bourne, 1997), which can be downloaded from http:// www.sdsc.edu/pb/wpdb, is a protein structural data resource. It is a Microsoft windows-based, data management and query system that can be used locally to interrogate the 3D structures of biomacromolecules found in PDB. The program provides a query ability to find structures and performs sequence alignment, property profile analyses, secondary structure assignment, 3D rendering, geometric calculation, and structure superposition (based on PDB atomic coordinates without computation). The accompanying WPDB loader (WPDBL) permits the user to build a subset of database from selected pdb files for use as tutorials on protein structures (Tsai, 2001).

The summaries and analyses of PDB structures can be accessed from PDBsum (Laskowski, 2001) at http://www.biochem.ucl.ac.uk/pdbsum. Search the structure by entering the four-character PDB code (e.g., 1lyz, this returns the summary page) or a string of the protein name (e.g., lysozyme, this returns a list of hits; you have to select the desired entry to open the summary page). The summary page (Figure 12.5) contains brief descriptions of the structure, hyperlinks to PDB, GRASS (Virtual graphic server at Colombia University), MMDB, CATH, SCOP, PROCHECK (Ramachandran plot statistics), PROMOTIF (summary of protein secondary structures at the site), and clickable presentations of CATH classification, secondary structures, PROMITIF, TOP, SAS (annotated FASTA alignment of related sequences in the PDB), PROSITE, and LIGAND (ligand molecules and ligand binding residues). In addition to the graphical representation of secondary structures

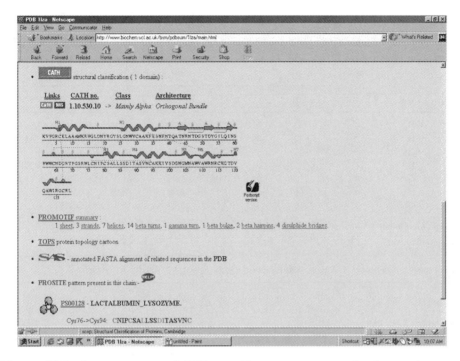

Figure 12.5. Summary page of PDBsum. The summary page for human lysozyme complex (1LZC) shows hyperlinks to various structure analyses.

along the sequence, the structural information can be viewed by selecting PROMO-TIF→sheet/strands/helices/beta turns/gamma turn/bulge/hairpins/disulfide bridges as exemplified in Table 12.4 for helical structure of human lysozyme. Click on Helical wheels and nets to view diagrams of helical wheels. The residues are color coded for hydrophobic (purple), polar (blue) and charged (red) amino acid types.

12.3.2. Secondary Structure Predictions

A number of Web sites offer services in the secondary structure predictions of proteins using different approaches with varied accuracy. The Secondary structure prediction of ExPASy Proteomic tools (http://www.expasy.ch/tools/) provides pointers to different Web servers for predicting secondary structures of proteins. The ProtScale of ExPASy Proteomic tools produces conformational profiles by plotting statistical scales of various parameters (e.g., Chou and Fasman's conformational propensities, Levitt's conformational parameters) against residue positions.

Numerous prediction methods including SOPM (Geourjon and Deleage, 1994), SOPAMA (Geourjon and Deleage, 1995), HNN, MLRC (Guermeur et al., 1999), DPM (Deleage and Roux, 1987), DSC (King and Sternberg, 1996), GOR I (Garnier et al., 1978), GOR III (Gibrat et al., 1987), GOR IV (Garnier et al., 1996), PHD, PREDATOR, and SIMPA96 are available at Network Protein Sequence Analysis (NPS@) (Combet et al., 2000). The server can be accessed via http://npsa-pbil.ibcp.fr. For example, the prediction method of Garnier, Osguthorpe, and Robson (GOR I,

TABLE 12.4. Secondary (Helical) Structure Descriptions from PROMOTIF

Individual Helices

Helix Number	Start	End	Type	Number of Residues	Unit Residues					
					Length	Rise	per Turn	Pitch	Deviation	Sequence
1	5	14	H	10	15.52	1.51	3.62	5.47	3.5	RCELAAAMKR
2	25	36	H	12	17.82	1.49	3.74	5.56	9.6	LGNWVCAAKFES
3	80	84	G	5	9.27	1.80	3.18	5.71	17.0	CSALL
4	89	99	H	11	17.05	1.51	3.65	5.51	9.1	TASVNCAKKIV
5	104	107	G	4	7.11	1.95	3.06	5.95	41.7	GMNA
6	109	114	H	6	8.66	1.43	3.57	5.09	8.9	VAWRNR
7	120	124	G	5	9.19	1.84	3.13	5.77	15.5	VQAWI

Helix Interactions Involving This Chain

Helix Numbers		Helix Types		Distance	Angle	Interaction Type		No. of Residues			
								Total	Helix 1	Helix 2	
1	2	H	H	9.6	123.5	I	I	9	5	4	Intrachain
1	7	H	G	7.7	−109.8	N	C	4	2	3	Intrachain
2	4	H	H	13.4	110.0	I	I	2	1	2	Intrachain
2	5	H	G	11.8	−4.7	C	C	3	3	1	Intrachain
2	6	H	H	8.2	118.7	I	I	8	5	4	Intrachain
2	7	H	G	6.9	77.2	I	I	11	7	3	Intrachain
3	4	G	H	6.1	−109.5	c	n	3	2	2	Intrachain
4	5	H	G	7.2	114.7	C	N	4	2	3	Intrachain
5	6	G	H	5.0	−120.3	c	n	5	3	3	Intrachain

Note: The helix types are H for α helix and G for 3_{10} helix. The helical geometry is followed by a measure of the deviation from an ideal helix (this value should be 0 for a perfect helix). For the interacting pairs, the interaction type describes where in each of the helices the distance of closest approach occurs (C, beyond the C terminus of the helix; N, beyond the N terminus of the helix; I, internal to the helix).

III, or IV) can be initiated by selecting GOR IV under Secondary structure prediction tool to open the query form. Paste the query sequence into the sequence box and click the Submit button. The result is returned showing the query sequence with the corresponding predicted conformations (c, h, e, and t for coil, helix, extended strand, and turn, respectively), summarized content (%) of conformations, and conformational profiles (Figure 12.6).

Click Secondary Structure Consensus Prediction enabling the simultaneous execution of a number of selected prediction methods. The predicted secondary structures including the consensus secondary structure (Sec. Cons.) is returned (Figure 12.7).

The BCM (Baylor College of Medicine) at http://www.hgsc.bcm.tmc.edu/Search/ Launcher/ offers segment-oriented prediction (PSSP/SSP) in which the most probable secondary structure segments (a for α helix, b for β strand, and c for the remainder) are assigned based on the probability for a, b, or c (0 to 9). The PredSS assignments of a and b are made. The nnPredict (http://www.cmpharm.ucsf.edu/ ~nomi/nnpredict.html) uses the neural network in the structure prediction and reports both the predicted secondary structures (H = helix, E = strand, and − = no prediction) and tertiary structure class (all-α, all-β, and α/β). The Protein Sequence

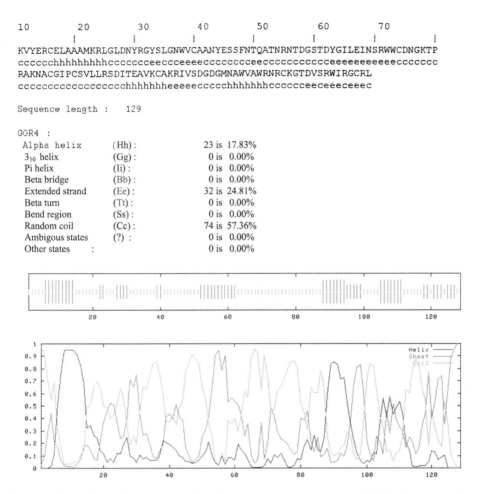

```
       10         20         30         40         50         60         70
        |          |          |          |          |          |          |          |
KVYERCELAAAMKRLGLDNYRGYSLGNWVCAANYESSFNTQATNRNTDGSTDYGILEINSRWWCDNGKTP
cccccchhhhhhhhhhcccccccceeccceeeeccccccccceecccccccccccceeeeeeeeeeeecccccccc
RAKNACGIPCSVLLRSDITEAVKCAKRIVSDGDGMNAWVAWRNRCKGTDVSRWIRGCRL
ccccccccccccccccchhhhhhheeeeeccccchhhhhhhccccccceeceeeceeec

Sequence length :   129

GOR4 :
   Alpha helix      (Hh):              23 is 17.83%
   3₁₀ helix        (Gg):               0 is  0.00%
   Pi helix         (Ii):               0 is  0.00%
   Beta bridge      (Bb):               0 is  0.00%
   Extended strand  (Ee):              32 is 24.81%
   Beta turn        (Tt):               0 is  0.00%
   Bend region      (Ss):               0 is  0.00%
   Random coil      (Cc):              74 is 57.36%
   Ambigous states  (?) :               0 is  0.00%
   Other states     :                   0 is  0.00%
```

Figure 12.6. Secondary structure prediction of duck lysozyme at NPS@. The predicted secondary structures of duck lysozyme at Network Protein Sequence Analysis (NPS@) with GOR IV method are depicted in different representations.

Analysis (PSA) server of BMERC predicts secondary structures and folding classes from a query sequence. On the PSA home page at http://bmerc-www.bu.edu/psa/index.html, select Submit a sequence analysis request to submit the query sequence and your e-mail address. The returned results include (a) probability distribution plots (conventional X/Y and contour plots) for strand, turn, and helix and (b) a list of structure probabilities for loop, helix, turn, and strand for every amino acid residues.

The method of Rost and Sander (Rost and Sander, 1993), which combines neural networks with multiple sequence alignments known as PHD, is available from the PredictProtein (Rost, 1996) server of Columbia University (http://cubic.bioc.columbia.edu/predictprotein/). This Web site offers the comprehensive protein sequence analysis and structure prediction (Figure 12.8). For the secondary structure prediction, choose Submit a protein sequence for prediction to open the submission form. Enter e-mail address, paste the sequence, choose options, and then click the

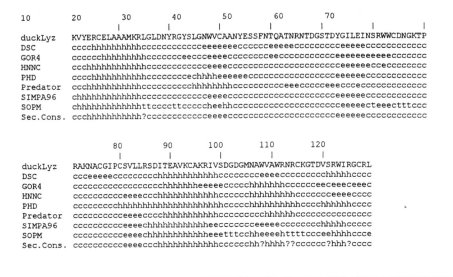

```
          10        20        30        40        50        60        70
           |         |         |         |         |         |         |                |
duckLyz   KVYERCELAAAMKRLGLDNYRGYSLGNWVCAANYESSFNTQATNRNTDGSTDYGILEINSRWWCDNGKTP
DSC       ccccHHHHHHHHHHHCCCCCCCCCCCCCeeeeeeeCCCCCCeeeeeCCCCCCCCeeeeeeCCCCCCCCCCCC
GOR4      ccccccHHHHHHHHHHccccccceeeeeeeecccccccceecccccccccceeeeeeeeeeeeccccccc
HNNC      ccchhhHHHHHHHHHHHHHccccccccccceeeeecccccccccccccccccccccceeeeeeeccecccccccc
PHD       ccccHHHHHHHHHHHHHcccccccceeHHHHeeeeeecccccccccccccccccccceeeeeeecccccccccccc
Predator  ccchhHHHHHHHHHHHccccccccccHHHHHHHHccccccccceeeccccccceeecccccccccccccccc
SIMPA96   cHHHHHHHHHHHHHHHHcccccccccccceeeeecccccccccccccccccccceeeeeeecccccccccccc
SOPM      cHHHHHHHHHHHHHHHttcccccttccccccheeHHccccccccccccccccccccccceeeeectccctttccc
Sec.Cons. cccHHHHHHHHHHHH?ccccccccccccceeeeecccccccccccccccccccceeeeeeecccccccccccc


                    80        90       100       110       120
                     |         |         |         |         |
duckLyz   RAKNACGIPCSVLLRSDITEAVKCAKRIVSDGDGMNAWVAWRNRCKGTDVSRWIRGCRL
DSC       cccceeeeecccccccccchHHHHHHHHHHhhhccccccccceeeecccccccccchHHHHhcccc
GOR4      cccccccccccccccccccchHHHHHHheeeeecccccccHHHHHHhcccccccceeceeecccc
HNNC      cccccccccceeeeecccchHHHHHHHHHHhhhcccccchHHHHHHHHHhcccccccceeeeccccc
PHD       cccccccchHHHHHHHHHHHHHHHHHHHHHHHccccchHHHHHHHHHhcccchHHHHHhcccc
Predator  ccccccccceeeecchHHHHHHHHHHhhhcccccccccchHHHHHHhcccccccccccchHHHHHcccc
SIMPA96   cccccccceeeeechHHHHHHHHHHhhheeccccccceeeeecccccccchHHHHHcccc
SOPM      cccccccccceeeeehHHHHHHHHHHhhheeeettcchHeeeeehttttccceeHHHHcccce
Sec.Cons. cccccccccceeeeecchHHHHHHHHHHhhhhccccccchh?hHHh??cccccc?hHH?cccc
```

Figure 12.7. Secondary structure consensus prediction at NPS@. Network Protein Sequence Analysis (NPS@) offers numerous methods for secondary structure prediction of proteins. The secondary structure consensus prediction (Sec.Cons.) of duck lysozyme is derived from simultaneous execution of predictions with more than one methods.

Submit/Run Prediction button. The predicted results for the secondary structure (PHDsec), solvent accessibility (PHDacc), and helical transmembrane regions (PHDhtm) are returned by e-mail (Figure 12.9). The prediction reports the multiple alignment (with statistical summary) of homologous proteins, and it reports the query sequence (AA) with

- PHD_sec: predicted H for helix and E for strand
- Rel_sec: reliability index for PHDsec prediction with 0 = low and 9 = high
- P_3_acc: predicted relative solvent accessibility in $b = 0-9\%$, $i = 9-36\%$, and $e = 36-100\%$
- Rel_acc: reliability index for PHDacc prediction with 0 = low and 9 = high

The prediction of the secondary structures can be made by the structure similarity search of PDB collection at the site. Several servers provide such prediction method. The Jpred, which aligns the query sequence against PDB library, can be accessed at http://jura.ebi.ac.uk:8888/index.html. To predict the secondary structures, however, check Bypass the current Brookhaven Protein Database box and then click Run Secondary Structure Prediction on the home page of Jped to open the query page (Figure 12.10). Upload the sequence file via browser or paste the query sequence into the sequence box. Enter your e-mail address (optional) and click the Run Secondary Structure Prediction button. The results with the consensus structures are returned either online (linked file) or via e-mail (if e-mail address is entered).

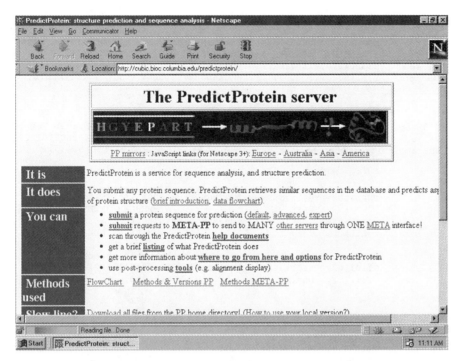

Figure 12.8. Home page of PredictProtein. PredictProtein server at Columbia University provides various protein function and structure analyses locally or remotely (via META-PP). The analyses include: signal peptide (SignalP), O-glycosylation sites (NetOglyc), phosphorylation sites (NetPhos), cleavage sites of piconarviral protease (NetPico), and transit peptides (ChloroP) for protein functions, and homology modeling (SWISS-MODEL CPHmodels, SDSCI), threading (FRSVR, Loopp, Sausage), membrane regions (DAS, TMHMM, TopPred), and secondary structure (Jpred, PHD, Predator, PROF, PSIpred, PSSP, SAM-T99, Sspro).

The 3D–1D compatibility algorithm (Ito et al., 1997) is applied to predict the secondary structures by threading at SSThread of DDBJ (http://www.ddbj.nig.ac.jp/ E-mail/ssthread/www_service.html). Paste the query sequence (fasta format) into the sequence box, enter your e-mail address, and click the Send button. The e-mail returns the threading result reporting the amino acid sequence with the predicted secondary structures (H for α helix, E for β strand, and C for coil or other).

12.3.3. Toward 3D Structure Prediction: Structural Similarity, Classification, and Folds

The structural similarity search can be conducted at Jpred if the Bypass on the current Brookhaven Protein Database box is unchecked. The search returns output for the pairwise alignments of the query sequence against the sequences of PDB files with the consensus sequence between them and the summary of the alignments. The G To P site (http://spock.genes.nig.ac.jp/ ~ genome/gtop.html) provides BLAST and PSI-BLAST searches of a query sequence against 3D structures (PDB or SCOP). Select either BLAST (Altschul et al., 1990) or PSI-BLAST to open the search form.

```
PHD results (normal):
        ....,....1....,....2....,....3....,....4....,....5....,....6
AA      MRSLLILVLCFLPLAALGKVFGRCELAAAMKRHGLDNYRGYSLGNWVCVAKFESNFNTQA
PHD_sec  EEEEEEE EEEEEEHHHHHHHHHHHHHHHHHH            HHHHHEEEE
Rel_sec 93320101010133110111137999999999944787774220346312453156745 22
SUB_sec L.................HHHHHHHHHH..LLLLL......H....E..LLL.L..


P_3_acc    bebbbbbbbbbbbbbbbebeebeebebbeebeeebbee ebbbbbebbbbbebebe ebeb
Rel_acc    1216668674652201132110161622161311300121043087846200210 1222
SUB_acc    ...bbbbbbbb...........b.b...b............b..bbbbb.........


        ....,....7....,....8....,....9....,....10...,....11...,....12
AA      TNRNTDGSTDYGILQINSRWWCNDGRTPGSRNLCNIPCSALLSSDITASVNCAKKIVSDG
PHD_sec       EEEEE                       HHHHHHH HHHHHHHHHHHHH
Rel_sec 2459999964346641651113899999888853343569884212899999999993 47
SUB_sec ..LLLLLLL...EE..LL....LLLLLLLLLLL....HHHHH....HHHHHHHHHHH..L

P_3_acc eeeeeeebeebbbbbbbbbebbbeeeeeebeebbebbbbbebbeeebeebbebbeebbeee
Rel_acc 012013021008671611214521041221211917190044111402143891252 311
SUB_acc ..........bbb.b....bb...e.......b.b.b..bb...b...b.bb..b....


        ....,....13...,....14...,....15
AA      NGMSAWVAWRNRCKGTDVQAWIRGCRL
PHD_sec    HHHHHHHHH      HHHHHH
Rel_sec 8615879999843788858866335 99
SUB_sec LL.HHHHHHHH..LLLLHHHHH..LLL

P_3_acc ebbebbbbbeeebeeeebeebbebbee
Rel_acc 40113405312033032112141032 0
SUB_acc e....b.b.............b.....
```

Figure 12.9. Protein structure prediction with PHD. The amino acid sequence of chicken lysozyme precursor (147 amino acids) is submitted to PredictProtein server for PHD structure predictions. The returned e-mail reports protein class based on secondary structures, predicted secondary structure composition (%H, %E, and %L), residue composition, data interpretation, and predicted data in two levels (brief and normal of which the normal is shown). Search for the database can be performed by making choice(s) from the list(s) of pop-up box(es).

Paste the query sequence and click the Search button. The search result (BLAST against PDB and PSI_BLAST againt SCOP) with summary of the search and multiple alignment is returned.

The Structural Classification of Proteins (SCOP) database (Section 12.2) is available at http://scop.mrc-lmb.cam.ac.uk/scop/. You may start the search by selecting (clicking) Top of the hierarchy to move down the classification hierarchy, that is, Root (scop)→Class→Fold→Superfamily→Family→Protein→Species. Alternatively, you may search at any stage of classification by going to http://scop.mrc-lamb.cam.ac.uk/scop/search.cgi and enter the PDB ID or the keyword at the Root (Figure 12.11). Click the Search button to view the list of related entries (matching the keyword). Choose the desired entry to display the lineage and PDB entry domains with pointers. The SCOP is augmented by the ASTRAL compendium (Brenner et al., 2000) at http://astral.stanford.edu/ that provides sequence databases corresponding to the domain structures.

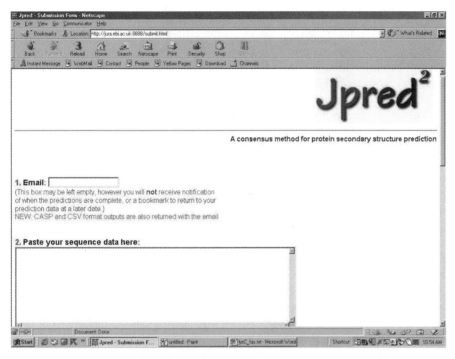

Figure 12.10. Query page of Jpred. Jpred aligns the input single sequence or unaligned multiple sequences to generate PSIBLAST profile and uses the profile for Jnet to produce prediction. Alternatively, the aligned multiple sequences can be supplied for Jnet prediction.

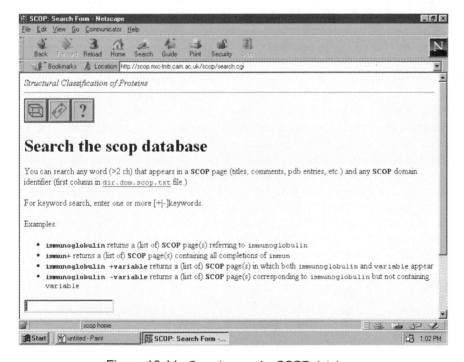

Figure 12.11. Search page for SCOP database.

The CATH database (http://www.biochem.ucl.ac.uk/bsm/cath_new/index.html) is another hierarchical protein structure classification (Pearl et al., 2000) that classified protein domains into four Classes. Each class is subclassified into several structural Architectures (Section 12.2), which is further classified into fold family (Topology) and then Homologous/evolutionary superfamily. The representative domain for each architecture description can be displayed by selecting Navigation from Top of the hierarchy. A search for the domain structure can be conducted by specifying PDB code or general text. To search the database, go to DHS (Dictionary of Homologous Superfamilies) at http://www.biochem.ucl.ac.uk/bsm/dhs/. You may click one of the four classes to initiate a search that returns the CATH table (classification) of the class. Alternatively, you may search category by entering PDB code (PDB ID) or CATH number (c.aa.tt.ss, where c stands for class, a for architecture, t for topology, and s for superfamily, e.g., 3.40.80.10 for lysozyme) or Enzyme classification (EC number) and click the Search button.

To request 3D structure analysis by multiservers via META-PP submission at the PredictProtein site, click Submit Requests to META-PP to open the submission form. Enter your e-mail address, paste the query sequence, and pick (check boxes) of the desired 3D analyses (homology modeling and threading) then click the Submit/Run prediction button. The analytical results are returned via e-mail. As mentioned above, the PredictProtein server also carries out the prediction of relative solvent accessibility (PHDacc), which is reported in three states (b, i, and e). The threading prediction of the query sequence against the structural library chosen from PDB at LIBRA I (http://www.ddbj.nig.ac.jp/E-mail/libra/LIBRA_I.html) reports a list of compatible structures and 3D–1D alignment results recording the secondary structures and accessibility (Figure 12.12).

The 3D-PSSM (Kelley et al., 2000) server at http://www.bmm.icnet.uk/~3dpssm/ offers online protein fold recognition. On the submission form, enter your e-mail address and a one-line description of the query protein, then paste the query sequence into the sequence box and click the Submit button. The query sequence is used to search the Fold library for homologues. You will be informed of the URL where the result is located for 4 days. The output includes a summary table (hits with statistics; models that can be viewed with RasMol; classifications; and links) and fold recognition by 3D-PSSM with a printout as exemplified in Figure 12.13. The alignment displays consensus sequence, secondary structures (C for coil, E for extended, and H for helix), and core score (0 for exterior to 9 for interior core).

The FUGUE site (Shi et. al., 2001) at http://www-cryst.bioc.cam.ac.uk/~fugue/prfsearch.html is a protein fold recognition server. On the request form, enter your e-mail address and the name of the query sequence (optional), upload or paste the query sequence, and select "search options" then click the Search button. The query sequence is used to search structurally homologous proteins in the Fold library for hits (representatives). You will be informed with a short summary of the result and the URL where the detailed result can be examined for 5 days. The fold recognition results (Figure 12.14) are summarized in a view ranking listed according to Z-score with the following recommended cutoffs.

Recommended cutoff :	ZSCORE> = 6.0 (CERTAIN 99% confidence)
Other cutoff :	ZSCORE> = 5.0 (LIKELY 95% confidence)
Other cutoff :	ZSCORE> = 4.7 (MARGINAL 90% confidence)
Other cutoff :	ZSCORE> = 3.5 (GUESS 50% confidence)
Other cutoff :	ZSCORE< 3.5 (UNCERTAIN)

```
Rk StrC  Protein                        Lsr Lal    Rsc     SD    Rs/N   ID%
 1 2ohxA ALCOHOL DEHYDROGENASE (HOLO FO  374 375 -211.8  -7.34 -0.565  53.3
 2 1kevA NADP-DEPENDENT ALCOHOL DEHYDRO  351 372 -148.0  -4.54 -0.398  21.8
 3 1bouB 4,5-DIOXYGENASE ALPHA CHAIN;    298 299 -109.5  -2.86 -0.366   9.4
 4 2cnd- NADH-DEPENDENT NITRATE REDUCTA  260 296 -107.8  -2.78 -0.364  10.8
 5 2bbkH METHYLAMINE DEHYDROGENASE (MAD  355 357 -106.6  -2.72 -0.298   8.7

    .   .   .   .   .   .          .   .   .    .   .   .     . .

3D-1D alignments are sequences (only 1 2ohxA structure vs codADH is shown):

2631138164 9768413133 9535293852 6413697949 6-69998889
egeleeBBBB BBBegeelee eBBBBBBBee eeelBBBBBB B-BBBeeAAA
STAGKVIKCK AAVLWEEKKP FSIEEVEVAP PKAHEVRIKM V-ATGICRSD
 : :::::::: ::: ::  ::    ::: ::  : : : :::
ATVGKVIKCK AAVAWEANKP LVIEEIEVDV PHANEIRIKI IATGVCHTDL


    .   .   .   .   .   .   .   .   .   .   .

9899986998 8966719249 8985262383 1892499754 7724244123
geAAAAAAAA AAAAglegBB BBBegegggA AAAAAAlegB BBegggegee
LGGVGLSVIM GCKAAGAARI IGVDINKDKF AKAKEVGATE CVNPQDYKKP
:: :::   : :: ::: :: : :: : ::: ::: :::  ::: :   :
LGAVGLAAVM GCHSAGAKRI IAVDLNPDKF EKAKVFGATD FVNPNDHSEP


8425942892 4698999896 5251884789 6996586868 9997683424
AAAAAAAAgl leegBBBBgg leAAAAAAAA AgeegggeBB BBgeeeeele
IQEVLTEMSN GGVDFSFEVI GRLDTMVTAL SCCQEAYGVS VIVGVPPDSQ
 :  ::  : :  ::::::: :   :   :  ::   :  ::: : ::
ISQVLSKMTN GGVDFSLECV GNVGVMRNAL ESCLKGWGVS VLVGWTDLHD


2718294775 7777995999 8695954892 3993484213 9481896564
eeeeegAAAA gleBBBBeel lleeAAAAAA AAAAAAAglg eegggBBBB
NLSMNPMLLL SGRTWKGAIF GGFKSKDSVP KLVADFMAKK FALDPLITHV
    :      ::::::: : :::: :: :: : :    ::    ::  :::
VATRPIQLI- AGRTWKGSMF GGFKGKDGVP KMVKAYLDKK VKLDEFITHR


5171494486 2674243679 78838-
BBgggAAAAA AAAAAlgegB BBBBe-
LPFEKINEGF DLLRSGESIR TILTF-

  : :  :     ::   :  :: : :

MPLESVNDAI DLMKHGKCIR TVLSLE
```

Figure 12.12. Protein structure prediction by LIBRA I server. The amino acid sequence of Baltic cod alcohol dehydrogenase is used for structure predictions against the structure library by LIBRA I. The results consist of a list (a partial list is shown) of compatible structures and aligned sequences with predicted structural features (only one set is shown). The abbreviations used for the list are Rk, rank position; StrC, structural code; Lsr, length of the structural template; Lal, length of the aligned region; Rsc, raw score of the structural template; SD, standardized score; Rs/N, raw score (Rsc) normalized by the alignment length (Lal); and ID%, sequence identity. The structural features in the aligned sequences are as follows. First line: accessibility of the aligned site with scores from 1 (exposed to solvent) to 9 (buried in protein); second line: local conformation of the aligned site with A(alpha), B(beta), gel(coils classified by the dihedral angles); third line: sequence of the template structure, fourth line: match site of the alignment and fifth line: query sequence.

```
                     Sawted
  Domain   E-value  Score  Class         Fold            Superfamily       Family
  ======   =======  =====  =====         ====            ===========       ======
: c1teha_  1.07e-05 0.00   no SCOP  3D-PSSM pseudo-family  founder id c1h1da_    -
: c1h1da_  3.73e-02 0.11   no SCOP  3D-PSSM pseudo-family  founder id c1h1da_    -
: c1cdoa_  5.42e-02 0.07   no SCOP  3D-PSSM pseudo-family  founder id c1h1da_    -
:
: d1teha2  6.04e+00 0.00   α and β  NAD(P)-bd Rossmann-fold  NAD(P)bd-Rossmann-fold  Alc/glc DH
: d1cdoa2  7.85e+00 0.07   α and β  NAD(P)-bd Rossmann-fold  NAD(P)bd-Rossmann-fold  Alc/glc DH
: d3btoa2  7.85e+00 0.11   α and β  NAD(P)-bd Rossmann-fold  NAD(P)bd-Rossmann-fold  Alc/glc DH
:
```

```
              150                                                200
fish_al_PSS  EEEEEEEEEE ECCCCCCCCE EEEECCHHHC CCEEEECCCC CCCCEEEEEC
fish_al_Seq  TVVNQIAVAK IDPSAPLDTV CLLGCGVSTG FGAAVNTAKV EPGSTCAVFG
-----------                       G+STG +GAAVNTAK+ EPGS-CAVFG
d1teha2_Seq  .......... .......... .....GISTG YGAAVNTAKL EPGSVCAVFG
d1teha2_SS   .......... .......... .....HHHHH HHHHHHCCCC CCCCEEEEEC
CORE         .......... .......... .....00013 0098001305 0001257932
```

```
              200                                                250
fish_al_PSS  CHHHHHHHHH HHHHHCCCEE EEEECCHHHH HHHHHCCCCC CCCHHHCCCC
fish_al_Seq  LGAVGLAAVM GCHSAGAKRI IAVDLNPDKF EKAKVFGATD FVNPNDHSEP
-----------  LG-VGLA--M GC++AGA+RI I+VD+N+DKF ++AK+FGAT+ ++NP+D+S+P
d1teha2_Seq  LGGVGLAVIM GCKVAGASRI IGVDINKDKF ARAKEFGATE CINPQDFSKP
d1teha2_SS   CCHHHHHHHH HHHHHCCCEE EEECCCHHHH HHHHHHCCCE EECHHCCCC
CORE         3003211960 2700104006 5560000000 0050012600 2501000000
```

```
              250                                                300
fish_al_PSS  HHHHHHCCCC CCCCEEEEEE CCHHHHHHHH HHHHCCCCEE EEEEEC.CHH
fish_al_Seq  ISQVLSKMTN GGVDFSLECV GNVGVMRNAL ESCLKGWGVS VLVGWT.DLH
-----------  I++VL++MT+ GGVD+S+EC+ GNV+VMR+AL E+C+KGWGVS V+VG++ +++
d1teha2_Seq  IQEVLIEMTD GGVDYSFECI GNVKVMRAAL EACHKGWGVS VVVGVAASGE
d1teha2_SS   HHHHHHHHHC CCCCEEEECC CCHHHHHHHH HCCCCCCCEE EECCCCCCCC
CORE         3001700110 0050038051 0000100020 0410000001 5080000000
```

```
              300                                                350
fish_al_PSS  HCCCCCEEEE CCCEEEEEEE ECCCCHHHHH HHHHHHHCCC CCCCCCEEEC
fish_al_Seq  DVATRPIQLI AGRTWKGSMF GGFKGKDGVP KMVKAYLDKK VKLDEFITHR
-----------  ++ATRP+QL+ +GRTWKG++F GG+K+
d1teha2_Seq  EIATRPFQLV TGRTWKGTAF GGWKS..... .......... ..........
d1teha2_SS   CCCCCHHHHH HCCEEEECCH HHCCH..... .......... ..........
CORE         0000000000 0000001300 20000..... .......... ..........
```

Figure 12.13. Protein fold recognition by 3D_PSSB. The summary fold recognition results of cod alcohol dehydrogenase analyzed by 3D_PSSB (secondary structure prediction by PsiPred is also included) are sorted by E value, most confident assignment first. E values below 0.05 are highly confident, and E values up to 1.0 are worthy of attention. Following the list (shown partial, abbreviations used are no SCOP, not in SCOP v1.50; NAD(P)-bd, NAD(P)-binding; and Alc/glc DH, Alcohol/glucose dehydrogenase) is the pairwise top 20 alignments (only partial of one example is shown).

The keys linking to the files for alignments with structures/fold assignments are available from the View Alignments table. The query sequence is aligned against the sequence(s)/structure(s) of either all representatives or a selection of the group according to the following alignment options (keys):

View Ranking (Click on a profile hit will bring you to the corresponding HOMSTRAD family)

Profile Hit	PLEN	RAWS	RVN	ZSCORE	ZORI	AL	
adh	375	1713	2043	76.58	77.55	00	CERTAIN
hsd1keva1	187	-2	210	11.11	12.08	02	CERTAIN
hsd1qora1	196	-35	185	9.11	10.08	02	CERTAIN
LRR	485	-382	26	3.58	4.54	00	GUESS
hs1d0va	346	-182	76	3.40	4.37	00	UNCERTAIN
ATP-synt	535	-503	82	3.33	4.30	00	UNCERTAIN
hsd1ayl	532	-485	63	2.50	3.47	00	UNCERTAIN
hsd1orda1	107	4	17	2.33	3.30	22	UNCERTAIN
hs1fbna	225	-147	64	2.32	3.28	02	UNCERTAIN
kex	498	-537	108	2.31	3.28	00	UNCERTAIN

View Alignments (aa option is shown)

```
                               10        20        30        40        50
3huda       (   1 )   stagkvikckAAVLwevkkpFsiedVeVappkayEVRIkMvAVGICrtDD
1cdoa       (   1 )   atvgkvIkckAAVAweankpLvieeIeVdvPhaneIRIkIiATGVChtDL
1teha       (   3 )    anevikckAAVAweagkpLsieeIeVapPkahEVRIkIiATAVChtDa
2ohxa       (   1 )   stagkvikckAAVLweekkpFsieeVeVapPkahEVRIkMvATGICrSDd
1dlta       (   1 )   gtagkvikckAAVLweqkqpFsieeIeVapPktkEVRIkIlATGICrtDD
gi|482344|            ATVGKVIKCKAAVAWEANKPLVIEEIEVDVPHANEIRIKIIATGVCHTDL
                      bbbbbbb       bbbbbb       bbbbbbbbbb   aaaa

                               60        70        80        90       100
3huda       (  51 )   hVvsgn-lvTplpVILGHEAAGiVesvgegVttVkpgdkVIPLFTPqCgk
1cdoa       (  51 )   YHLfegkhkdGFpVVLGHEGAGiVesvGpgVteFqpgekVIPLfisqcge
1teha       (  51 )   yTlsgadpegcfpVILGHEGAGiVesvGegVtkLkagdTVIPLyiPgCge
2ohxa       (  51 )   hVvsgt-lvTplpVIAGHEAAGiVesigegVttVrpgdkVIPLFTPqCgk
1dlta       (  51 )   hVikgt-mvSkfpVIVGHEATGiVesigegVttVkpgDkVIPLfLPQcre
gi|482344|            YHLFEGKHKDGFPVVLGHEGAGIVESVGPGVTEFQPGEKVIPLFISQCGE
                      aaa          bbbbbbbb        bbbb

                              110       120       130       140       150
3huda       ( 100 )   crvCknpesnyClkndlgnprGtLqdgtrrFtCrgkpIhHFLgTSTFSqy
1cdoa       ( 101 )   crfCqspktnqCvkgwanespdvmSpkeTrFtCkgrkVlQFLgTSTFSqy
1teha       ( 100 )   ckfClnpktNlCgkirvtggkglMpdgtSRFtCkgktIlHymgTSTFSey
2ohxa       ( 100 )   crvCkhpegnfClkNdlsmprGtMqdgtsRFtCrgkpIhHFLgTSTFSqy
1dlta       ( 100 )   cnaCrnpdgnlCirsDit-grgvLadgttRFtCkgkpVHHFlnTSTFTey
gi|482344|            CRFCQSPKTNQCVKGWANESPDVMSPKETRFRFTCKGRKVLQFLGTSTFSQY
                                                bb        bb          b

                              160       170       180       190       200
3huda       ( 150 )   TVVdenaVAkIdaaSpLekVCLIGCGFSTGyGSAvnvAkVtpgStCAVFG
1cdoa       ( 151 )   TVVnqiaVAkIdpsApldTVCLLGCGVSTGfGAAvntAkVepgstCAVFG
1teha       ( 150 )   TVVadiSVAkIdplApldkVCLLGCGISTGYGAAvntAkLepgsvCAVFG
2ohxa       ( 150 )   TVVdeiSVAkIdaaSpLekVCLIGCGFSTGyGSAvkvAkVtqgstCAVFG
1dlta       ( 150 )   TVVdesSVAkIddaAppekVCLIGCGFStGyGAAvktGkVkpgstCVVFg
```

Figure 12.14. Protein fold recognition by FUGUE. The recognition results for cod alcohol dehydrogenase performed by FUGUE are summarized in view ranking and view alignment tables that provide links to pertaining structural information. The view ranking lists information on: PLEN, profile length; RAWS, raw alignment score; RVN, (Raw score)–(Raw score for NULL model); ZSCORE, Z-score normalized by sequence divergence; ZORI, original Z-score (before normalization); and AL, alignment algorithm used for Z-score/alignment calculation and comments on the confidence of the analysis. The view alignments provide links to sequence/structure information of the compatible (hit) structures with keys (aa, ma, mh, and hh) for alignment options of which the aa structure alignment is presented.

```
gi|482344|                        TVVNQIAVAKIDPSAPLDTVCLLGCGVSTGFGAAVNTAKVEPGSTCAVFG
                                  bbbb333bbb       aaaa3333 aaaaaaaaaa          bbbbb

                                      210       220       230       240       250
3huda      ( 200 )                LGGVGLSAVmGCkaagAarIIAVDinkdkfakAkelgAteCinpgdykkp
1cdoa      ( 201 )                LGAVGLAAVMGChsagAkrIIAVDlnpdKfekAkvFgAtdfvnpndhsep
1teha      ( 200 )                LGGVGLAVIMGCkvagAsrIIGVDinkdkfarAkeFgAteCinpqdfskp
2ohxa      ( 200 )                LGGVGLSVImGCkaagAarIIGVDinkdkfakAkevGAteCvnpqdykkp
1d1ta      ( 200 )                LGGvGLSVImGCksagAsrIIGIdlnkdkfekAmaVgAteCiSpkdstkp
gi|482344|                        LGAVGLAAVMGCHSAGAKRIIAVDLNPDKFEKAKVFGATDFVNPNDHSEP
                                  aaaaaaaaaaaa     bbbbb    333aaaaaaa    bbb 333

                                      260       270       280       290       300
3huda      ( 250 )                iqevLkemtdgGVdfSFEViGrldtMmaSLlCCheacGtsSViVGvppasq
1cdoa      ( 251 )                isqvLskmtngGVdfSLECvGnvgvMrnALeSClkgwGvSVlVGwT-dlh
1teha      ( 250 )                iqevLiemtdgGVdySFECiGnvkvMrAALeAChkgwGvSVvVGvaasge
2ohxa      ( 250 )                iqevLtemSngGVdfSFEVIGrldtMvtALsCCqeayGvSViVGvppdsq
1d1ta      ( 250 )                isevLsemTgnnVgyTFEViGhletMidALaSChmnyGtsSVvVgvppsak
gi|482344|                        ISQVLSKMTNGGVDFSLECVGNVGVMRNALESCLKGWGVSVLVGW-TDLH
                                  aaaaaaaa      bbbb     aaaaaaaa       bbbb

                                      310       320       330       340       350
3huda      ( 300 )                nlsinpmllltgRtwkgavYGgfkSkegIpklVadfmakkfsLdaLIthv
1cdoa      ( 300 )                dvatrpiqliagrtwkgSmFGgfkGkdgVpkmVkayldkkvkLdeFIthr
1teha      ( 300 )                eiaTrpfqlvtgrtwkgtaFGgwkSvesVpklLVseymskkIkVdeFVthn
2ohxa      ( 300 )                nlsmnpmlllsgrtwkgaiFGgfkSkdsVpklVadfmakkfaLdpLithv
1d1ta      ( 300 )                mltydpmllftgrtwkgCvFGgLkSrddVpklVteflakkfdLdqLIthv
gi|482344|                        DVATRPIQLIAGRTWKGSMFGGFKGKDGVPKMVKAYLDKKVKLDEFITHR
                                  aaaa   bbbb  333  aaaaaaaaaaaaaa      333bbbb

                                      360       370
3huda      ( 350 )                lpFekIneGFdlLhsgkSiRTVLtf
1cdoa      ( 350 )                mpLesVndAIdlMkhgkCIRTVLsl
1teha      ( 350 )                LsFdeInkAfelmhsgksIRTVVki
2ohxa      ( 350 )                lpFekIneGFdlLrsgeSIRTILtf
1d1ta      ( 350 )                lpfkkIseGfellnsgqSIRTVLtf
gi|482344|                        MPLESVNDAIDLMKHGKCIRTVLSLE
                                  bb333aaaaaaaaaa      bbbbb
```

Figure 12.14. Continued.

aa: query sequences (including PSI-BLAST homologues) aligned against all the representative structures from a HOMSTRAD family

ma: master sequence aligned against all the representative structures from a HOMSTRAD family

mh: master sequence aligned against a single structure of highest sequence identity from a HOMSTRAD family

hh: single sequence/structure pair with highest sequence identity in 'aa'

Sequences of the representative proteins are displayed in JOY protein sequence/ structure representation (http://www-cryst.bioc.cam.ac.uk/~joy/). The representations are uppercase for solvent inaccessible, lowercase for solvent accessible, red for α helix, blue for β strand, maroon for 3_{10} helix, **bold** for hydrogen bond to main chain amide, underline for hydrogen bond to main chain carbonyl, cedilla for disulfide bond, and *italic* for positive ϕ angle. The query sequence is displayed in all capital letters. The consensus secondary structure (a for α helix, b for β strand, and 3 for 3_{10} helix) as defined, if greater than 70% of the residues in a given position in that particular conformation, is given underneath.

12.3.4. Post-translational Modifications and Functional Sites

The annotated amino acid sequence files normally provide information concerning the sites and types for post-translational modifications and functions. These descriptions can be found from Features line(s) in all of the three sequence formats, namely, PIR, GenPept, and Swiss-Prot. The ExPASy Proteomics tools (http://www.expasy.ch/tools/) provide numerous links for such services.

The prediction of signal peptide cleavage site is performed at CBS prediction server, SignalP (http://www.cbs.dtu.dk/services/SignalP/). On the submission form, paste the query sequence, enter the sequence name, select one of the organism groups, and request Postscript graphics then click the Submit Sequence button. The SignalP returns three scores (C for raw cleavage site, S for signal peptide, and Y for combined cleavage site) between 0 to 1 for each residue position in a table and a profile graph (Figure 12.15)

The CBS prediction server also provides services for predicting O-glycosylation sites (NetGly) in mammalian proteins (http://www.cbs.dtu.dk/services/NetOGly-2.0/) and phosphorylation sites (NetPhos) in eukaryotic proteins (http://www.cbs.dtu.dk/services/NetPhos/). Paste the query sequence and click the Submit Sequence button to receive the predicted results. NetOGly returns tables of potential *versus* threshold assignments for threonine and serine residues as well as a plot of O-glycosylation potential *versus* sequence position. NetPhos returns tables of context (nanopeptides, [S,T,Y] $\pm$ 4 residues) and scores for serine/threonine/tyrosine predictions.

The ProSite (http://www.expasy.ch/sprot/prosite.html) is the collection of sequence information for biologically significant patterns, motifs, signatures, and

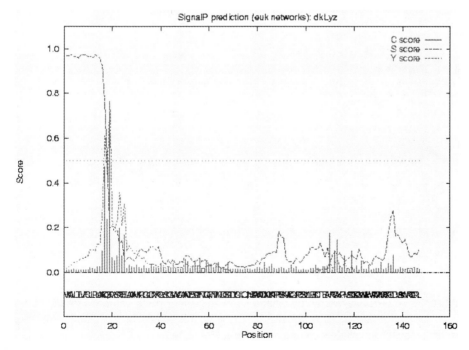

Figure 12.15. Prediction of signal peptide cleavage site. The profile plot shows the prediction of signal peptide cleavage site of duck lysozyme by SignalP.

```
KESPAKKFQR QHMDPDSSSS NSSNYCNLMM SRRNMTQGRC KPVNTFVHES LADVQAVCSQ
INVNCKNGQT NCYQSNSTMH ITDCRQTGSS KYPNCAYKAS QEQKHIIVAC EGNPPVPVHF
DASV
```

```
[1] PDOC00001  PS00001          ASN_GLYCOSYLATION
N-glycosylation site
Number of matches:    3
        1       21-24 NSSN
        2       34-37 NMTQ
        3       76-79 NSTM
 [2] PDOC00005  PS00005    PKC_PHOSPHO_SITE
Protein  kinase C  phosphorylation site
Number of matches:     2
        1       31-33 SRR
        2       89-91 SSK
[3]  PDOC00006   PS00006          CK2_PHOSPHO_SITE
Casein kinase II phosphorylation site
            50-53 SLAD
[4]  PDOC00008   PS00008          MYRISTYL
N-myristoylation site
            68-73 GQTNCY
[5]  PDOC00118   PS00127          RNASE_PANCREATIC
Pancreatic ribonuclease family signature
            40-46 CKPVNTF
```

```
PPSearch output  :
Sequence /net/nfs0/vol1/production/w3nobody/tmp/132934.ppsearchres (124 residues):


Matching pattern     PS00001  ASN_GLYCOSYLATION:
   21:  NSSN
   34:  NMTQ
   76:  NST M
Total matches:   3
Matching pattern     PS00005  PKC_PHOSPHO_SITE:
   31:  S RR
   89:  SS K
Total matches:   2
Matching pattern     PS00006  CK2_PHOSPHO_SITE:
   50:  SLA D
Total matches:   1
Matching pattern     PS00008  MYRISTYL:
   68:  GQTNCY
Total matches:   1
Matching pattern     PS00127  RNASE_PANCREATIC:
   40:  CKPVNT F
Total matches:   1
Total no of hits in this sequence:      8
```

Figure 12.16. ProSite search by different servers. The ProSite searches for Pig ribonuclease are performed by ScanProsite of ExPASy, PPSearch of EBI, and ProfileScan.
Identical results with slightly different outputs are obtained.

ProfileScan output:
```
pigRNase, 124 bases, BB4B412D checksum.

KESPAKKFQRQHMDPDSSSSNSSN YCNLMM SRRNMTQGRCKPVNTF VHESLAD VQAVCSQ  60
INVNCKN GQTNCY QSNSTMHITDCRQTG SSKYPNCAYKASQEQKHI IVACEGNPPVPVHF 120
DASV
```

```
Sequence length: 124 res.
? 0.1000    0 pos.   21 -  24 PS00001 | ASN_GLYCOSYLATION  N-glycosylation site [document],InterPro]
? 0.1000    0 pos.   34 -  37 PS00001 | ASN_GLYCOSYLATION  N-glycosylation site [document],InterPro]
? 0.1000    0 pos.   76 -  79 PS00001 | ASN_GLYCOSYLATION  N-glycosylation site [document],InterPro]
? 0.1000    0 pos.   31 -  33 PS00005 | PKC_PHOSPHO_SITE  Protein kinase C phosphorylation site [doc],IPro]
? 0.1000    0 pos.   89 -  91 PS00005 | PKC_PHOSPHO_SITE  Protein kinase C phosphorylation site [doc],IPro]
? 0.1000    0 pos.   50 -  53 PS00006 | CK2_PHOSPHO_SITE  Casein kinase II phosphorylation site [doc],IrPro]
? 0.1000    0 pos.   68 -  73 PS00008 | MYRISTYL  N-myristoylation site [documentation],InterPro]
! 1.0000    1 pos.   40 -  46 PS00127 | RNASE_PANCREATIC  Pancreatic RNasefamily signature [doc],IPro]
Significant matches are labeled with a red " !" in the first column
```

Figure 12.16. Continued.

fingerprints, each of which associated with an accession number (PDOCxxxxx). Some of the descriptors are: Post-translational modification sites, Domain profiles/ sequences, DNA or RNA associated protein profiles/signatures, Enzyme active sites/signatures (for oxidoreductases, transferases, hydrolases, lyases, isomerases, and ligases), Electron transport protein signatures, Structural protein signatures, Cytokine and growth factor signatures, Hormone and active peptide signatures, Toxin signatures, Inhibitor signatures, and Secretive protein and chaperone signatures.

The ProtSite search can be conducted at ScanProsite tool (selecting Scan a sequence for the occurrence of PROSITE patterns) of ExPASy Proteomic tools (http://www.expasy.ch/tools/scnpsite.html), or PPSearch of EBI (http://www2.ebi.ac. uk/ppsearch/) or ProfileScan server (http://www.isrec.isb-sib.ch/software/PFSCAN_ form.html). On the search form, paste the query sequence, select options (e.g., display in PPSearch and databases in ProfileScan) and click the Run/Start the Scan button. The search results are returned (after clicking SEView applet button in ProfileScan) with different outputs as shown in Figure 12.16.

To search amino acid sequence against the BLOCKS database at http:// www.blocks.fhcrc.org/, select Block Searcher to open the search form. Choose the options, paste the query sequence, and click the Perform Search button. The search returns lists of matched blocks (I.D. descriptions, number of blocks, and combined *E* values), locations (amino acid positions of individual blocks and its E value), and pairwise alignment of the blocks with the matched query sequences (Figure 12.17).

The protein domain families database, Pfam, contains curated multiple sequence alignments for each family with linked functional annotation and profile hidden Markow models (HMMs) for finding the domains in the query sequences (Bateman et. al. 2000). The database can be accessed interactively at http://www.sanger.ac.uk/ Pfam/ by Keyword search (enter keywords, e.g., enzyme name), Protein search (paste or upload query sequence) or DNA search (paste query sequence). Select Protein search to paste or upload the query sequence and click the Search Pfam button. Pfam returns tabulated search results and alignments of the matched domains (Figure 12.18). Select the hyperlinked match (domain) to view the functional description, linked database, and references for the domain. Each family provides two multiple alignments that can be viewed/saved by clicking the option button. The

```
Query=Unknown (cod alcohol dehydrogenase)
Size=375 Amino Acids
Blocks Searched=11117
Alignments Done=        4440928
Cutoff combined expected value for hits=  5
Cutoff block expected value for repeats/other=  5
```

```
                                                            Combined
Family                                       Strand  Blocks  E-value
IPB002328  Zinc-containing alcohol dehydrogena   1   5 of 5  1.2e-92
IPB002364  Quinone oxidoreductase/zeta-crystal   1   2 of 4  4.9e-12
IPB000464  Fumarate reductase / succinate dehy   1   1 of 10   0.033
IPB002162  D-isomer specific 2-hydroxyacid deh   1   1 of 4    0.3
IPB001625  Glutamate/leucine/phenylalanine/val   1   1 of 5    0.42
PR01211    Synuclein signature                   1   1 of 5    0.45
BP00240    COLLAGEN GLYCOPROTEIN PRECUR          1   1 of 2    0.89
IPB001557  L-lactate dehydrogenase               1   1 of 6    1.2
PR00302    Lupus La protein signature            1   1 of 5    1.7
PF01873    Domain found in IF2B/IF5              1   1 of 4    2.1
DM01023    2 GLYCOSYL HYDROLASES FAMILY 5.       1   1 of 6    2.3
PF00718    Polyomavirus coat protein             1   1 of 8    3.1
IPB000171  Bacterial-type phytoene dehydrogena   1   1 of 5    3.2
IPB001636  SAICAR synthetase                     1   1 of 7    3.9
PR01259    Na+/Ca2+ exchanger signature          1   1 of 4    4.9
```

```
>IPB002364 2/4 blocks Combined E-value= 4.9e-12: Quinone oxidoreductase/zeta-
  crystallin
Block      Frame    Location (aa)       Block E-value
IPB002364B  0          63-88             2.6e-11
IPB002364D  0          260-271             53
Up to 1 repeats expected:
Other reported alignments:

                           |--- 2364 amino acids---|
           IPB002364 AB:..................C........................................D
           Unknown  :B::D


IPB002364B          <->B   (52,4627):62
QOR_PSEAE|P43903     58    PSGLGSEGAGEVEAVGSEVTRFKVGD
                           |  || |||| || ||   || |  |
Unknown              63    PVVLGhEGAGIVESVGPGVTeFqPGE


IPB002364D          B<->D  (110,5614):171
QOR_LEIAM|P42865    212    PKGVDVVYECVG
                           |||   ||||
Unknown             260    nGGVDfsLECVG
```

Figure 12.17. Search a sequence against Blocks database. The search of cod alcohol dehydrogenase sequence against Blocks database returns a list of hits and pairwise alignments of blocks (only one set is shown).

seed alignment that contains a relatively small number of representative family members is used to build profile HMM. The family alignment contains all family members in the database. Select Browse Pfam on the home page to search/view the functional description (linked databases, references, alignments, and organization) of Pfam domains.

The PANAL/MetaFam server (http://mgd.ahc.umn.edu/panal/run_panal.html) analyzes protein sequence for Prosite patterns, Prosite profiles, BLOCKS, PRINTS, and Pfam. Check the options, paste the query sequence, and click the Submit button. The analytical results are returned with a graphical sketch of the predicted features,

Trusted matches- domains scoring higher than the gathering threshold

Domain	Start	End	Bits	Evalue	Alignment
pyr_redox	47	368	335.90	4.5e-97	Align
pyr_redox_dim	392	501	205.50	8.2e-58	Align

Alignments of Pfam-A domains to HMMs

```
Alignment of pyr_redox vs UserSeq/47-368

                  *->dvvIIGgGpAGlaAAiraaragflGrkrlkvalvEkepplkrgtlGG
                  d+++IGgGp+G++AAir a+   lG   lk+++vEk+         gtlGG
     UserSeq   47 DLCVIGGGPGGYVAAIRGAQ---LG---LKTICVEKR-----GTLGG 82


                  TClNvgciprkpllkaalvgeeakd.d.f..i.vggpeldlkpleqykek
                  TClNvgcip+k+ll++++++++++k++ + ++i+v+g+++l +++++k+
     UserSeq   83 TCLNVGCIPSKALLNNSHIYHTVKHdTkRrgIdVSGVSVNLSQMMKAKDD 132


        .       .       .       .       .       .       .       .       .       .

                  DlVlvAiGrrPntellglegaGiGleldrrGkGgIvVDeylrTsgtrTsv
                  D++lvAiGr+P te+lgl+++G  + +d+   ++++ D      +r    T++
     UserSeq  310 DVLLVAIGRVPYTEGLGLDKLG--ISMDKS--NRVIMDSEYR-----TNI 35        0


                  pgIyAaGDvaggplrlahv<-*
                  p+I ++GD + g ++lah+
     UserSeq  351 PHIRVIGDATLG-PMLAHK    368


Alignment of pyr_redox_dim vs UserSeq/392-501

                  *->vPsvvftdPEiAsVGLTEeeAkekGgdednvkvgkfpFaangrAlay
                  +P+v +t+PE+A+VG TE+ Ake+G++    +++g+f F+an+rA+++
     UserSeq  392 IPAVMYTHPEVAWVGITEQKAKESGIK---YRIGTFGFSANSRAKTN 435


                  gegateGfvKlvadaetgriLGaHivGpnAgEklIqeaalAikmGaTvee
                  +  ++G+vK+++daet+r+LG+H++Gp AgE lI+ea+lA+++Ga++e
     UserSeq  436 MD--ADGLVKVIVDAETDRLLGVHMIGPMAGE-LIGEATLALEYGASAE- 48        1


                  dlantihaHPTlsEalkeaa<-*
                  d+a+ +haHPTlsEa kea
     UserSeq  482 DVARVCHAHPTLSEATKEAM    501
```

Figure 12.18. Output of Pfam search results. Pfam search is performed with amino acid sequence derived from lipoamide dehydrogenase (*Schizosaccharomyces pombe*). The table for the trusted matches from Pfam-A for pyr_redox (pyridine nucleotide disulfide oxidoreductase) and pyr_redox_dim (pyridine nucleotide disulfide oxidoreductase, dimerization) domains and their alignments (partial) to HMMs (*→) are shown. The trusted matches from Pfam-B, the potential matches (Thi4 for thiamine biosynthetic enzyme domain), and the bead-on-a-string sketches are not shown. Select the linked domain name to view the functional description of the domain. The HMM alignments are followed by an option button (Align to seed or Align to family) that enables the user to view/save the multiple alignment of each matched family.

TABLE 12.5. Summary List of PANAL for Cod Alcohol Dehydrogenase

Program	Domain	Start	End	Motifs	E Value	Description
PROSITE	PS00013	201	211			Membrane lipprot.lipid attachment site
PROSITE	PS00059	66	80			Copper fist DNA binding domain profile
BLOCKS	IPB002328	20	243	5 of 5	4.592e-91	Zinc-containing alcohol dehydrogena
BLOCKS	IPB002364	63	271	2 of 4	1.875e-10	Quinone oxidoreductase/zeta-crystal
PFAM	adh_zinc	20	374		2.3e-134	Zinc-binding dehydrogenases

and a summary list is shown in Table 12.5. The detailed listings of the results are also available by clicking Full Report.

The PEST search can be conducted at http://www.icnet.uk/LRITu/projects/pest/runpest.html. Paste the query sequence, enter options, and click Submit button. Candidate PEST sequences with hydrophobicity index and PEST score are returned.

12.4. WORKSHOPS

1. Conduct the structural similarity search for a protein with the following sequence:

```
MSQDENIVKAVEESAEPAQVILGEDGKPLSKKALKKLQKEQEKQRKKEERALQLEAEREAREKKAAAEDT
AKDNYGKLPLIQSRDSDRTGQKRVKFVDLDEAKDSDKEVLFRARVHNTRQQGATLAFLTLRQQASLIQGL
VKANKEGTISKNMVKWAGSLNLESIVLVRGIVKKVDEPIKSATVQNLEIHITKIYTISETPEALPILLED
ASRSEAEAEAAGLPVVNLDTRLDYRVIDLRTVTNQAIFRIQAGVCELFREYLATKKFTEVHTPKLLGAPS
EGGSSVFEVTYFKGKAYLAQSPQFNKQQLIVADFERVYEIGPVFRAENSNTHRHMTEFTGLDMEMAFEEH
YHEVLDTLSELFVFIFSELPKRFAHEIELVRKQYPVEEFKLPKDGKMVRLTYKEGIEMLRAAGKEIGDFE
DLSTENEKFLGKLVRDKYDTDFYILDKFPLEIRPFYTMPDPANPKYSNSYDFFMRGEEILSGAQRIHDHA
LLQERMKAHGLSPEDPGLKDYCDGFSYGCPPHAGGGIGLERVVMFYLDLKNIRRASLFPRDPKRLRPCGV
PSFPPNLSARVVGGEDARPHSWPWQISLQYLKDDTWRHTCGGTLIASNFVLTAAHCISNTWTYRVAVGKN
NLEVEDEEGSLFVGVDTIHVHKRWNALLLRNDIALIKLAEHVELSDTIQVACLPEKDSLLPKDYPCYVTG
WGRLWTNGPIADKLQQGLQPVVDHATCSRIDWWGFRVKKTMVCAGGDGVISACNGDSGGPLNCQLENGSW
EVFGIVSFGSRRGCNTRKKPVVYTRVSAYIDWINEKMQL
```

2. Predict the secondary structure of the following protein by GOR IV method.

```
CGVPSFPPNLSARVVGGEDARPHSWPWQISLQYLKDDTWRHTCGGTLIASNFVLTAAHCISNTWTYRVAV
GKNNLEVEDEEGSLFVGVDTIHVHKRWNALLLRNDIALIKLAEHVELSDTIQVACLPEKDSLLPKDYPCY
VTGWGRLWTNGPIADKLQQGLQPVVDHATCSRIDWWGFRVKKTMVCAGGDGVISACNGDSGGPLNCQLEN
GSWEVFGIVSFGSRRGCNTRKKPVVYTRVSAYIDWINEKMQL
```

3. Perform the consensus secondary structure prediction of pea alcohol dehydrogenase at NSP@ by selecting six different methods.

```
>gi|81891|pir||S00912 alcohol dehydrogenase (EC 1.1.1.1) 1 - garden pea
MSNTVGQIIKCRAAVAWEAGKPLVIEEVEVAPPQAGEVRLKILFTSLCHTDVYFWEAKGQTPLFPRIFGH
EAGGIVESVGEGVTHLKPGDHALPVFTGECGECPHCKSEESNMCDLLRINTDRGVMLNDNKSRFSIKGQP
VHHFVGTSTFSEYTVVHAGCVAKINPDAPLDKVCILSCGICTGLGATINVAKPKPGSSVAIFGLGAVGLA
AAEGARISGASRIIGVDLVSSRFELAKKFGVNEFVNPKEHDKPVQQVIAEMTNGGVDRAVECTGSIQAMI
```

```
SAFECVHDGWGVAVLVGVPSKDDAFKTHPMNFLNERTLKGTFYGNYKPRTDLPNVVEKYMKGELELEKFI
THTVPFSEINKAFDYMLKGESIRCIIKMEE
```

4. List SCOP lineages for the following proteins:
   a. Pig D-amino acid oxidase, N-terminal domain
   b. Human carbonic anhydrase
   c. Human cyclin A
   d. Nuclear estrogen receptor, DNA-binding domain
   e. Goose lysozyme, active site domain
   f. HIV-1 reverse transcriptase, palm domain
5. Furnish CATH architecture descriptions for the following proteins:
   a. Tyrosyl-tRNA synthetase
   b. Serine/threonine protein phosphatase
   c. ATP Synthase, domain 1
   d. Galactose oxidase, domain 1
   e. Glysosyl transferase
   f. Phenylalanine-tRNA synthetase
   g. Carbonic anhydrase II
   h. Dihydrolipoamide transferase
   i. Inorganic pyrophosphatase
   j. Dihydrofolate reductase, subunit A

6. Submit the following sequence to a structure prediction by displaying the secondary structure and core index in the multiple alignment of the query sequence, and with the representative sequences of proteins whose 3D structures are known.

```
KVYERCELAAAMKRLGLDNYRGYSLGNWVCAANYESSFNTQATNRNTDGSTDYGILEINSRWWCDNGKTP
RAKNACGIPCSVLLRSDITEAVKCAKRIVSDGDGMNAWVAWRNRCKGTDVSRWIRGCRL
```

Retrieve the PDB file of the protein which is structurally the most homologous to the query protein.

7. Submit the following sequence to structure prediction by the multiple alignment of the query sequence with the reference sequences of proteins whose 3D structures are known in Joy sequence/structure representation.

```
RTDCYGNVNRIDTTGASCKTAKPEGLSYCGVPASKTIAERDLKAMDRYKTIIKKVGEKLCVEPAVIAGII
SRESHAGKVLKNGWGDRGNGFGLMQVDKRSHKPQGTWNGEVHITQGTTILTDFIKRIQKKFPSWTKDQQL
KGGISAYNAGAGNVRSYARMDIGTTHDDYANDVVARAQYYKQHGY
```

Retrieve the PDB file of the protein which is structurally most homologous to the query protein.

8. What is the most likely structural domain/motif of a protein with the following sequence?

```
MIKLAKFPRDFVWGTATSSYQIEGAVNEDGRTPSIWDTFSKTEGKTYKGHTGDVACDHYHRYKEDVEILK
EIGVKAYRFSIAWPRIFPEEGKYNPKGMDFYKKLIDELQKRDIVPAATIYHWDLPQWAYDKGGGWLNRES
IKWYVEYATKLFEELGDAIPLWITHNEPWCSSILSYGIGEHAPGHKNYREALIAAHHILLSHGEAVKAFR
EMNIKGSKIGITLNLTPAYPASEKEEDKLAAQYADGFANRWFLDPIFKGNYPEDMMELYSKIIGEFDFIK
EGDLETISVPIDFLGVNYYTRSIVKYDEDSMLKAENVPGPGKRTEMGWEISPESLYDLLKRLDREYTKLP
MYITENGAAFKDEVTEDGRVHDDERIEYIKEHLKAAAKFIGEGGNLKGYFVWSLMDNFEWAHGYSKRFGI
VYVVDYTTQKRILKDSALWYKEVILDDGIED
```

9. Identify the signal peptide and family signature sequences of pig pancreatic ribonuclease with the given amino acid sequence:

```
>gi|67344|pir||NRPG pancreatic ribonuclease (EC 3.1.27.5) - pig
KESPAKKFQRQHMDPDSSSSNSSNYCNLMMSRRNMTQGRCKPVNTFVHESLADVQAVCSQINVNCKNGQT
NCYQSNSTMHITDCRQTGSSKYPNCAYKASQEQKHIIVACEGNPPVPVHFDASV
```

10. Identify the ProSite residues and protein domain families for the following proteins of given amino acid sequences:

```
>gi|11373714|pir||CCZP cytochrome c - Schizosaccharomyces pombe
MPYAPGDEKKGASLFKTRCAQCHTVEKGGANKVGPNLHGVFGRKTGQAEGFSYTEANRDKGITWDEETLF
AYLENPKKYIPGTKMAFAGFKKPADRNNVITYLKKATSE
```

```
>gi|7430909|pir||S52316 flavodoxin - Escherichia coli
MNMGLFYGSSTCYTEMAAEKIRDIIGPELVTLHNLKDDSPKLMEQYDVLILGIPTWDFGEIQEDWEAVWD
QLDDLNLEGKIVALYGLGDQLGYGEWFLDALGMLHDKLSTKGVKCVGYWPTEGYEFTSPKPVIADGQLFV
GLALDETNQYDLSDERIQSWCEQILNEMAEHYA
```

```
>gi|2318115|gb|AAB66483.1| serine protease )Homo sapiens]
MKKLMVVLSLIAAAWAEEQNKLVHGGPCDKTSHPYQAALYTSGHLLCGGVLIHPLWVLTAAHCKKPNLQV
FLGKHNLRQRESSQEQSSVVRAVIHPDYDAASHDQDIMLLRLARPAKLSELIQPLPLERDCSANTTSCHI
LGWGKTADGDFPDTIQCAYIHLVSREECEHAYPGQITQNMLCAGDEKYGKDSCQGDSGGPLVCGDHLRGL
VSWGNIPCGSKEKPGVYTNVCRYTNWIQKTIQAK
```

```
>gi|7489456|pir||T03296 beta-glucosidase (EC 3.2.1.21), rice chloroplast
MADLFTEYADFCFKTFGNRVKHWFTFNEPRIVALLGYDQGTNPPKRCTKCAAGGNSATEPYIVAHNFLLS
HAAAVARYRTKYQAAQQGKVGIVLDFNWYEALSNSTEDQAAAQRARDFHIGWYLDPLINGHYSQIMQDLV
KDRLPKFTPEQARLVKGSADYIGINQYTASYMKGQQLMQQTPTSYSADWQVTYVFAKNGKPIGPQANSNW
LYIVPWGMYGCVNYIKQKYGNPTVVITENGMDQPANLSRDQYLRDTTRVHFYRSYLTQLKKAIDEGANVA
GYFAWSLLDNFEWLSGYTSKFGIVYVDFNTLERHPKASAYWFRDMLKH
```

```
>gi|6319698|ref|NP_009780.1| beta subunit of pyruvate dehydrogenase -Sacchar-
omyces cerevisiae
MFSRLPTSLARNVARRAPTSFVRPSAAAAALRFSSTKTMTVREALNSAMAEELDRDDDVFLIGEEVAQYN
GAYKVSKGLLDRFGERRVVDTPITEYGFTGLAVGAALKGLKPIVEFMSFNFSMQAIDHVVNSAAKTHYMS
GGTQKCQMVFRGPNGAAVGVGAQHSQDFSPWYGSIPGLKVLVPYSAEDARGLLKAAIRDPNPVVFLENEL
LYGESFEISEEALSPEFTLPYKAKIEREGTDISIVTYTRNVQFSLEAAEILQKKYGVSAEVINLRSIRPL
DTEAIIKTVKKTNHLITVESTFPSFGVGAEIVAQVMESEAFDYLDAPIQRVTGADVPTPYAKELEDFAFP
DTPTIVKAVKEVLSIE
```

REFERENCES

Altschul, S. F., Gish, W., Miller, W., Myers, E. W., and Lipman, D. J. (1990) *J. Mol. Biol.* **215**:403–410.

Anfinsen, C. B. (1973) *Science* **181**:223–230.

Apweiler, R., Biswas, M., Fleischmann, W., Kanapin, A., Karavidopoulou, Y., Kersey, P., Kriventseva, E. V., Mittard, V., Mulder, N., Phan, I., and Zdobnov, E. (2001) *Nucleic Acids Res.* **29**:44–48.

Bateman, A., Birney, E., Durbin, R., Eddy, S. R., Howe, K. L., and Sonnhammer, E. L. L. (2000) *Nucleic Acids Res.* **28**:263–266.

Berman, H., Westbrook, J., Feng, Z., Gilliland, G., Bhat, T. N., Weissig, H., Shindyalov, I. N., and Bourne, P. E. (2000) *Nucleic Acids Res.* **28**:235–242.

Bernstein, F. C., Koetzle, T. F., Williams, G. J. B., Meyer, E. F. Jr., Brice, M. D., Rogers, J. R., Kennard, O., Shimanouchi, T., Tasumi, M. (1977) *J. Mol. Biol.* **112**:535–542.

Bharath, R. (1994) *Neural Network Computing*, McGraw-Hill, New York.

Brenner, S. E., Chothia, C., Hubbard, T. J. P., and Murzin, A. G. (1996) *Meth. Enzymol.* **266**:635–643.

Brenner, S. E., Koehl, P., and Levitt, M. (2000) *Nucleic Acids Res.* **28**:254–256.

Chou, P. Y., and Fasman, G. D. (1974) *Biochemistry* **13**:211–222

Combet, C., Blanchet, C., Geourjon, C., and Delage, G. (2000) *Trends Biochem. Sci.* **25**:147–150.

Conte, L. L., Ailey, B., Hubbard, T. J. P., Brenner, S. E., Murzin, A. G., and Chothia, C. (2000) *Nucleic Acids Res.* **28**:257–259.

Deleage, G., and Roux, B. (1987) *Protein Eng.* **1**:289–294.

Fasman, G. D. Ed. (1989) *Prediction of Protein Structure and Principles of Protein Conformation.* Plenum Press, New York.

Frishman, D., and Argos, P (1996) *Protein Eng.* **9**:133–142.

Garnier, J., Osguthorpe, J. D., and Robson, B. (1978) *J. Mol. Biol.* **120**:97–120.

Garnier, J., Gibrat, J-F., and Robson, B. (1996) *Meth. Enzymol.* **266**:540–553.

Geourjon, C., and Deleage, G. (1994) *Protein Eng.* **7**:157–164.

Geourjon, C., and Deleage, G. (1995) *Comput. Appl. Biosci.* **11**:681–684.

Gibrat, J. F., Garnier, J., and Robson, B. (1987) *J. Mol. Biol.* **198**:425–443.

Guermeur, Y., Geourjon, C., Gallinari, P., and Deleage, G. (1999) *Bioinformatics* **15**:413–421.

Gupta, R., Birch, H., Rapacki, K., Brunak, S., and Hansen, J. E. (1999) *Nucleic Acids Res.* **27**:370–372.

Henikoff, S., and Henikoff, J. G. (1994) *Genomics* **19**:97–107.

Hofmann, K., Bucher, P., Falquet, L., and Bairoch, A. (2000) *Nucleic Acids Res.* **28**:215–219.

Ito, M., Matsuo, Y., and Nishikawa, K. (1997) *CABIOS* **13**:415–423.

Johnson, M. S., May, A. C. W., Rodionov, M. A., and Overington, J. P. (1996) *Meth. Enzymol.* **266**:575–598.

Jones, D. T., Taylor, W. R., and Thornton, J. M. (1992) *Nature* **358**:86–89.

Kelley, L. A., MacCallum, R. M., and Sternberg, M. J. E. (2000) *J. Mol. Biol.* **299**:501–522.

King, R. D., and Sternberg, M. J. (1996) *Protein Sci.* **5**:2298–2310.

Kneller, D. G., Cohen, F. E., and Langridge, R. (1990) *J. Mol Biol.* **214**:171–182.

Kreegipuu, A., Blom, N., and Brunak, S. (1999) *Nucleic Acids Res.* **27**:237–239.

Laskowski, R. A. (2001) *Nucleic Acids Res.* **29**:221–222.

Lesk, A. M. (1991) *Protein Architecture: A Practical Approach.* IRL/Oxford Press, Oxford.

Lesk, A. M., and Chothia, C. (1984) *J. Mol. Biol.* **174**:175–191.

Levin, J. M., Robson, B., and Garnier, J. (1986) *FEBS Lett.* **15**:303–308.

Lim, V. I. (1974) *J. Mol. Biol.* **88**:873–894.

Martin, J., and Hartl, F. U. (1997) *Curr. Opin. Struct. Biol.* **7**:41–62.

Nielsen, H., Brunak, S., and von Heijne, G. (1999) *Protein Eng.* **12**:3–9.

Orengo, C. A. (1999) *Protein Sci.* **7**:233–242.

Pearl, F. M. G., Lee, D., Bray, J. E., Sillitoe, I., Todd, A. E., Harrison, A. P., Thornton, J. M., Orengo, C. A. (2000) *Nucleic Acids Res.* **28**:277–282.

Pittsyn, O. B., and Finkelstein, A. (1983) *Biopolymers* **22**:15–25.

Richardson, J. S. (1981) *Adv. Protein Chem.* **34**:167–339.

Rodgers, S., Wells, R., and Rechsteiner, M. (1986) *Science* **234**:364–368.

Rost, B. (1996) *Meth. Enzymol.* **266**:525–539.

Rost, B., and Sander, C. (1993) *J. Mol. Biol.* **232**:584–599.

Schiffer, M., and Edmundson, A. (1967) *Biophys. J.* **7:**121–135.

Shi, J., Blundell, T. L., and Mizuguchi, K. (2001) *J. Mol. Biol.* **310:**243–257.

Shindyalov, I. N., and Bourne, P. E. (1997) *Comput. Appl. Biosci. (CABIOS)* **13:**487–496.

Tsai, C. S. (2001) *J. Chem. Ed.* **78:**837–839.

van Rooij, A. J. F., Jain, L. C., and Johnson, R. P. (1996) *Neural Network Training Using Genetic Algorithms.* World Scientific, River Edge, NJ.

Wang, Y., Anderson, J. B., Chen, J., Geer, L. Y., He, S., Hurwitz, D. I., Liebert, C. A., Madej, T., Marchler, G. H., Marchler-Bauer, A., Panchenko, A. R., Shoemaker, B. A., Song, J. S., Thiessen, P. A., Yamashita, R. A., and Bryant, S. H. (2002) *Nucleic Acids Res.* **30:**249–252.

Westbrook, J., Feng, Z., Jain, S., Bhat, T. N., Thanki, N., Ravichandran, V., Gilliland, G. L., Bluhm, W., Weissig, H., Greer, D. S., Bourne, P. E. and Berman, H. (2002) *Nucleic Acids Res.* **30:**245–248.

Zvelebil, M. J., Barton, G. J., Taylor, W. R., and Sternberg, M. J. E. (1987) *J. Mol. Biol.* **195:**957–961.

13

PHYLOGENETIC ANALYSIS

Phylogenetic analysis is the means of inferring or estimating evolutionary relation-ships. Nucleotide sequences of DNA or RNA and amino acid sequences of proteins are the most popular data used to construct phylogenetic trees. Methods of phylogenetic analysis using sequence data are introduced and performed with a software package, PHYLIP locally and online.

13.1. ELEMENTS OF PHYLOGENY

Systematics is the science of comparative biology. The primary goal of systematics is to describe taxic diversity and to reconstruct the hierarchy, or phylogenetic relationships, of those taxa. Molecular biology has offered systematicists an almost endless array of characters in the forms of DNA, RNA, and protein sequences with different structural/functional properties, mutational/selectional biases, and evol-utionary rates. The tremendous flexibility in their resolving power ensures the importance and widespread acceptance of DNA/RNA and protein sequences in phylogenetic research (Miyamoto and Cracroft, 1991). Phylogenetic analysis is the means of inferring or estimating evolutionary relationships. The physical events yielding a phylogeny happened in the past, and they can only be inferred or estimated. Phylogenetic analysis of biological sequences assumes that the observed differences between the sequences are the result of specific evolutionary processes. These assumptions are:

1. The evolutionary divergences are strictly bifurcating, so that the observed data can be represented by a treelike phylogeny.

2. The sequence is correct and originates from a specific source.
3. If the sequences analyzed are homologous, they are descended from a shared ancestral sequence.
4. Each multiple sequences included in a common analysis has a common phylogenetic history with the others.
5. The sampling of taxa is adequate to resolve the problem of interest.
6. The sequence variability in the sample contains phylogenetic signal adequate to resolve the problem of interest.
7. Each position in the sequence evolved homogeneously and independently.

The evolutionary history inferred from phylogenetic analysis is usually depicted as branching (treelike) diagrams. The correct evolutionary tree is a rooted tree that graphically depicts the cladistic relationships that exist among the operational taxonomic units (OTUs)—for example, contemporary sequences. A tree consists of nodes connected by lines, and a rooted tree starts from the root, which is the node ancestral to all other nodes. Each ancestral node gives rise to two descendant nodes (in bifurcating trees). Terminal or exterior nodes having no further descendants correspond to OTUs. All nonroot, nonexterior nodes are interior nodes. They correspond to ancestral sequences and may be inferred from the contemporary sequences. The connecting lines between two adjacent nodes are branches of the tree. A branch represents the topological relationship between nodes. Branches are also divided into external and internal ones. An external branch connects an external node and an internal node, whereas an internal branch connects two internal nodes. The number of unmatched sites between the aligned sequences of the two adjacent nodes is the length of that branch. The number of these sequence changes counted over all branches constitutes the length of the tree.

Calculation of tree length is simplified by removing the root from the tree. Such an unrooted tree will retain the interior nodes and the exterior nodes (OTUs). A tree of N exterior nodes has $N - 2$ interior nodes and $2N - 3$ links. Thus a tree of N OTUs can be converted to $2N - 3$ dendrograms. The phylogenetic procedure can reconstruct ancestral sequences for each interior node of a tree but cannot determine which interior node or which pair of adjacent interior nodes is closer to the root. However, after the search for the optimal tree is terminated, one can use the full weight of evidence on the cladistic relationships of the OTUs to assign a root to the optimal tree or, at least, to some of its branches. In the *out-group* method of rooting, one (or more) sequence that is known to be an out-group to the N sequences of the unrooted tree is added. This converts the unrooted tree of N sequences to the recalculated rooted tree of $N + 1$ sequences.

The number of possible tree topologies rapidly increases with an increase in the number (N) of OTUs. The general equation (Miyamoto and Cracroft, 1991) for the possible number of topologies for bifurcating unrooted trees (T_N) with n ($\geqslant 3$) OTUs (taxa) is given by

$$T_N = (2N - 5)! / [2^{N-3}(N - 3)!]$$

It is clear that a search for the true phylogenetic tree by phylogenetic analysis of a large number of sequences is computationally demanding and difficult.

13.2. METHODS OF PHYLOGENETIC ANALYSIS

The five basic steps in phylogenetic analysis of sequences (Hillis et al., 1993) organized as shown in Figure 13.1 consist of:

1. Sequence alignment
2. Assessing phylogenetic signal
3. Choosing methods for phylogenetic analysis
4. Construction of optimal phylogenetic tree
5. Assessing phylogenetic reliability

The sequence under study must be aligned so that positional homologues may be analyzed. Most methods of sequence alignment are designed for pairwise comparison. Two approaches to sequence alignments are used: a global alignment (Needleman and Wunsch, 1970) and a local alignment (Smith and Waterman, 1981). The former compares similarity across the full stretch of sequences, whereas the latter searches for regions of similarity in parts of the sequences (Chapter 11). Modifications of these approaches can also be used to align multiple sequences. Because the multiple alignment is inefficient with sequences if INDELs are common and substitution rates are high, most studies restrict comparisons to regions in which alignments are relatively obvious. Unless the number of taxa is few and INDELs are

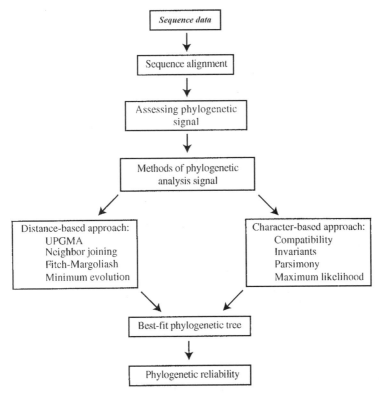

Figure 13.1. Strategy in phylogenetic analysis of sequences.

uncommon, it is not feasible to ensure that the optimum alignment has been achieved. For this reason, regions of questionable alignments are often removed from consideration prior to phylogenetic analysis. In general, substitutions are more frequent between nucleotide/amino acids that are biochemically similar (e.g., Table 11.1). In the case of DNA, the *transitions* between purine→purine and pyrimidine→pyrimidine are usually more frequent than the *transversion* between purine→pyrimidine and pyrimidine→purine. Such biases will affect the estimated divergence between two sequences. Specification of relative rates of substitution among particular residues usually takes the form of a square matrix; the number of rows/columns is 4 for DNA and 20 for proteins and 61 for codons. The diagonal elements represent the cost of having the same nucleotide/amino acid in different sequences. The off diagonal elements of the matrix correspond to the relative costs of going from one nucleotide/amino acid to another. A clustal program such as ClustalW (Higgins et al., 1996), which aligns sequences according to an explicitly phylogenetic criterion, is the most commonly used program for the multiple alignment of biochemical sequences.

For most sequence data, some positions are highly conserved, whereas other positions are randomized with respect to phylogenetic history. Thus assessment of phylogenetic signal is often required. In general, pairwise comparisons of the sequences are performed to evaluate the potential phylogenetic importance of the data. For example, the transition/transversion ratios for sequence pairs of DNA can be compared to those expected for the sequences at equilibrium, given the observed base compositions. DNA sequences that are largely free of homoplasy (convergence, parallelism, and reversal) will have transition/transversion ratios greater than those for sequences that are saturated by change, but similar to those observed for closely related taxa which remain highly structured. Similarly, pairwise divergences have been used to assess the potential phylogenetic value of DNA sequences, by plotting percent divergence against time. In such plots, regions of sequences or classes of character state transformations that are saturated by change do not show a significant positive relationship with time. Both the transition/transversion ratio and sequence divergence are influenced by homoplasy, thus both can provide insights into the potential phylogenetic value of sequence data. Another way of detecting the presence of a phylogenetic signal in a given data set is to examine the shape of the tree-length distribution (Fitch, 1979). Data sets with skewed tree-length distributions are likely to be phylogenetically informative. The data sets that produce significantly skewed tree-length distributions are also likely to produce the correct tree topology in phylogenetic analysis.

The following considerations may guide a choice of methods for phylogenetic analysis, namely, (1) assumptions about evolution, (2) parameters of sequence evolution, (3) the primary goal of analysis, (4) size of the data set, and (5) the limitation of computer time. Numerous methods are available, and each makes different assumptions about the molecular evolutionary process. It is unrealistic to assume that any one method can solve all problems, given the complexity of genomes/proteomes and their evolution. Phylogenetic analyses of sequences can be conducted by analyzing discrete characters such as aligned nucleotides and amino acids of the sequences (i.e., character-based approach) or by making pairwise comparisons of whole sequences (i.e. distance-based approach). Deciding whether to use a distance-based or a character-based method depends on the assumptions and the goals of the study.

Distance-Based Approaches. Distance-based methods use the amount of the distance (dissimilarity) between two aligned sequences to derive phylogenetic trees. A distance method would reconstruct the true tree if all genetic divergence events were accurately recorded in the sequence. Distance matrix methods simply count the number of differences between two sequences. This number is referred to as the evolutionary distance, and its exact size depends on the evolutionary model used. The actual tree is then computed from the matrix of distance values by running a clustering algorithm that starts with the most similar sequences or by trying to minimize the total branch length of the tree. Both the type and the position of a mutation have been incorporated into the parameters of divergence values. The simplest of the divergence measures is the two-parameter model of Kimura (Kimura, 1980). This model is designed to estimate divergence as well as the more complicated algorithms, over a broad range of divergences, without the need for additional specifics about the evolutionary process.

Methods for clustering distance data can be divided into those that provide single topology by following a specific series of steps (single-tree algorithms) and those that use optimality criterion to compare alternative topologies and to select the tree (multiple-tree algorithms). The representative single-tree algorithms are (a) the unweighed pair group method using arithmetic means (UPGMA) and (b) the neighbor joining method (NJ). The NJ (Saitou and Nei, 1987) has become a popular approach for analyzing sequence distances because of its ability to handle unequal rates, its connection to minimum-length trees, and its ease of calculation with regard to both tree topology and branch length. Several criteria of optimality are used for building distance trees. Each approach permits unequal rates and assumes additivity. The Fitch–Margoliash method (Fitch and Margoliash, 1967) minimizes the deviation between the observed pairwise distances and the path-length distances for all pairs of taxa on a tree. The best phylogenetic tree maximizes the fit of observed (original) versus tree-derived (patristic) distances, as measured by % standard deviation. Branch lengths are determined by linear algebraic calculations of observed distances among three different taxa interconnected by a common node. Minimum evolution (Saitou and Imanishi, 1989) aims to find the shortest tree that is consistent with the path lengths measured in a manner similar to that of Fitch–Margoliash approach without using all possible pairwise distances and all possible associated tree path lengths. Rather, it chooses the location of internal tree nodes based on the distance to external nodes, and then it optimizes the internal branch length according to the minimum measured error between these observed points. The phylogenetic tree is the one with minimal total overall length.

Character-Based Approaches. The individual aligned sites of biochemical sequences are equivalent to *characters*. The actual nucleotide or amino acid occupying a site is the *character state*. The character-based approaches treat each substitution separately rather than reducing all of the individual variation to a single divergence value. The relationships among organisms by the distribution of observed mutations are determined by counting each mutation event. These methods are preferred for studying character evolution, for combining multiple data sets, and for inferring ancestral genotypes. All sequence information is retained through the analyses. No information is lost in the conversion to distances. The approaches include parsimony, maximum likelihood, and method of invariants.

The principle of parsimony searches for a tree that requires the smallest number of changes to explain the differences observed among the taxa under study (Czelusnlak et al., 1990). The method minimizes the number of evolutionary events required to explain the original data. In practical terms, the parsimony tree is the shortest and the one with the fewest changes (least homoplasy). Parsimony remains the most popular character-based approach for sequence data due to its logical simplicity, its ease of interpretation, its prediction of both ancestral character states and amount of change along branches, the availability of efficient programs for its implementation, and its flexibility of conducting character analyses.

The maximum likelihood method (Felsenstein, 1981) calculates the probability of a data set, given the particular model of evolutionary change and specific topology. The likelihood of changes in the data can be determined by considering each site separately with respect to a particular topology and model of molecular evolution. Therefore the maximum likelihood method depends largely on the model chosen and on how well it reflects the evolutionary properties of the macromolecule being studied. In practice, the maximum likelihood is derived for each base position in an alignment. An individual likelihood is calculated in terms of the probability that the pattern of variation produced at a site by a particular substitution process with reference to the overall observed base frequencies. The likelihood becomes the sum of the probabilities of each possible reconstruction of substitutions under a particular substitution process. The likelihoods for all the sites are multiplied to give an overall likelihood of the tree. Because of questions about the accuracy of the models, coupled with the computational complexities of the approach, maximum likelihood methods have not received wide attention.

When large amounts of homoplasy exist among distantly related branches, one can avoid the abundant homoplasy while recognizing the phylogenetic signal by relying on a few specific patterns of nucleotide variation that represent the most conservative changes (Lake, 1987). For example, three possible topologies for four taxa called operator invariants are calculated. These calculations are based on the variable positions with two purines and two pyrimidines. Zero-value invariants represent cases in which random multiple mutation events have canceled each other out, and as such a chi-squared (χ^2) or binomial test is used to identify the correct topology as the one with an invariant significantly greater than zero. There are 36 patterns of transitions/transversions for four taxa with three possible topologies. Different methods of phylogenetic inference rely on various combinations of these components, with 12 used to calculate the three operator invariants of evolutionary parsimony. When more than four taxa are considered, all possible groups of four are typically analyzed and a composite tree constructed from the individual results.

Once a method and appropriate software have been selected, the best tree under the selected optimality criterion must be estimated (Saitou, 1996). The number of distinct tree topologies is very large (Felsenstein, 1978), even for a modest number of taxa (e.g., over 2.8×10^{74} distinct, labeled, bifurcating trees for 50 taxa). For relatively few taxa (up to as many as 20 or 30) it is possible to use exact algorithms (algorithmic approach) that will find an optimal tree. For greater numbers of taxa, one must rely on heuristic algorithms that approximate the exact solutions but may not give the optimal solution under all conditions. The exact algorithm searches exhaustively through all possible tree topologies for the best solution(s). This method is computationally simple for 9 or fewer taxa and becomes impractical for 13 or more taxa. An alternative exact algorithm known as the branch-and-bound algorithm (Hendy and Penny, 1982) can be applied to speed up the analysis of data set consisting of less than 10–11 taxa. If the exact algorithms are not feasible for a given data set, various heuristic approaches can be attempted. Heuristic procedures (Stagle, 1971) are a computer simulation and programming philosophy suited to finding a problem solution exploiting any empirical trial, strategy, or shortcut by which the computer acquires knowledge of the structure of the problem space beyond its pure abstract definition. The heuristic methodology does not, however, guarantee (in contrast to the algorithmic approach) the discovery of all possible goals in an absolute sense. Most heuristic techniques start by finding a reasonably good estimate of optimal tree(s) and then attempting to find a better solution by examining structurally related trees. The initial tree is usually found by a stepwise addition of taxa sequentially to a tree at the optimal place in the growing tree. The subsequent improvement is achieved by examining related topologies by a series of procedures known as branch swapping. All involve rearranging branches of the initial tree to search for a shorter alternative.

To assess confidence of phylogenetic results, it is generally recommended that subsets of taxa be examined from within the more complex topologies by focusing on specific major questions targeted before the analysis. Alternatively, one could limit the comparisons to just those topologies considered plausible for biological reasons. The reliability in phylogenetic results can be assessed by analytical methods or resampling methods. Analytical techniques (Felsenstein, 1988; Lake, 1987) for testing phylogenetic reliability operate by comparing the support for one tree to that for another under the assumption of randomly distributed data. Resampling techniques estimate the reliability of a phylogenetic result by bootstrapping or jackknifing the characters of the original data set. The bootstrapping approach (Felsenstein, 1985; Hillis and Bull, 1993) creates a new data set of the original size by sampling the available characters with replacement, whereas the jackknife approach (Penny and Hendy, 1986) randomly drops one or more data points or taxa at a time, creating smaller data sets by sampling without replacement. The ultimate criterion for determining phylogenetic reliability rests on tests of congruence among independent data sets representing both molecular and non-molecular information (Hillis, 1987). Different character types and data sets are unlikely to be exposed to the same evolutionary biases, and as such, congruent results supported by each are more likely to reflect convergence onto the single, correct tree. Therefore congruence analysis provides an important mechanism with which to evaluate the reliability of different methods for constructing phylogenetic trees. Given a common data set, those approaches leading to the congruent result should be preferred over those that do not. Studies of congruence can also provide insights into the limitations and assumption of different tree-constructing algorithms. Congruence analysis further permits an evaluation of the reliability of different weighting schemes of character transformations used in a phylogenetic analysis.

The catalog of software programs for phylogenetic analyses is available at http://evolution.gentics.washington.edu/philip/software.html.

13.3. APPLICATION OF SEQUENCE ANALYSES IN PHYLOGENETIC INFERENCE

13.3.1. Phylogenetic Analysis with Phylip

Phylip (Phylogenetic Inference Package) is a software package comprising about 30 command-line programs that cover almost any aspect of phylogenetic analysis

TABLE 13.1. Phylogenetic Methods Available from Phylip

Phylogenetic Methods	Exe. Program	Sequences	Description
Distance-based:			
Distance measure	DnaDist	DNA	Compute distances to distance matrix
Distance measure	ProtDist	Protein	Compute distances to distance matrix
Fitch-Margoliash (FM)	Fitch	DNA/Protein	Fitch-Margoliash method of analysis
FM with molecular cloak	Kitsch	DNA/Protein	FM method with molecular cloak
Neighbor joining	Neighbor	DNA/Protein	Neighbor joining analysis
Character-based:			
Compatibility	DnaComp	DNA	Search with compatibility criterion
Invariant	DnaInvar	DNA	Lake's phylogenetic invariants
Parsimony	DnaPars	DNA	Parsimony method
Parsimony	ProtPars	Protein	Parsimony method
Parsimony/compatibility	DnaMove	DNA	Interactive parsimony or compatibility
Maximum likelihood	DnaML	DNA	Maximum likelihood without molecular cloak
Maximum likelihood	DnaMLK	DNA	Maximum likelihood with molecular cloak
Search/reliability:			
Exact search	DnaPenny	DNA	Brance-and-bound search
Exact search	Penny	DNA/Protein	Branch-and-bound search 10–11 taxa or less
Resampling	SeqBoot	DNA/Protein	Multiple data sets from bootstrap resampling
Statistic	Contrast	DNA/Protein	Independent contrast for multivariate statistics
Phylogenetic tree:			
Drawing	DrawGram	DNA/Protein	Plot rooted tree
Drawing	Draw/Tree	DNA/Protein	Plot unrooted tree
Consensus tree	Consensus	DNA/Protein	Compute consensus tree by majority rule
Editing	ReTree	DNA/Protein	Reroot, flip branches and renaming

(Felsenstein, 1985; Felsenstein, 1996). The software can be downloaded from http://evolution.genetics.washington.edu/phylip.html. The package consists of a diverse collection of programs, including routines for calculating estimates of divergence and programs for both distance-based and character-based phylogenetic analyses. Some of the executable files (.exe) are listed in Table 13.1. The general user's manual can be found in main.doc, and detailed user's instruction to each executable file is in the separate documentation (.doc) file.

The sequence data, in Phylip format, should be in an input file called *infile* within the package. Use ReadSeq at http://dot.imgen.bcm.tmc.edu:9331/seq-util/Options/readseq.html to convert retrieved sequences in any formats (e.g., fasta format) to Phylip3.2 format. Copy and paste the sequence set, select Phylip3.2 as the output format, and click the Perform conversion button. From the output, copy only the identifier (number of species and number of characters), species name, and characters (sequences) into the infile (Figure 13.2).

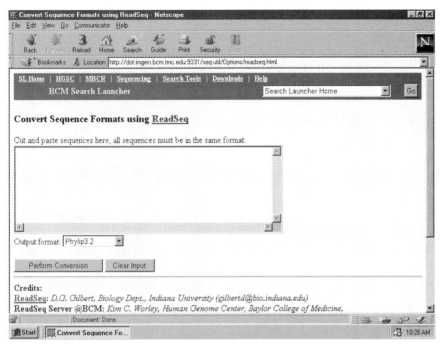

Figure 13.2. Conversion of sequence format with ReadSeq.

With Phylip, DnaDist computes a distance matrix from nucleotide sequences. Phylogenetic trees are generated by anyone of the available tools utilizing the distance matrix programs (Neighbor, Fitch, or Kitsch). DnaDist allows the user to choose between three models of nucleotide substitution. The Kimura two-parameter model allows the user to weigh transversion more heavily than transitions. Phylip also comprises DnaPars and DnaML to estimate phylogenetic relationships by the parsimony method and the maximum likelihood methods from nucleotide sequences, respectively. ProtDist is a program that computes a distance matrix for an alignment of protein sequences. It allows the user to choose between one of three evolutionary models of amino acid replacements. The simplest, fastest model assumes that each amino acid has an equal chance of turning into one of the other 19 amino acids. The second is a category model in which the amino acids are distributed among different groups and transitions are evaluated differently depending on whether the change would result in an amino acid switching in the group. The third (default) method uses a table of empirically observed transitions between amino acids (the Dayhoff PAM 001 matrix, Chapter 11). ProtPars is the parsimony program for protein sequences.

To perform distance-based analysis, start Dnadist/Protdist program by double-clicking Dnadist.exe/Protdist.exe. The execution of the program is indicated by a return of the option menu. Because the Phylip3.2 format derived from ReadSeq is sequential, the input option needs to be changed. Type *i* to change from Yes to *No, sequential* in response to "Input sequences interleaved?," and then type *Y* to accept the remaining default options. The program automatically searches for infile and writes the distance matrix into *outfile*. Copy/save outfile as otherfile for later

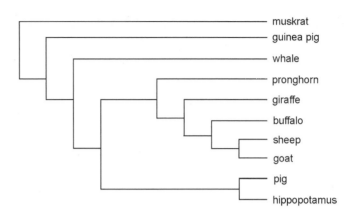

Figure 13.3. The TreeView is used to draw a phylogenetic tree from treefile of Phylip in radial, slanted cladogram, rectangular cladogram, or phytogram. The rectangular cladogram for treefile derived from amino acid sequences of ribonucleases is drawn with TreeView.

reference and rename outfile to infile to be analyzed by one of the distance matrix programs (e.g., Neighbor, Fitch, or Kitsch). The analytical results are recorded in *outfile* and *treefile*.

The character-based analysis of sequence data can be initiated by double-clicking the appropriate executable file (e.g., DnaPars, DnaML, or ProtPars). The acceptance of the input sequences is indicated by a return of the option menu. Type *i* to change the input option and accept the subsequent default options by typing *Y*. The program automatically searches and executes the *infile* and records the results in *outfile* and *treefile*. The treefile can be further processed with TreeView (Page, 1996), which allows the user to manipulate the tree and save the file in commonly used graphic formats. The tree-drawing software, TreeView, can be downloaded from http://taxanomy.zoology.gla.ac.uk/rod/treeview.html. Launch TreeView and open the treefile from Phylip by selecting File→Phylip (*.). Choose the tree view (radial, slanted cladogram, rectangular cladogram or phytogram) as shown in Figure 13.3, and save graphic file as treename.wmf.

To search for the consensus tree by bootstrapping technique, Seqboot accepts an input from the infile and multiplies it in a user-specified number of times (enter odd number *n* to yield a total of *n* + 1 data sets). The resulting outfile, after renaming to infile, is subjected to either distance-based or character-based analysis. Copy/save treefile to otherfile (for later reference) and rename treefile to infile. The resulting trees are reduced to the single tree by the use of Consense program, which returns the consensus tree with the bootstrap values as numbers on the branches. The topology of the consensus tree of the outfile can be viewed with any text editor.

13.3.2. Phylogenetic Analysis Online

Clustal is the common program for executing multiple sequence alignment (Higgins et al., 1996). After the alignment, the program constructs neighbor joining trees with bootstrapping. ClustalW can be accessed at EBI (http://www.ebi.ac.uk/), BCM (http://dot.imgen.bcm.tme.edu..9331/multi-align/multi-align.html), and DDBJ

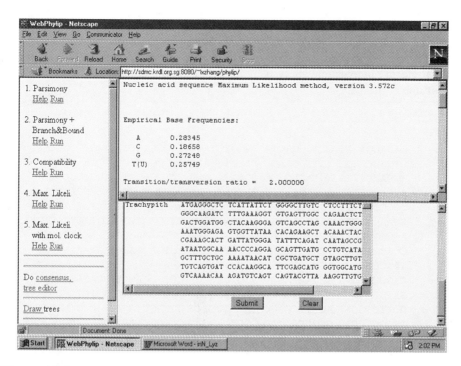

Figure 13.4. Phylogenetic analysis with WebPhylip. The phylogenetic analysis of DNA sequences encoding lysozyme precursors is performed with WebPhylip by maximum likelihood method.

(http://www.ddbj.nig.ac.jp/). The tree file can be requested in Phylip format, which can be saved as alignname.ph and displayed with TreeView. The output alignment file can be saved as alignname.aln which can be further analyzed using WebPhylip at http://sdmc.krdl.org.sg:8080/~lxzhang/phylip/ (Lim and Zhang, 1999).

The home page of WebPhylip consists of three windows. The left window is the command window for selecting and issuing the execution (Run) of the analysis/operation. The analysis results appear in the upper window. The lower window is the query window with options, a query box for pasting the query sequences (in Phylip format), and Submit/Clear buttons. To perform phylogenetic analysis, select DNA/Protein for Phylogeny method in the left window to open the available analytical methods (parsimony, parsimony + branch & bound, compatibility, maximum likelihood, and maximum likelihood with molecular clock for DNA whereas only parsimony for protein). Click Run (under the desired method) to open the query window with options. Select the appropriate Input type (either interleaved or sequential), paste the query sequences and click the Submit button. The analytical results are displayed in the upper window (Figure 13.4).

Select Do Consensus and click Run of Consensus tree (left window) to open the request form (lower window). Choose Yes to "Use tree file from last stage?" and click the Submit button to display the consensus tree (upper window) and save the tree file. Select Draw trees and click Run of cladograms/phenograms/phylogenies. Choose Yes to "Use tree file from last stage?" and click the Submit button to save the drawing treename.ps.

13.4. WORKSHOP

1. The nucleotide sequences of gastrin precursors in Phylip format are given below. Preform phylogenetic analysis by neighbor joining method using the Kimura two-parameter model.

```
    6 447
Human     CAGACGAGAT GCAGCGACTA TGTGTGTATG TGCTGATCTT TGCACTGGCT
          CTGGCCGCCT TCTCTGAAGC TTCTTGGAAG CCCCGCTCCC AGCAGCCAGA
          TGCACCCTTA GGTACAGGGG CCAACAGGGA CCTGGAGCTA CCCTGGCTGG
          AGCAGCAGGG CCCAGCCTCT CATCATCGAA GGCAGCTGGG ACCCCAGGGT
          CCCCCACACC TCGTGGCAGA CCCGTCCAAG AAGCAGGGAC CATGGCTGGA
          GGAAGAAGAA GAAGCCTATG GATGGATGGA CTTCGGCCGC CGCAGTGCTG
          AGGATGAGAA CTAACAATCC TAGAACCAAG CTTCAGAGCC TAGCCACCTC
          CCACCCCACT TCAGCCCTGT CCCCTGAAAA ACTGATCAAA AATAAACTAG
          TTTCCAGTGG ATC------- ---------- ---------- -------
Bovine    CAGGCTCCAG GCCAACAGCA GCGTCACCCA CCTCCCAGCT TCACAGACAA
          GATGCAGCGA CTGTGTGCAC ATGTGCTGAT CTTGGTGCTG GCTCTGGCCG
          CCTTCTGCGA AGCTTCTTGG AAGCCCCACT CCCACCTGCA AGACGCGCCC
          GTCGCTCCAG GGGCCAATAG GGGCCAGGAG CCGCTCAGGA TGAACCGGCT
          GGGCCCAGCC TCGAACCCCC GGAGGCAGCT GGGGCTTCAG GATCCGCCAC
          ACATGGTCGC AGACCTGTCC AAGAAGCAGG GGCCATGGGT GGAGGAAGAG
          GAAGCAGCAT ATGGATGGAT GGACTTCGGC CGCCGCAGTG CTGAGGAAGG
          GGACCAACAT CCCTGAGAAC AAGCTTTGGA GCCCAGTCAC CTCCCAGCCC
          AGCCCAGCCC CATGAAACAC ACATCAAAAT AAACTAGCTT CCAATGG
Horse     CCATCCGCAG CTTCTCCCAC GTCCCATCTC TGCAGACAAG ATGCGGCGAC
          TGTGTGTGTA TGTGCTGATC TTTGCGCTGG CTCTGGCCGC CTTCTCCGAA
          GCTTCTTGGA AGCCCTGCTC CCAGCTGCAG GACGCACCCT CAGGTCCAGG
          GGCCAATAAG GGCCTGTACC CGCATTGGCC AGACCAGCTG GACAGGCTGG
          GCCCAGCCTC TCATCACCGA AGGCAGCTGG GGCTCCAGGG TTCCCCACAC
          TTGGTAGCAG ACCTGTCCAA GAAGCAAGGG CCATGGCTGG AAAAAGAAGA
          AGCAGCCTAC GGATGGATGG ACTTTGGCCG CCGCAGCGCT GAGGAAGGGG
          ATCAAAGTCC TTAGAAAAAG CTTGGGAGCC CAGCCGGCTC CCACCCAGCC
          CAGCCCTGCC CTGTGAAAAA CCAACCAAAA TAAACTAGCT TCGGATG
Chicken   AAAGTGCGGA CGGAGCGCAG GGAGAGGTGC GGAGCCCCCG GAGCAGCAGC
          GTGAGCCATG AAGACGAAGG TGTTCCTCGG CCTCATCCTC AGCGCGGCGG
          TGACCGCCTG TCTGTGCCGG CCGGCAGCGA AGGCCCCGGG GGGCTCCCAC
          CGCCCCACCT CCAGCCTGGC CCGGCGGGAT TGGCCCGAGC CCCCGTCCCA
          GGAGCAGCAG CAGCGCTTCA TCTCCCGCTT CCTGCCCCAC GTCTTCGCAG
          AGCTGAGCGA CCGCAAAGGC TTCGTGCAGG GGAACGGGGC GGTAGAGGCC
          CTGCACGACC ACTTCTACCC CGACTGGATG GACTTCGGCC GCCGGAGCAC
          AGAGGATGCG GCCGATGCCG CGTAGCCCGC GCAGCGCCCC GACCCTCTCA
          GCACATCTCT GGTCCCGCAA TAAAGCTTTG GCACTCCC-- -------
Mouse     AGAGCAGAGC TGACCCAGCG CCACAACAGC CAACTATTCC CCAGCTCTGT
          GGACAAGATG CCTCGACTGT GTGTGTACAT GCTGGTCTTA GTGCTGGCTC
          TAGCTACCTT CTCGGAAGCT TCTTGGAAGC CCCGCTCCCA GCTACAGGAT
          GCATCATCTG GACCAGGGAC CAATGAGGAC CTGGAACAGC GCCAGTTCAA
          CAAGCTGGGC TCAGCCTCTC ACCATCGAAG GCAGCTGGGG CTCCAGGGTC
          CTCAACACTT CATAGCAGAC CTGTCCAAGA AGCAGAGGCC ACGGATGGAG
          GAAGAAGAAG AGGCCTACGG ATGGATGGAC TTTGGCCGCC GCAGTGCTGA
          GGAAGACCAG TAGGACTAGC AACACTCTTC CAGAGCCCAG CCATCTCCAG
          CCACCCCTCC CCCAGCTCCG TCCTTACAAA ACATATTAAA AATAAGC
```

```
Rat         TCCAGGCTCT GAAGGATCAC TGCAACAGAA ATGGACAAGA AGGTTTGTGT
            TAGTATTCTG CTAGCCATGC TTGCCATAGC AGCCTTGTGC AGGCCAATGA
            CAGAGCTAGA ATCAGCAAGA CACGGAGCTC AAAGGAAGAA CTCCATTTCT
            GATGTCAGCA GACGGGACCT CCTAGCATCC CTGACCCATG AACAGAAGCA
            GCTGATCATG TCACAGCTTC TCCCCGAGCT ACTTTCAGAA CTCTCCAATG
            CTGAGGACCA TCTTCACCCC ATGCGTGACC GGGACTATGC TGGATGGATG
            GATTTTGGCC GCAGAAGTTC TGAAGTCACA GAATCTTAAA TCCTTTCCTT
            TACCATCTCT GCCTATTTTA CTGCTGTGAA CGTCCCAATG GGGATTAATG
            ATGCTAATAA ATTATTGGTC TGAT------ ---------- -------
```

2. The amino acid sequences of ribonuclease in Phylip format are given below. Perform phylogenetic analyses by Neighbor joining method using the Dayfoff PAM 001 matrix (default).

```
       10 124
buffalo       KETAAAKFQR QHMDSSTSSA SSSNYCNQMM KSRNMTSDRC KPVNTFVHES
              LADVQAVCSQ ENVACKNGQT NCYQSYSTMS ITDCRETGSS KYPNCAYKTT
              QANKHIIVAC EGNPYVPVHY DASV
giraffe       KESAAAKFER QHIDSSTSSV SSSNYCNQMM TSRNLTQDRC KPVNTFVHES
              LADVQAVCSQ KNVACKNGQT NCYQSYSAMS ITDCRETGNS KYPNCAYQTT
              QAEKHIIVAC EGNPYVPVHY DASV
goat          KESAAAKFER QHMDSSTSSA SSSNYCBZMM KSRNLTQDRC KPVNTFVHES
              LADVZAVCSZ KNVACKNGZT BCYQSYSTMS ITBCRZTGSS KYPNCAYKTT
              QAEKHIIVAC ZGBPYVPVHF DASV
guinea pig    AESSAMKFER QHVDSGGSSS SNANYCNEMM KKREMTKDRC KPVNTFVHEP
              LAEVQAVCSQ RNVSCKNGQT NCYQSYSSMH ITECRLTSGS KFPNCSYRTS
              QAQKSIIVAC EGKPYVPVHF DNSV
hippopotamus  KETAAEKFQR QHMDTSSSLS NDSNYCNQMM VRRNMTQDRC KPVNTFVHES
              EADVKAVCSQ KNVTCKNGQT NCYQSNSTMH ITDCRETGSS KYPNCAYKTS
              QLQKHIIVAC EGDPYVPVHY DASV
muskrat       KETSAQKFER QHMDSTGSSS SSPTYCNQMM KRREMTQGYC KPVNTFVHEP
              LADVQAVCSQ ENVTCKNGNS NCYKSRSALH ITDCRLKGNS KYPNCDYQTS
              QLQKQVIVAC EGSPFVPVHF DASV
pig           KESPAKKFQR QHMDPDSSSS NSSNYCNLMM SRRNMTQGRC KPVNTFVHES
              LADVQAVCSQ INVNCKNGQT NCYQSNSTMH ITDCRQTGSS KYPNCAYKAS
              QEQKHIIVAC EGNPPVPVHF DASV
pronghorn     KETAAAKFER QHIDSNPSSV SSSNYCNQMM KSRNLTQGRC KPVNTFVHES
              LADVQAVCSQ KNVACKNGQT NCYQSYSTMS ITDCRETGSS KYPNCAYKTT
              QAKKHIIVAC EGNPYVPVHY DASV
sheep         KESAAAKFER QHMDSSTSSA SSSNYCNQMM KSRNLTQDRC KPVNTFVHES
              LADVQAVCSQ KNVACKNGQT NCYQSYSTMS ITDCRETGSS KYPNCAYKTT
              QAEKHIIVAC EGNPYVPVHF DASV
whale         RESPAMKFQR QHMDSGNSPG NNPNYCNQMM MRRKMTQGRC KPVNTFVHES
              LEDVKAVCSQ KNVLCKNGRT NCYESNSTMH ITDCRQTGSS KYPNCAYKTS
              QKEKHIIVAC EGNPYVPVHF DNSV
```

3. Retrieve the amino acid sequences of liver alcohol dehydrogenase from six organisms to perform phylogenetic analysis. Compare phylogenetic results from Fitch–Margoliash methods without (Fitch) versus with (Kisch) molecular clock using the Dayhoff PAM 001 matrix.

4. Perform phylogenetic analysis by parsimony approach of ribonucleases with amino acid sequences given in Exercise 2.

5. The DNA nucleotide sequences encoding tRNA specific for lysine are given below. Perform phylogenetic analysis by maximum likelihood approach.

```
>Squid tRNA-Lys(UUU)
GCCTCCATAGCTCAGTCGGTAGAGCATCAGACTTTTAANCTGAGGGTCTGGGGTTCGAGTCCCCATGTGG
GCTCCA
>Rat liver transfer RNA-Lys
GCCCGGATAGCTCAGTCGGTAGAGCATCAGACTTTTATCTGAGGGTCCAGGGTTCAAGTCCCTGTTCGGG
CGCCA
>Mouse Lys-tRNA
AGGCCCTTAGCTCAGCAGGTAGAGCAAATGACTTTTAATCATTGGGTCAGAGGTTCAAATCCTCTAGGGC
GTACC
>Fruitfly Lys-tRNA
GCCCGGCTAGCTCAGTCGGTAGAGCATGAGACTCTTAATCTCAGGGTCGTGGGTTCGAGCCCCACGTTGG
GCGCCA
```

6. Perform parsimony analysis with bootstrapping technique on gastrin precursors with nucleotide sequences given in Exercise 1.

7. The amino acid sequences for cytochrome C are given below. Draw cladogram of the consensus tree after conducting parsimony analysis at WebPhylip.

```
>bovine
GDVEKGKKIFVQKCAQCHTVEKGGKHKTGPNLHGLFGRKTGQAPGFSYTDANKNKGITWGEETLMEYLEN
PKKYIPGTKMIFAGIKKKGEREDLIAYLKKATNE
>chicken
MGDIEKGKKIFVQKCSQCHTVEKGGKHKTGPNLHGLFGRKTGQAEGFSYTDANKNKGITWGEDTLMEYLE
NPKKYIPGTKMIFAGIKKKSERVDLIAYLKDATSK
>dogfish
GDVEKGKKVFVQKCAQCHTVENGGKHKTGPNLSGLFGRKTGQAQGFSYTDANKSKGITWQQETLRIYLEN
PKKYIPGTKMIFAGLKKKSERQDLIAYLKKTAAS
>duck
GDVEKGKKIFVQKCSQCHTVEKGGKHKTGPNLHGLFGRKTGQAEGFSYTDANKNKGITWGEDTLMEYLEN
PKKYIPGTKMIFAGIKKKSERADLIAYLKDATAK
>hippopotamus
GDVEKGKKIFVQKCAQCHTVEKGGKHKTGPNLHGLFGRKTGQSPGFSYTDANKNKGITWGEETLMEYLEN
PKKYIPGTKMIFAGIKKKGERADLIAYLKQATNE
>horse
GDVEKGKKIFVQKCAQCHTVEKGGKHKTGPNLHGLFGRKTGQAPGFTYTDANKNKGITWKEETLMEYLEN
PKKYIPGTKMIFAGIKKKTEREDLIAYLKKATNE
>human
MGDVEKGKKIFIMKCSQCHTVEKGGKHKTGPNLHGLFGRKTGQAPGYSYTAANKNKGIIWGEDTLMEYLE
NPKKYIPGTKMIFVGIKKKEERADLIAYLKKATNE
>mouse
MGDVEKGKKIFVQKCAQCHTVEKGGKHKTGPNLHGLFGRKTGQAAGFSYTDANKNKGITWGEDTLMEYLE
NPKKYIPGTKMIFAGIKKKGERADLIAYLKKATNE
>ostrich
GDIEKGKKIFVQKCSQCHTVEKGGKHKTGPNLDGLFGRKTGQAEGFSYTDANKNKGITWGEDTLMEYLEN
PKKYIPGTKMIFAGIKKKSERADLIAYLKDATSK
>penguin
GDIEKGKKIFVQKCSQCHTVEKGGKHKTGPNLHGIFGRKTGQAEGFSYTDANKNKGITWGEDTLMEYLEN
PKKYIPGTKMIFAGIKKKSERADLIAYLKDATSK
>pig
GDVEKGKKIFVQKCAQCHTVEKGGKHKTGPNLHGLFGRKTGQAPGFSYTDANKNKGITWGEETLMEYLEN
PKKYIPGTKMIFAGIKKKGEREDLIAYLKKATNE
```

```
>pigeon
GDIEKGKKIFVQKCSQCHTVEKGGKHKTGPNLHGLFGRKTGQAEGFSYTDANKNKGITWGEDTLMEYLEN
PKKYIPGTKMIFAGIKKKAERADLIAYLKQATAK
>rabbit
GDVEKGKKIFVQKCAQCHTVEKGGKHKTGPNLHGLFGRKTGQAVGFSYTDANKNKGITWGEDTLMEYLEN
PKKYIPGTKMIFAGIKKKDERADLIAYLKKATNE
>rat
MGDVEKGKKIFVQKCAQCHTVEKGGKHKTGPNLHGLFGRKTGQAAGFSYTDANKNKGITWGEDTLMEYLE
NPKKYIPGTKMIFAGIKKKGERADLIAYLKKATNE
>tuna
GDVAKGKKTFVQKCAQCHTVENGGKHKVGPNLWGLFGRKTGQAEGYSYTDANKSKGIVWNENTLMEYLEN
PKKYIPGTKMIFAGIKKKGERQDLVAYLKSATS
>turkey
GDIEKGKKIFVQKCSQCHTVEKGGKHKTGPNLHGLFGRKTGQAEGFSYTDANKNKGITWGEDTLMEYLEN
PKKYIPGTKMIFAGIKKKSERVDLIAYLKDATSK
>whale
GDVEKGKKIFVQKCAQCHTVEKGGKHKTGPNLHGLFGRKTGQAVGFSYTDANKNKGITWGEETLMEYLEN
PKKYIPGTKMIFAGIKKKGERADLIAYLKKATNE
```

8. Retrieve the nucleotide sequences encoding type C lysozyme precursors from eight organisms and draw their phylogenies of the consensus tree after performing compatibility analysis at WebPhilip.

9. The DNA nucleotide sequences encoding tRNA specific for phenylalanine are given below. Draw a cladogram of the consensus tree after performing parsimony analysis + branch & bound at WebPhylip.

```
>Saccharomyces Cerevisia tRNA-Phe
NCGGATTTANCTCAGNNGGGAGAGCNCCAGANTNAANANNTGGAGNTCNTGTGNNCGNTCCACAGAATTC
GCACCA
>Mycoplasma sp. tRNA-Phe (GAA)
GGTCGTGTAGCTCAGTCGGTAGAGCAGCAGACTGAAGCTCTGCGTGTCGGCGGTTCAATTCCGTCCACGA
CCACCA
>Neurospora crassa tRNA-Phe (GAA)
GCGGGTTTAGCTCAGTTGGGAGAGCGTCAGACTGAAGATCTGAAGGTCGTGTGTTCGATCCACACAAACC
GCACCA
>X.leavis Phe-tRNA
GCCGAAATAGCTCAGTTGGGAGAGCGTTAGACTGAAGATCTAAAGGTCCCTGGTTCGATCCCGGGTTTCG
GCACCA
>R.rubrum Phe-tRNA
GCCCGGGTAGCTCAGCTGGTAGAGCACGTGACTGAAAATCACGGTGTCGGTGGTTCGACTCCGCCCCCGG
GCACCA
>E.coli Phe-tRNA
GCCCGGATAGCTCAGTCGGTAGAGCAGGGGATTGAAAATCCCCGTGTCCTTGGTTCGATTCCGAGTCCGG
GCACCA
>D.melanogaster Phe-tRNA
GCCGAAATAGCTCAGTTGGGAGAGCGTTAGACTGAAGATCTAAAGGTCCCCGGTTCAATCCCGGGTTTCG
GCACCA
>cyanobacterium Phe-tRNA
GCCAGGATAGCTCAGTTGGTAGAGCAGAGGACTGAATATCCTCGTGTCGGCGGTTCAATTCCGCCTCCCG
GCACCA
>B.subtilis Phe-tRNA
GGCTCGGTAGCTCAGTTGGTAGAGCAACGGACTGAAAATCCGTGTGTCGGCGGTTCGATTCCGTCCCGAG
CCACCA
>B.stearothermophilus Phe-tRNA
GGCTCGGTAGCTCAGTCGGTAGAGCAAAGGACTGAAAATCCTTGTGTCGGCGGTTCGATTCCGTCCCGAG
CCACCA
```

10. Retrieve the amino acid sequences of type C lysozymes from twelve biological sources. Perform ClustalW multiple sequence alignment and WebPhylip parsimony analysis to draw a cladogram of the consensus tree.

REFERENCES

Czelusnlak, J., Goodman, M., Moncrief, N. D., and Kehoe, S. M. (1990) *Meth. Enzymol.* **183**:601–615.

Felsenstein, J. (1978) *Syst. Zool.* **27**:27–33.

Felsenstein, J. (1981) *J. Mol. Evol.* **17**:368–376.

Felsenstein, J. (1985) *Evolution* **39**:783–791.

Felsenstein, J. (1988) *Annu. Rev. Genet.* **22**:521–565.

Felsenstein, J. (1996) *Meth. Enzymol.* **266**:418–427.

Fitch, W. M. (1979) *Syst. Zool.* **28**:375–379.

Fitch, W. M., and Margoliash, E. (1967) *Science* **155**:279–284.

Hendy, M. D., and Penny, D. (1982) *Math. Biosci.* **59**:277–290.

Higgins, D. G., Thompson, J. D., and Gibson, T. J. (1996). *Meth. Enzymol.* **266**:383–402.

Hillis, D. M. (1987) *Annu. Rev. Ecol. Syst.* **18**:23–42.

Hillis, D. M., and Bull, J. J. (1993) *Syst. Biol.* **42**:182–192.

Hillis, D. M., Allard, M. W., and Miyamotao, M. M. (1993) *Meth. Enzymol.* **224**:456–487.

Kimura, M. (1980) *J. Mol. Evol.* **16**:111–120.

Lake, J. A. (1987) *Mol. Biol. Evol.* **4**:167–191.

Lim, A., and Zhang, L. (1999) *Bioinformatics*, **15**:1068–1069.

Miyamoto, M. M., and Cracroft, J. Ed. (1991) *Phylogenetic Analysis of DNA Sequences.* Oxford University Press, Oxford.

Needleman, S. B., and Wunsch, C. D. (1970) *J. Mol. Biol.* **48**:443–453.

Page, R. D. M. (1996) *Comput. Appl. Biosci.* **12**:357–358.

Penny, D., and Hendy, M. (1986) *Mol. Biol. Evol.* **3**:403–417.

Saitou, N., and Imanishi, T. (1989) *Mol. Biol. Evol.* **6**:514–525.

Saitou, N., and Nei, M. (1987) *Mol. Biol. Evol.* **4**:406–425.

Saitou, N. (1996) *Meth. Enzymol.* **266**:427–449.

Smith, T. F., and Waterman, M. S. (1981) *J. Mol. Biol.* **147**:195–197.

Stagle, J. R. (1971) *Artificial Intellligence: The Heuristic Programming Approach.* McGraw-Hill, New York.

14

MOLECULAR MODELING: MOLECULAR MECHANICS

Molecular modeling can be defined as an application of computers to generate, manipulate, calculate and predict realistic molecular structures and associated properties. The computational approaches, in particular molecular mechanics, to obtain energetic and structural information for biomolecules are presented. Methodologies for energy calculation/geometry optimization, dynamic simulation and conformational search are discussed. The applications of molecular modeling packages, Chem3D and HyperChem, are described.

14.1. INTRODUCTION TO MOLECULAR MODELING

Molecular modeling can be considered as a range of computerized techniques based on theoretical chemistry methods and experimental data that can be used either to analyze molecules and molecular systems or to predict molecular, chemical, and biochemical properties (Höltje and Folkeis, 1997; Leach, 1996; Schlecht, 1998). It serves as a bridge between theory and experiment to:

1. Extract results for a particular model.
2. Compare experimental results of the system.
3. Compare theoretical predictions for the model.
4. Help understanding and interpreting experimental observations.

5. Correlate between microscopic details at atomic and molecular level and macroscopic properties.

6. Provide information not available from real experiments.

Thus molecular modeling can be defined as the generation, manipulation, calculation, and prediction of realistic molecular structures and associated physicochemical as well as biochemical properties by the use of a computer. It is primarily a mean of communication between scientist and computer, the imperative interface between human-comprehensive symbolism, and the mathematical description of the molecule. The endeavor is made to perceive and recognize a molecular structure from its symbolic representations with a computer. Thus functions of the molecular modeling include:

- *Structure retrieval or generation:* Crystal structures of organic compounds can be found in the Cambridge Crystallographic Datafiles (http://www.ccdc.cam. ac.uk/). Those that does not exist may be generated by 3D rendering software. The 3D structural coordinates of biomacromolecules can be retrieved from Protein Data Bank (http://www.rcsb.org/pdb/).

- *Structural visualization:* Computer graphics is the most effective means for visualization and interactive manipulation of molecules and molecular systems. Numerous software programs (e.g., Cn3D, RasMol and KineMage, Chapter 4) are available for visualization, management, and manipulation of molecular structures.

- *Energy calculation and minimization:* One of the fundamental properties of molecules is their energy content and energy level. Three major theoretical computational methods of their calculation include empirical (molecular mechanics), semiempirical, and *ab initio* (quantum mechanics) approaches. Energy minimization results in geometry optimization of the molecular structure.

- *Dynamics simulation and conformation search:* Solving motion of nuclei in the average field of the electrons is called *quantum dynamics*. Solution to the Newton's equation of motion for the nuclei is known as *molecular dynamics*. Integration of Newton's equation of motion for all atoms in the system generates molecular trajectories. *Conformation search* is carried out by repeating the process by rotating reference bonds (dihedral angles) of the molecule under investigation for finding lowest energy conformations of molecular systems.

- *Calculation of molecular properties:* Methods of estimating or computing properties (i.e., interpolating properties, extrapolating properties, and computing properties). Some computing properties are boiling point, molar volume, solubility, heat capacity, density, thermodynamic quantities, molar refractivity, magnetic susceptibility, dipole moment, partial atomic charge, ionization potential, electrostatic potential, van der Waals surface area, and solvent accessible surface area.

- *Structure superposition and alignment:* Computing activities and properties of molecules often involves comparisons across a homologous series. Such techniques require superposition or alignment of structures.

- *Molecular interactions, docking:* The intermolecular interaction in a ligand–receptor complex is important and requires difficult modeling exercises. Usually the receptor (e.g., protein) is kept rigid or partially rigid while the conformation of ligand molecule is allowed to change.

The advent of high-speed computers, availability of sophisticated algorithms, and state-of-the-art computer graphics have made plausible the use of computationally intensive methods such as quantum mechanics, molecular mechanics, and molecular dynamics simulations to determine those physical and structural properties most commonly involved in molecular processes. The power of molecular modeling rests solidly on a variety of well-established scientific disciplines including computer science, theoretical chemistry, biochemistry, and biophysics. Molecular modeling has become an indispensable complementary tool for most experimental scientific research.

Computational biochemistry and computer-assisted molecular modeling have rapidly become a vital component of biochemical research. Mechanisms of ligand–receptor and enzyme–substrate interactions, protein folding, protein–protein and protein–nucleic acid recognition, and *de novo* protein engineering are but a few examples of problems that may be addressed and facilitated by this technology.

14.2. ENERGY MINIMIZATION, DYNAMICS SIMULATION, AND CONFORMATIONAL SEARCH

14.2.1. Energy Calculation

Computational approaches to potential energy may be divided into two broad categories: quantum mechanics (Hehre et al., 1986) and molecular mechanics (Berkert and Allinger, 1982). The basis for this division depends on the incorporation of the Schrödinger equation or its matrix equivalent. It is now widely recognized that both methods reinforce one another in an attempt to understand chemical and biological behavior at the molecular level. From a purely practical standpoint, the complexity of the problem, time constraints, computer size, and other limiting factors typically determine which method is feasible.

Quantum mechanics (QM) can be further divided into *ab initio* and semiempirical methods. The *ab initio* approach uses the Schrödinger equation as the starting point with post-perturbation calculation to solve electron correlation. Various approximations are made that the wave function can be described by some functional form. The functions used most often are a linear combination of Slater-type orbitals (STO), $\exp(-ax)$, or Gaussian-type orbitals (GTO), $\exp(-ax^2)$. In general, *ab initio* calculations are iterative procedures based on self-consistent field (SCF) methods. Self-consistency is achieved by a procedure in which a set of orbitals is assumed and the electron–electron repulsion is calculated. This energy is then used to calculate a new set of orbitals, and these in turn are used to calculate a new repulsion energy. The process is continued until convergence occurs and self-consistency is achieved.

On the other hand, the term semiempirical is usually reserved for those calculations where families of difficult-to-solve integrals are replaced by equations and parameters that are fitted to experimental data. Semiempirical methods describe molecules in terms of explicit interactions between electrons and nuclei and are based on the principles:

- Nuclei and electrons are distinguished from each other.
- Electron–electron and electron–nuclear interaction are explicit.
- Interactions are governed by nuclear and electron charges — that is, potential energy and electron motions.
- Interactions determine the space distribution of nuclei and electrons and their energies.

For the best result, the molecule being computed should be similar to molecules in the database used to parameterize the method. However, if the molecule being computed is significantly different from anything in the parameterization set, erratic results may be obtained. Semiempirical calculations have been successful in dealing with organic compounds.

For large biomolecules, semiempirical calculations cannot be applied effectively; in these cases the methods referred to as molecular mechanics (MM) can be used to model their structures and behaviors. Molecular mechanics utilizes simple algebraic expressions for the total energy of a compound without computing a wave function or total electron density (Boyd and Lipkowitz, 1982; Brooks et al., 1988). The fundamental assumption of MM or its tool, empirical force field (EFF or simply force field, FF), is that data determined experimentally for small molecules can be extrapolated to larger molecules. It is aimed at quickly providing energetically favorable conformations for large systems. Molecular mechanical method is based on the following principles:

1. Nuclei and electrons are lumped together and treated as unified atom-like particles.
2. Atom-like particles are treated as spherical balls.
3. Bonds between particles are viewed as springs.
4. Interactions between these particles are treated using potential functions derived from classical mechanics.
5. Individual potential functions are used to describe different types of interactions.
6. Potential energy functions rely on empirically derived parameters that describe the interactions between sets of atoms.
7. The potential functions and the parameters used for evaluating interactions are termed a force field.
8. The sum of interactions determines the conformation of atom-like particles.

Therefore, MM energies have no meaning as absolute quantities. They are used for comparing relative strain energy between two or more conformations.

In molecular mechanical calculations, the force fields generally take the form

$$E_{\text{total}} = E_r + E_\theta + E_\phi + E_{\text{nb}} + [\text{special terms}]$$

in which the successive terms, expressing the total energy (E_{total}), are energies associated with bond stretching (E_r), bond angle bending (E_θ), bond torsion (E_ϕ), nonbond interactions (E_{nb}), plus specific terms such as hydrogen bonding (E_{hb}) in biochemical systems. Most MM equations are similar in the types of terms they contain. There are some differences in the forms of the equations that can affect the choice of FF and parameters for the systems of interest. Examples are (a) MM2/3 for organic compounds (Altona and Faber, 1974) and (b) AMBER (Weiner et al., 1984; Weiner et al., 1986) or CHARMm (Brooks et al., 1983) for biological molecules.

For most FF, the internal energy terms are similar, namely,

$$E_r = \sum K_r (r - r_0)^2$$
$$E_\theta = \sum K_\theta (\theta - \theta_0)^2$$

and

$$E_\phi = \sum K_\phi [1 + \cos(n\phi - \phi_0)]$$

where K_r, K_ϕ, and K_θ are force constants for bond, angle, and dihedral angle, respectively. r_0, θ_0, and ϕ_0 define the equilibrium distance, equilibrium angle, and phase angle for the given type. n is the periodicity of the Fourier term. These parameter values are derived from model molecules and vary among different FF. However, different potential functions may be used by different FF for E_{nb} and special terms such as E_{hb}. Lennard-Jones 6-12 potential is the most commonly used for van der Waals interactions (E_{vdW}) in E_{nb}, such that

$$E_{vdW} = \Sigma\Sigma(A_{ij}/r_{ij}^{12} - B_{ij}/r_{ij}^6)$$

Because biochemical molecules are often charged, an electrostatic energy (E_{elec}) term is added to E_{nb}:

$$E_{elec} = \Sigma\Sigma(q_i q_j)/(Dr_{ij})$$

or separately to take into account the interactions between nonbound but interacting atoms i and j with the distance of r_{ij}. A_{ij} and B_{ij} are van der Waal parameters. D is a molecular dielectric constant (vary from 1 *in vaccuo* to 80 in water) that accounts for the environmental attenuation of electrostatic interaction between the two atoms with the point charge q_i and q_j. The hydrogen bonding E_{hb} term differs. Of the two most commonly used force fields in biochemistry, AMBER introduces the 10–12 potential, that is,

$$E_{hb} = \Sigma\Sigma(C_{ij}/r_{ij}^{12} - D_{ij}/r_{ij}^{10})$$

while CHARMm consider both the distance and angle of the hydrogen bond interactions among three atoms (A for acceptor, H for hydrogen, and D for donor). You can choose to calculate all nonbonded interactions or to truncate (cut off) the nonbonded interaction calculations using a switched or shifted function. Useful guidelines for nonbonded interactions are as follows:

- Calculate all nonbonded interactions for small and medium-sized molecules.
- Use either switched function or shifted function to decrease computing time for macromolecules such as proteins and nucleic acids.

- Switched function is a smooth function, applied from the inner radius (R_{on}) to the outer radius (R_{off}), that gradually reduces nonbonded interaction to zero. The suggested outer radius is approximately 14 Å, and the inner radius is approximately 4 Å less than the outer radius.
- Shifted function is smooth function, applied over a whole nonbonded distance, from zero to outer radius, that gradually reduces nonbonded interaction to zero.

Thus different FF are designed for different systems and purposes. The databases used also differ. The users of MM software should be aware of these differences because a particular FF may work extremely well within one molecular structure class but may fail when applied to other types of structures.

Molecular mechanics is an extremely widely used method for generating molecular models for a multitude of purposes in chemistry and biochemistry. The first major reason for the popularity of MM is its speed, which makes it computationally feasible for routine usage. The alternative methods for generating molecular geometries, such as *ab initio* or semiempirical molecular orbital calculations, consume much larger amounts of computer time, making them much more expensive to use. The economy of MM makes studies of relatively large molecules such as biomacromolecules feasible on a routine basis. Thus, MM has become a primary tool of computational biochemists. MM is relatively simple to understand. The total strain energy is broken down into chemically meaningful components that correspond to an easily visualized picture of molecular structure. Molecular mechanics also has some limitations. The potential pitfall to be aware of is that MM routines will generate a conformation for which the strain energy is minimized. However, the minimum found during a calculation may not be the global minimum. It is relatively easy for the procedure to become trapped in a local energy minimum. There are schemes for minimizing the risk of such entrapment in local minima. For example, the calculation can be done a number of times starting from different initial geometries to see if the final geometry found remains the same. Another obvious drawback of the MM approach is that it cannot be used to study any molecular system where electronic effects are dominant. Here, quantum mechanical approaches that explicitly account for the electrons in molecules must be used.

To study electronic behavior in biomolecules, QM and MM are combined into one calculation (QM/MM) (Gogonea et al., 2001; Warshel, 1991) that models a large molecule (e.g., enzyme) using MM and one crucial section of the molecule (e.g., active site) with QM. This is designed to give results that have good speed where only the region needs to be modeled quantum mechanically.

Normally, a single-point energy calculation is for stationary point on a potential energy surface. The calculation provides an energy and the gradient of that energy. The gradient is the root-mean-square (RMS gradient) of the derivative of the energy with respect to Cartesian coordinates, that is,

$$\text{RMS Gradient} = (3N)^{-1}[\Sigma(\delta E/\delta X)^2 + (\delta E/\delta Y)^2 + (\delta E/\delta Z)^2]^{1/2}$$

At a minimum, the gradient is zero. Thus the size of the gradient can provide qualitative information to assess if a structure is close to a minimum.

14.2.2. Energy Minimization and Geometry Optimization

The basic task in the computational portion of MM is to minimize the strain energy of the molecule by altering the atomic positions to optimal geometry. This means minimizing the total nonlinear strain energy represented by the FF equation with respect to the independent variables, which are the Cartesian coordinates of the atoms (Altona and Faber, 1974). The following issues are related to the energy minimization of a molecular structure:

- The most stable configuration of a molecule can be found by minimizing its free energy, G.
- Typically, the energy E is minimized by assuming the entropy effect can be neglected.
- At a minimum of the potential energy surface, the net force on each atom vanishes, therefor the stable configuration.
- Because the energy *zero* is arbitrary, the calculated energy is relative. It is meaningful only to compare energies calculated for different configurations of chemically identical systems.
- It is difficult to determine if a particular minimum is the *global minimum*, which is the lowest energy point where force is zero and second derivative matrix is positive definite. *Local minimum* results from the net zero forces and positive definite second derivative matrix, and *saddle point* results from the net zero forces and at least one negative eigenvalue of the second derivative matrix.

The most widely used methods fall into two general categories: (1) steepest descent and related methods such as conjugate gradient, which use first derivatives, and (2) Newton–Raphson procedures, which additionally use second derivatives.

The steepest descent method (Wiberg, 1965) depends on (1) either calculating or estimating the first derivative of the strain energy with respect to each coordinate of each atom and (2) moving the atoms. The derivative is estimated for each coordinate of each atom by incrementally moving the atom and storing the resultant strain energy change. The atom is then returned to its original position, and the same calculation is repeated for the next atom. After all the atoms have been tested, their positions are all changed by a distance proportional to the derivative calculated in step 1. The entire cycle is then repeated. The calculation is terminated when the energy is reduced to an acceptable level. The main problem with the steepest descent method is that of determining the appropriate step size for atom movement during the derivative estimation steps and the atom movement steps. The sizes of these increments determine the efficiency of minimization and the quality of the result. An advantage of the first-derivative methods is the relative ease with which the force field can be changed.

The conjugate gradient method is a first-order minimization technique. It uses both the current gradient and the previous search direction to drive the minimization. Because the conjugated gradient method uses the minimization history to calculate the search direction and contains a scaling factor for determining step size, the method converges faster and makes the step sizes optimal as compared to the steepest descent technique. However, the number of computing cycles required for a conjugated gradient calculation is approximately proportional to the number of atoms (N), and the time per cycle is proportional to N^2. The Fletcher–Reeves approach chooses a descent direction to lower energy by considering the current gradient, its conjugate, and the gradient for the previous step. The Polak–Ribiere algorithm improves on the Fletcher–Reeves approach by additional consideration of the previous conjugate and tends to converge more quickly.

The Newton–Raphson methods of energy minimization (Berkert and Allinger, 1982) utilize the curvature of the strain energy surface to locate minima. The computations are considerably more complex than the first-derivative methods, but they utilize the available information more fully and therefore converge more quickly. These methods involve setting up a system of simultaneous equations of size $(3N - 6)$ $(3N - 6)$ and solving for the atomic positions that are the solution of the system. Large matrices must be inverted as part of this approach.

The general strategy is to use steepest descents for the first 10–100 steps (500–1000 steps for proteins or nucleic acids) and then use conjugate gradients or Newton–Raphson to complete minimization for convergence (using RMS gradient or/and energy difference as an indicator). For most calculations, RMS gradient is set to 0.10 (you can use values greater than 0.10 for quick, approximate calculations). The calculated minimum represents the potential energy closest to the starting structure of a molecule. The energy minimization is often used to generate a structure at a stationary point for a subsequent single-point calculation or to remove excessive strain in a molecule, preparing it for a molecular dynamic simulation.

14.2.3. Dynamics Simulation

Molecules are dynamic, undergoing vibrations and rotations continually. The static picture of molecular structure provided by MM therefore is not realistic. Flexibility and motion are clearly important to the biological functioning of proteins and nucleic acids. These molecules are not static structures, but exhibit a variety of complex motions both in solution and in the crystalline state. The most commonly employed simulation method used to study the motion of protein and nucleic acid on the atomic level is the *molecular dynamics* (MolD) method (McCammon and Harvey, 1987). It is a simulation procedure consisting of the computation of the motion of atoms in a molecule according to Newton's laws of motion. The forces acting on the atoms, required to simulate their motions, are generally calculated using molecular mechanics force fields. Rather than being confined to a single low-energy conformation, MolD allows the sampling of a thermally distributed range of intramolecular conformation. Molecular dynamics calculations provide information about possible conformations, thermodynamic properties, and dynamic behavior of molecules according to Newtonian mechanics. A simulation first determines the force on each atom (F_i) as the function of time, equal to the negative gradient of the potential energy (V) with respect to the position (x_i) of atom i:

$$F_i = \delta V / \delta x_i$$

The acceleration a_i of each atom is determined by

$$a_i = F_i / m_i$$

The change in velocity v_i is equal to the integral of acceleration over time. In MolD, one numerically and iteratively integrates the classical equations of motion for every explicit atom N in the system by marching forward in time by tiny time increments, Δt. A number of algorithms exist for this purpose (Brooks et al., 1988; McCammon and Harvey, 1987), and the simplest formulation is shown below:

$$x_i(t + \Delta t) = x_i(t) + v_i(t)\Delta t$$

$$v_i(t + \Delta t) = v_i(t) + a_i(t)\Delta t = v_i(t) + \{F(x_1 \ldots x_N, t)/m\}\Delta t$$

The kinetic energy (K) is defined as

$$K = 1/2\Sigma m_i v_i$$

The total energy of the system, called the Hamiltonian (H), is the sum of the kinetic (K) and potential (V) energies:

$$H(r, p) = K(p) + V(r)$$

where p is the momenta of the atoms and r is the set of Cartesian coordinates.

The time increment must be sufficiently small that errors in integrating $6N$ equations ($3N$ velocities and $3N$ positions) are kept manageably small, as manifested by conservation of the energy. As a result, Δt must kept on the order of femtoseconds (10^{-15} s). Furthermore, because the forces F must be recalculated for every time step, MolD is a computation intensive task. Thus, the overall time scale accessible to MolD calculations is on the order of picoseconds ($1\,\text{ps} = 1 \times 10^{-12}\,\text{s}$). The molecular simulation approximates the condition in which the total energy of the system does not change during the equilibrium simulation. One way to test for success of a dynamics simulation and the length of the time step is to determine the change in kinetic and potential energies between time steps. In the microcanonical ensemble (constant number, volume, and energy), the change in kinetic energy should be of the opposite sign and exact magnitude as the potential energy.

The MolD normally consists of three phases: heating, equilibration, and cooling. To perform MolD, the structure is submitted to a minimization procedure to relieve any strain inherent in the starting positions of the atoms. The next step is to assign velocities to all the atoms. These velocities are drawn from a low-temperature Maxwellian distribution. The system is then equilibrated by integrating the equations of motion while slowly raising the temperature and adjusting the density. The temperature is raised by increasing the velocities of all of atoms. There is a simple analytical function expressing the relationship between kinetic energies of the atoms and the temperature of the system:

$$T(t) = 1/\{k_B(3N - n)\} \sum_{i=1}^{N} m_i|v_i|$$

where $T(t)$ = temperature of the system at time t

$(3N - n)$ = number of degrees of freedom in the system

v_i = velocity of atom i at time t

k_B = Boltzmann constant

m_i = mass of atom i

N = number of atoms in the system

This process of raising the temperature of the system will cover a time interval of 10–50 ps. The period of heating to the temperature of interest is followed by a period of equilibration with no temperature changes. The stabilization period will cover another time interval of 10–50 ps. The mean kinetic energy of the system is monitored; and when it remains constant, the system is ready for study. The structure is in an equilibrium state at the desired temperature.

The MolD experiment consists of allowing the molecular system to run free for a period of time, saving all the information about the atomic positions, velocities, and other variables as a function of time. This (voluminous) set of data is called *trajectory*. The length of time that can be saved during a trajectory sampling is limited by the computer time available and the speed of the computer. Once a trajectory has been calculated, all the equilibrium and dynamic properties of the system can be calculated from it. Equilibrium properties are obtained by averaging over the property during the time of the trajectory. Plots of the atomic positions as a function of time schematically depict the degree to which molecules are moving during the trajectory. The RMS fluctuations of all of the atoms in a molecule can be plotted against time to summarize the aggregate degree of fluctuation for the entire structure. The methods of MolD are becoming an important component for the study of protein structures in an effort to rationalize structural basis for protein activity and function.

Molecular dynamics simulations are efficient for searching the conformational space of medium-sized molecules, especially ligands in free and complexed states. Quenched dynamics is a combination of high-temperature molecular dynamics and energy minimization. For a conformation in a relatively deep local minimum, a room temperature molecular dynamics simulation may not overcome the barrier. To overcome barriers, conformational searches use elevated temperature (>600 K) at constant energy. To search conformational space adequately, simulations are run for 0.5–1.0 ps each at high temperature. For a better estimate of conformations the quenched dynamics should be combined with simulated annealing, which is a cooling simulation. Cooling a molecular system after heating or equilibration can (1) reduce stress on molecules caused by a simulation at elevated temperatures and take high-energy conformational states toward stable conformations and (2) overcome potential energy barriers and force a molecule into a lower energy conformation. Quenched dynamics can trap structures in local minimum. The molecular system is heated to elevated temperatures to overcome potential energy barriers and then cooled slowly to room temperature. If each structure occurs many times during the search, one is assured that the potential energy surface of that region has been adequately sought.

The molecular dynamics is useful for calculating the time-dependent properties of an isolated molecule. However, molecules in solution undergo collisions with other molecules and experience frictional forces as they move through the solvent. Langevin dynamics simulates the effect of molecular collisions and the resulting dissipation of energy that occur in real solvents without explicitly including solvent molecules by adding a frictional force (to model dissipative losses) and a random force (to model the effect of collisions) according to the Langevin equation of motion:

$$a_i = F_i/m_i - \gamma v_i + R_i/m_i$$

where γ is the friction coefficient of the solvent and R_i is the random force imparted to the solute atom by the solvent. The friction coefficient determines the strength of the viscous drag felt by atoms as they move through the medium, and γ is the friction coefficient related to the diffusion constant (D) of the solvent by $\gamma = k_B T/mD$. At low values of the friction coefficient, the dynamic aspects dominate and Newtonian mechanics is recovered as $\gamma \to 0$. At high values of γ, the random collisions dominate and the motion is diffusion-like.

Monte Carlo simulations are commonly used to compute the average thermodynamic properties of a molecule or a molecular system especially the structure and equilibrium properties of liquids and solutions (Allen and Tildesley, 1967). They have also been used to conduct conformational searches under nonequilibrium conditions. Unlike MolD or Langevin dynamics, which calculate ensemble averages by calculating averages over time, Monte Carlo calculations evaluate ensemble averages directly by sampling configurations from the statistical ensemble. To generate trajectories that sample commonly occurring configurations, the Metropolis method (Metropolis et al., 1953) is generally employed. Thermodynamically, the probability of finding a system in a state with ΔE above the ground state is proportional to $\exp(-\Delta E/kT)$. Thus, if the energy change associated with the random movement of atoms is negative, the move is accepted. If the energy change is positive, the move is accepted with probability $\exp(-\Delta E/kT)$.

14.2.3. Conformational Search

Conformational search is a process of finding low-energy conformations of molecular systems by varying user-specified dihedral angles. The method involved variation of dihedral angles to generate new structures and then energy minimizing each of these angles. Low-energy unique conformations are stored while high-energy duplicate structures are discarded. Because molecular flexibility is usually due to rotation of unhindered bond dihedral with little change in bond lengths or bond angles, only dihedral angles are considered in the conformational search. Its goal is to determine the global minimum of the potential energy surface of a molecular system. Several approaches have been applied to the problem of determining low-energy conformations of molecules (Howard and Kollman, 1988). These approaches generally consist of the following steps with differences in details:

1. *Selection of an initial structure:* The initial structure is the most recently accepted conformation (e.g., energy minimized structure) and remains unchanged during the search. This is often referred to as a *random walk* scheme in Monte Carlo searches. It is based on the observation that low-energy conformations tend to be similar, therefore starting from an accepted conformation tends to keep the search in a low-energy region of the potential surface. An alternative method, called the *usage-directed method*, seeks to uniformly sample a low-energy region by going through all previously

accepted conformations while selecting each initial structures (Chang et al., 1989). Comparative studies have found the usage-directed scheme to be superior for quickly finding low-energy conformations.

2. *Modification of the initial structure by varying geometric parameters:* The variations can be either systematic or random. Systematic variations can search the conformational space exhaustively for low-energy conformations. However, the number of variations becomes prohibitive except for the simplest systems. One approach to reduce the dimensionality of systematic variation is to first exhaust variations at a low resolution, then exhaust the new variations allowed by successively doubling the resolution. Random variations choose a new value for one or more geometric parameters from a continuous range or from sets of discrete values. To reduce the number of recurring conformations, several random variations have some sort of quick comparison with the sets of previous structures prior to performing energy minimization of the new structure.

3. *Geometry optimization of the modified structure to energy-minimized conformations:* The structures generated by variations in dihedral angles are energy-minimized to find a local minimum on the potential surface. Although the choice of optimizer (minimizer) has a minor effect on the conformational search, it is preferable to employ an optimizer that converges quickly to a local minimum without crossing barrier on the potential surface.

4. *Comparison of the conformation with those found previously:* The conformation is accepted if it is unique and its energy satisfies a criterion. Two types of criteria are used to decide an acceptance of the conformation. Firstly, geometric comparisons are made with previously accepted conformations to avoid duplication. Conformations are often compared by the maximum deviation of torsions or RMS deviation for internal coordinates, interatomic distances, or least-squares superposition of conformers. Because geometry optimization can invert chiral centers, the chiral centers of the modified structures should be checked after the energy minimization. Secondly, the energetic test for accepting a new conformer may be carried out by a simple cutoff relative to the best energy found so far or a Metropolis criterion where higher-energy structures are accepted with a probability determined by the energy difference and a temperature, for example, $\exp(-\Delta E/kT)$.

14.3. COMPUTATIONAL APPLICATION OF MOLECULAR MODELING PACKAGES

Published FF parameters for MM (Jalaie and Lipkowitz, 2000; Osawa and Lipkowitz, 1995) and software for molecular modeling (Boyd, 1995) have been compiled. Some of the MM programs applicable to molecular modeling of biomolecules are listed in Table 14.1.

All of these programs are available for Unix operating system. Most of the Windows versions are incorporated into commercial molecular modeling packages such as MM in Chem3D of CambridgeSoft (http://www.camsoft.com), AMBER and CHARMm in HyperChem of HyperCube (http://www.hyper.com), and SYBYL

TABLE 14.1. Molecular Modeling Programs of Biochemical Interest

MM/FF program	Source	Reference
AMBER	University of California, San Francisco/HyperChem	Weiner et al. (1984)
CHARMM	Harvard University/Accelrys, Inc.	Brooks et al. (1983)
ECEPP	Cornell University	Nemathy et al. (1983)
GROMOS	University of Groningen/Biomos	Herman et al. (1984)
SYBYL	Tripos, Inc.	Clark et al. (1989)
MM2/3	University of Georgia/Chem3D	Allinger et al. (1989)
MACROMODEL	Columbia University	Chemistry, Columbia University
OPLS	Yale University/HyperChem	Jorgensen and Tirado-Rives (1988)

in PC Spartan of WaveFunction (http://www.wavefun.com). Online servers such as SWEET (http://dkfz-heidelberg.de/spec/sweet2/doc/mainframe.html) performs MM2/3, AMBER server (http://www.amber.ucsf.edu/amber/amber.html) conducts *in vacuo* minimization with AMBER 5.0 and then electrostatic solvation with AMBER 6.0. B server (http://www.scripps.edu/nwhite/B/indexFrames.html) implements AMBER, and Swiss-Pdb Viewer (http://www.expasy.ch/spdbv/mainpage.html) executes GROMOS.

14.3.1. Carbohydrate Modeling at SWEET

The online Sweet (http://dkfz-heidelberg.de/spec/sweet2/doc/mainframe.html) is a program for constructing 3D models of saccharides (Bohne et al. 1998; Bohne et al. 1999) with an extension to peptides and other simple biomolecules. To perform energy minimization of monosaccharides such as α-D-mannose, α-L-fructose, β-D-galactose, α-D-glucose, β-D-glucosamine, and their oligomers:

- Select Work page of beginner version to open the request form.
- Select monosaccharide unit(s) and glycosidic linkage(s) from the pop-up lists.
- Click the Send button to open the Result table.
- Choose desired tools to view the saccharide structure and save the structure as PDB file.
- Select Method options with Full MM3(96) parameters and Gradient with 1.0 (default) of Optimize tool.
- Click the Optimize button.

To perform energy minimization of custom-made saccharides, peptides and other biomolecules (a list of templates is available from http://www.dkfz-heidleberg.

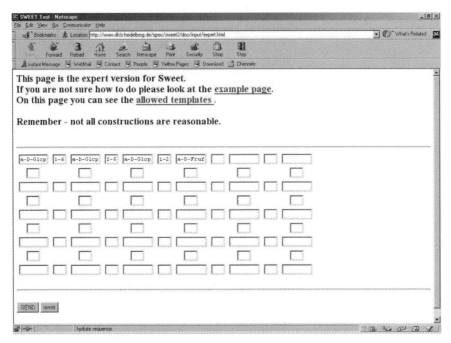

Figure 14.1. Construction of query saccharide structure for energy minimization with SWEET. Three-letter codes (IUBMB) for monosaccharides and amino acids are employed to construct oligosaccharide or oligopeptide chains. For monosaccharides, α- and β-anomers are prefixed with *a* and *b*, respectively. Pyranose and furanose rings are denoted as *p* and *f*, respectively.

de/spec/sweet2/doc/input/liblist.html, but not all templates have parameters for minimization):

- Select expert version to open the request form (Figure 14.1).
- Enter the compound name (e.g., a-D-Glcp for α-D-glucopyranose, b-D-Fruf for β-D-fructofuranose, three-letter symbols for amino acids) and the linkage (e.g., 1–4 for 1,4-glycosidic and N–C for peptide linkages).
- Click the Send button to access the Result table.
- From the Optimize tool, select Full MM3(96) parameters from Method option and 1.0 (default) from Gradient option.
- Click the Optimize button to access the Result table from which you may view and save the structure.

14.3.2. Folding of Nucleic Acids at mfold

RNA molecules form secondary structure by folding their polynucleotide chains via hydrogen bond formations between AU pairs and GC pairs. The thermochemical stability of forming such hydrogen bonds provides useful criterion for deducing the cloverleaf secondary structure of tRNAs; that is, tRNA molecules are folded into DH arm, φTC arm, anticodon arm, and extra loop (for some) with the hydrogen bonded stems. Some of the unpaired bases in these arms are hydrogen-bonded in the tertiary structures. The nucleotide sequence of RNA in fasta format can be submitted to RNA mfold server (http://bioinfo.math.rpi.edu/~mfold/rna/form1.cgi) with up to 500 bases for an immediate job and 3000 for batch submission (Mathew et al., 1999).

Enter the sequence name, paste the sequence in the query box, select or accept default options, and click the Submit button. The results (thermochemical data in text and plot files, and structure files in various formats including PostScrip) are returned (Figure 14.2). Analogous folding prediction for DNA is available at http://bioinfo.math.rpi.edu/~mfold/dna/form1.cgi.

14.3.3. Application of Chem3D

Chem3D is the molecular modeling software for desktop computer marketed by CambridgeSoft Corp. (http://www.camsoft.com) as a component of the ChemOffice suite (ChemDraw, Chem3D and ChemFinder). It converts 2D structures to 3D renderings and imports PDB, MSI ChemNote, Mopac, and Gaussian files. The program displays molecular surfaces, orbitals, electrostatic potential, and charge densities, and it performs energy calculation with MM2 force field. The *ultra* version includes MOPAC, which calculates transition state geometries and physical properties using AM1, PM3, and MNDO's. A conformational search program, Conformer (Princeton Simulations), can be integrated with Chem3D to perform conformational search and analysis. User's guides for ChemDraw and Chem3D should be consulted.

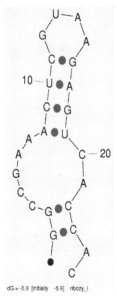

Loop Free-Energy
Decomposition
Structure 1
Initial dG = -5.90

Structural element	$\delta\delta$G	Information
External loop	-1.70	2 ss bases & 1 closing helices
Stack	-3.30	External closing pair is G^1-C^{23}
Helix	-3.30	2 base pairs
Interior loop	4.40	External closing pair is G^2-C^{22}
Stack	-2.20	External closing pair is A^8-U^{19}
Stack	-2.10	External closing pair is C^9-G^{18}
Stack	-2.40	External closing pair is U^{10}-A^{17}
Helix	-6.70	4 base pairs
Hairpin loop	1.40	Closing pair is C^{11}-G^{16}

Figure 14.2. RNA fold to form secondary structure. The nucleotide sequence of ribozyme (B chain of URX057 from NDB) is submitted to RNA mfold server for fold analysis. The output includes computed structure (as shown) and thermodynamic data in text (as shown) and dot plot (not shown).

The opening window of Chem3D consists of the workspace (display window where 3D structures are displayed with rotation bar, slider knob, and action buttons), the menu bar (File, Edit, View, Tools, Object, Analyze, MM2, Gaussian, MOPAC, and Window menus), tool pallette (action icons for the cursor), and replacement text box (element, label, or structure name typed in this box is converted to chemical structure). Structure file in *.mol, *.pdb, or *.sml can be opened and saved from the File menu. (*Note:* PDB files saved from Chem3D do not contain residue IDs.) The accompanying program, ChemDraw, draws 2D structures (.cdx) that are converted into 3D models (.c3d) by Chem3D. The molecular sketches from ISIS Draw (*.skc) have to be converted to *.cdx with ChemDraw for the 3D conversion.

To draw biomolecules using ready-made substructures of Chem3D, go to the View menu (to view tables of topologies, parameters, force fields used in the program, and substructures for constructing 3D) and then to the substructure table (*.TBL) to select the desired substructures. Copy and paste the substructures on the display window for subsequent modeling. Alternatively, 3D models can be built by entering substructure names into the Replacement text box (upper left-hand corner of display window), such as HSerThreAlaAsnLeuGluTyrOH for heptapeptide, STANLEY. Invoke Tools→Clean up structure to quickly correct unrealistic bond lengths and angles (Figure 14.3).

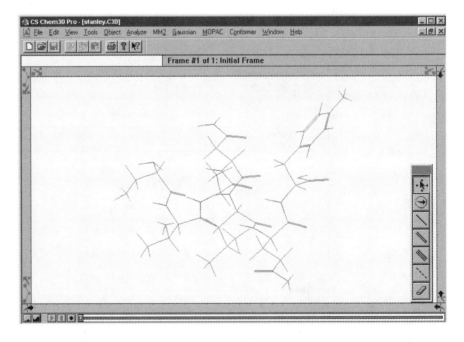

Figure 14.3. Display of 3D structure with Chem3D. Heptapeptide, STANLEY, is constructed by entering HSerThreAlaAsnLeuGluTyrOH into the replacement box (left side, under icons). Alternatively an ISIS Draw, stanldy.skc can be converted to stanley.cdx (Chem-Draw) and then imported into Chem3D. The atomic coordinates file is saved as stanley.pdb and can be edited for protein modeling. (*Note:* The file does not contain ATOM columns with residue ID.)

The Tools menu contains commands for manipulating the displayed structures such as Fit, Move, Reflect, Invert, Dock, and Overlay. To dock two interacting structures,

- Copy the first structure into clipboard.
- Open the second structure on the display window and paste the first structure upon the second.
- Select two atoms (one each from the two structures) as the first interaction pair, and set the interaction distance between them (Open the measurement table via Object→Set distance, and enter the desired interaction distance in the Optimal cell).
- Repeat the procedure (setting interaction distance between the two interacting atoms) for, at least, four interaction pairs in order to achieve reasonable dock.
- Invoke Dock command from the Tools menu.
- Set the values of the minimum RMS error and minimum RMS gradient to 0.01.
- Click Start to initiate docking computation, which terminates when either RMS error or RMS gradient becomes less than the set values.
- Compute the MM2 energy (MM2→Minimize energy) and save the file (Figure 14.4).

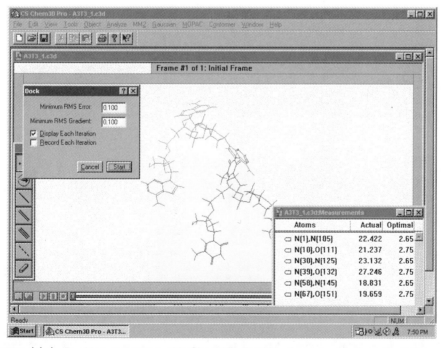

Figure 14.4. Docking structures with Chem3D. Docking (merging files) of dApApAp and dTpTpTp is accomplished by specifying distances between hydrogen bonding N_1 and N_6 of A with N_1 and O_6 of T, respectively, as demonstrated in the lower right inset. The upper left inset shows the dock dialog box.

Follow the similar procedures to overlay two structures for structural comparison. Specify, at least, three similarity atom pairs in order to obtain the acceptable
overlay. It is not necessary to give the bond distance between the two atoms in the
pair (assume to be zero) for the overlay. Choose Overlay command from the Tools
menu to set both RMS error and gradient to 0.01 and click start to initiate the
overlay computation.

The Analyze menu provides commands for measuring geometries. The MM2,
MOPAC, and Gaussian (if Gaussian program is installed) menus contain related
commands to execute energy computations. These include Run MM2/MOPAC job
(for single point energy calculation), Minimize energy, Molecular dynamics, and
Compute properties (dipole, charge, solvation, electrostatic potential, polarizability,
etc. by Mopac) commands. For energy minimization,

- Choose Minimize energy command from the MM2 menu to open the dialog
  box.
- With Job type tab active, select minimize energy, display options and set
  minimum RMS to 0.100.
- Click Run to initiate minimization.
- At the end of run, various energies (stretch, bend, torsion, van der Waals, and
  dipole–dipole) and total energy is displayed (Figure 14.5).
- Save the record and structure as emstruct.c3d.

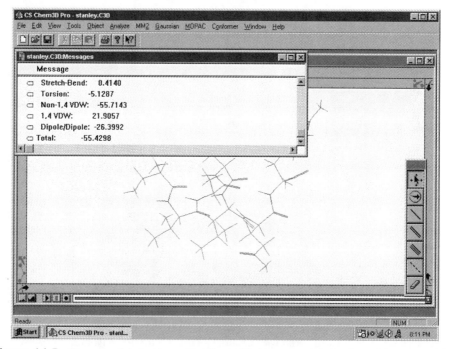

Figure 14.5. Energy minimization with Chem3D. Energy minimization of heptapeptide,
STANLEY, is carried out with Chem3D using MM2 force field. The inset (Message window)
shows the minimization result.

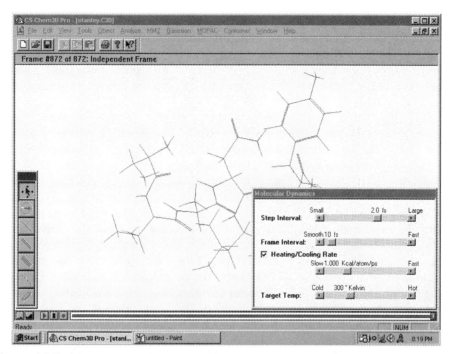

Figure 14.6. Molecular dynamics with Chem3D. Molecular dynamics of heptapeptide, STANLEY, is performed with Chem3D using MM2 force field. The inset shows the following conditions: 2.0 fs for step interval, 10 fs for frame interval, 1.0 kcal/atom/ps for heating/cooling rate, and 300 K as the target temperature. The display is taken midway through the dynamic simulation.

For dynamic simulation,

- Choose Molecular dynamics command from the MM2 menu to open the dialog box.
- With Dynamics tab active, enter simulation parameters (default settings: step interval, 2.0 fs; frame interval, 10 fs; terminate after 1000 steps, heating/cooling rate, 1.0 kcal/atom/ps; and target temperature, 300 K).
- Click Run button to initiate the dynamic simulation (Figure 14.6).
- The simulation is terminated when the target temperature is reached. You can view the energetics of the simulation from the message window and the trajectories (structural frames) by moving the slider knob at the bottom of the model window.
- Save the simulation record and trajectories as dynamics.c3d.
- To save an individual trajectory, pick the desired frame, choose Edit→Select all, and save as (File→Save as) trajname.pdb.

14.3.4. Application of HyperChem
HyperChem is the PC-based molecular modeling and simulation software package marketed by Hypercube Inc. (http://www.hyper.com). The program provides molecular mechanics (with MM+, AMBER, BIO+ (CHARMm), and OPLS force

fields), semiempirical (extented Hückel, CNDO, INDO, MINDO3, MNDO, AM1, PM3, and ZINDO), and *ab initio* quantum mechanics calculations. Computations are carried out for single-point energy, geometry optimization (energy minimization), molecular dynamics, Langevin dynamics, Monte Carlo simulation, and conformational search. The accompanied manuals should be consulted for various applications of HyperChem.

The HyperChem window displays the menu bar (File, Edit, Build, Select, Display, Databases, Setup, Compute, Annotation, Cancel, and Scipt menus), tool bar (draw, select, display, move and shortcut icons), Workspace, and status line. The program supports various 2D and 3D files including HyperChem (*.hin), PDB (pdb files in .ent), MDL (*.mol), Tripos (*.mlz) files. ISISDraw (*.skc) can also be opened or saved from the File menu. The 2D sketch can be converted into a 3D structure (and calculation of atom types for MM) by selecting Model build from the Build menu. The Build menu also provides tools for building the structure. United atoms tool simplifies a molecular structure and calculations by including hydrogen atoms in the definition of carbon atoms. HyperChem uses atom types for molecular mechanical calculations. The atom types can be calculated (Calculate types) or changed (Set atom type). The Edit menu contains items for manipulating displayed structure while the Display menu determines the appearance of molecules in the workspace (e.g., Show selection, Rendering, Overlay, Show isosurface, Show periodic box, Show multiple bonds, Show hydrogen bonds, Recompute H bonds, add Labels, and change Color). Rendering displays model rendering in sticks (stereo, ribbons, and wedges), balls, balls and cylinders, overlapping spheres, dots, and sticks and dots (Figure 14.7). Two selected molecules can be placed on top of one another by using Overlay tool of the Display menu. The Select menu enables the selection of Atoms, Residues, Molecules, and Spheres that encompasses all atoms within a 3D sphere plus all atoms with/without 2D rectangle.

The Setup and Compute menus contain tools for carrying out chemical calculations. For molecular mechanics energy computation,

- Select atoms or residues to be included in the calculation (default is the whole molecule).
- Choose Start Log command from the File menu if you want to store energy calculations in a log file (chem.log as default).
- Choose force field (MM +, AMBER, BIO + (CHARMm), or OPLS) from the Setup menu.
- Click the Options button to open the option dialog box.
- For MM + (energy calculations of small biomolecules or ligands): Choose either Bond dipoles or Atomic charges (assigned via Build→Set charge) for use in the calculations of nonbounded Electrostatic interactions. Select None (calculate all nonbonded interactions recommended for small molecules), Switched or Shifted for Cutoffs (for large molecules).
- For AMBER, BIO +, or OPLS (energy calculation of biomacromolecules): Choose Constant (for systems in a gas phase or in an explicit solvent) or Distance dependent (to approximate solvent effects in the absence of an explicit solvent) and set Scale factor for Dielectric permittivity ($\geqslant 1.0$ with the default of 1.0 being applicable to most systems). Select either Switched or Shifted for Cutoffs and set Electrostatic (the range is 0 to 1; use 0.5 for

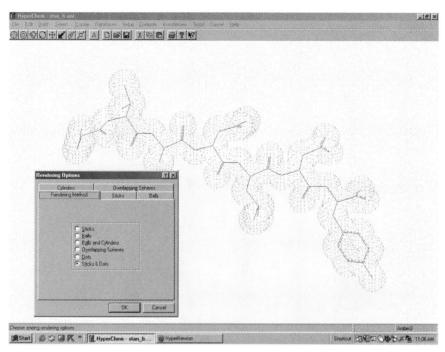

Figure 14.7. Display of molecular structure with HyperChem. Heptapeptide, STANLEY, is displayed in sticks and dots representation. The inset shows the dialog box for rendering options.

AMBER and OPLS, and use 0.4, 0.5, or 1.0 for BIO+) and van der Waals (the range is 0 to 1; use 0.5 for AMBER, 1.0 for BIO+, and 0.125 for OPLS). Different parameter sets are available for the AMBER force field (AMBER 2, 3, 94, or 96), which can be selected from Setup/Parameter of Force Field options (Figure 14.8).

- After choosing Setup menu options, execute chemical calculations from Compute menu commands to perform one of the followings.

   (a) Single point: calculates the total energy and the RMS gradient.

   (b) Geometry optimization: executes energy minimization, and calculates an optimum molecular structure with lowest energy and smallest RMS gradient.

   (c) Molecular dynamics: calculates the motion of selected atoms over picosecond intervals to search for stable conformations.

   (d) Langevin dynamics: calculates the motion of selected atoms over picosecond intervals using frictional effects to simulate the presence of a solvent.

   (e) Monte Carlo: calculates ensemble averages for selected atoms.

Additional commands are available for semiempirical and *ab initio* calculations. These include:

   (a) Vibration (calculates the vibrational motions of selected atoms)

   (b) Transition state (searches for transition states of reactant or product atoms)

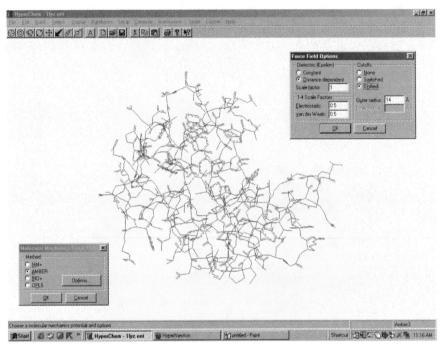

Figure 14.8. Molecular mechanics calculation with HyperChem. Setup for MM calculation of lysozyme (pdb1lyz.ent) includes selection of the method and options. The lower left inset displays dialog box for MM methods and the upper right inset shows dialog box for force field options.

(c) Plot molecular properties (displays the electrostatic potential, total spin density or total charge density)

(d) Orbitals (analyzes and displays orbitals and their energy levels)

(e) Vibrational spectrum (analyzes and displays the vibrational frequencies)

(f) Electronic spectrum (analyzes and displays the ultraviolet-visible spectrum)

For energy minimization,

- Select Compute→Geometry optimization to open Molecular mechanics optimization dialog box (Figure 14.9).
- Choose algorithm such as Steepest descent, Fletcher–Reeves (conjugate gradient), or Polak–Ribiere (conjugate gradient, default of HyperChem), and choose options for termination condition such as RMS gradient (e.g., 0.1 kcal/mol Å) or number of maximum cycles.
- Click OK to close the dialog box and start optimization.

To apply periodic boundary conditions for solvation,

- Select Setup→Periodic box to open Periodic box options box.
- Referring to the dimension given for the smallest box enclosing solute, assign the dimension for the Periodic box size (e.g., twice of the largest dimension for

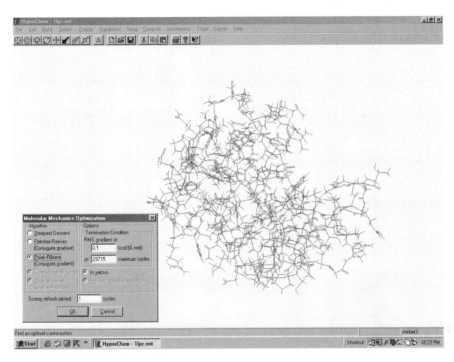

Figure 14.9. Geometry optimization with HyperChem. Setup for geometry optimization includes selection of algorithm and options. The inset shows the dialog box for these selections. Initially, *in vacuo* condition is chosen.

the given smallest box or a cube of 18.70 Å on the side recommended by HyperChem but not exceeding 56.10 Å on the side), maximum number of water molecules (for information), and minimum distance between solvent and solute atoms (practical range: 1–5 with the default of 2.3 Å).

- Click OK to close the option box and start optimization.

For molecular dynamics calculation,

- Select Compute→Molecular dynamics to open Molecular dynamics options dialog box.
- Set heat time (e.g., 5 ps), run time (at equilibrium, e.g. 5 ps), step size (e.g., 0.005 ps), starting temperature (e.g., 0 K), simulation or final temperature (e.g., 300 K), and temperature step (e.g., 30 K).
- Choose *In vacuo* (for the system not in a periodic box) or Periodic boundary conditions (for the system in a defined periodic box),
- Set Bath relaxation time (suggested range: step size to the default of 0.1 ps) and assign any number between − 32,768 and 32,768 for Random seed as the starting point for the random number generator used for the simulations. Friction coefficient (any positive value) is needed only for the Langevin dynamics.
- Click the Snapshots button if you want playback of MolD trajectories (saved in a movie file .avi).

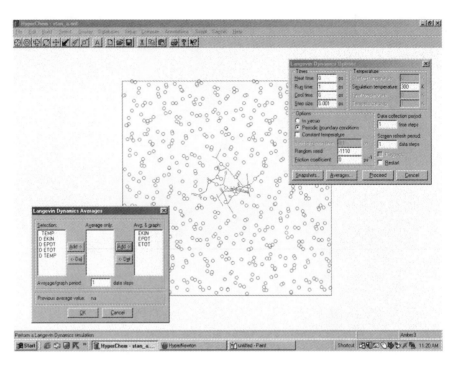

Figure 14.10. Molecular dynamics with HyperChem. Setup for MD under periodic boundary conditions includes defining the periodic boundary box, selecting molecular dynamics options, and setting dynamics average output. The setup for Langevin dynamics of heptapeptide (STANLEY) is illustrated (note that the periodic boundary conditions radio is checked). The upper right inset shows the dialog box for Langevin dynamics options, and the lower left dialog box depicts the dialog box (opened by clicking Average button of the MD options dialog box) for dynamics average output. Molecular dynamics calculation is initiated by clicking Proceed button of the MD options dialog box.

- Click the Average button to select average values of kinetic energy (EKIN), potential energy (EPOT), total energy (ETOT), and their RMS deviations and named selections (user selected interatomic distances, angles or torsion angles) to save (default, chem.csv) and plot after the simulation.
- Analogous procedures are applied to the Langevin dynamics (via Compute→ Langevin dynamics) as shown in Figure 14.10 and Monte Carlo simulation (via Compute→Monte Carlo).

The biopolymer modeling of HyperChem includes Building polynucleotides, polypeptides and polysaccharides, Amino acid sequence (fasta format) editing, Mutations, Overlapping by RMS fit, and Merging structures. To facilitate manipulation of protein structures, there is often a need to display the protein backbone only as follows.

- Open PDB file (*.ent) from the File menu.
- Set the select level to Molecule and use selection tool to click on the protein molecule.

- Change the selection to water molecules by choosing Complement selection on the Select menu.
- Choose Clear command on the Edit menu to remove water molecules and to display the protein structure.
- Turn off Show Hydrogens on the Display menu.
- Choose Select Backbone on the Select menu and the Show Selection Only on the Display menu to display the backbone of the polypeptide chain.

The Databases menu provides tools for building polypeptides (Amino Acids, Make Zwitterion, Sequence Editor), polynucleotides (Nucleic Acids), polysaccharides (Saccharides), and organic polymers (Polymers) from residues (monomer units) as exemplified for DNA in Figure 14.11.

To build protein structure,

- Select Amino Acids from the Databases menu to open the amino acid dialog box.

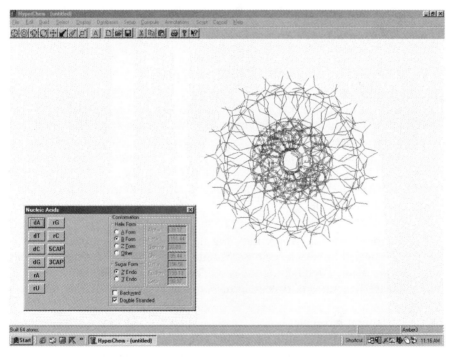

Figure 14.11. Construction of biopolymer with HyperChem. Two menus are available for creating 3D structure models in HyperChem. The Build menu provides tools for creating organic molecules. Use the Drawing tool to sketch atoms in a molecule and connect them with covalent bonds. Invoke the Model builder to create a 3D structure from the 2D sketch. The Databases menu offers tools for creating biopolymers from residues with user specified linkages and conformations—that is, polysaccharides from monosaccharides, polypeptides form amino acids, and polynucleotides from nucleotides. A double-stranded DNA chain, for example, is constructed from nucleotide residues in a desired conformation (the inset).

- Choose chain conformation (Alpha Helix or Beta Sheet or Other) and isomer (L or D), and pick amino acids from the N-terminus.
- Close the dialog box to complete the chain.

To construct nucleic acids,

- Choose Nucleic Acids on the Databases menu.
- From the dialog box, choose the helical conformation of nucleic acid (A, B, Z, or other form) and add nucleotides (dA, dT, dG, and dC for DNA; rA, rU, rG, and rC for RNA) in the direction of 5′ to 3′ (default) or a choice of Backward and Double stranded. Both termini can be capped (5′ Cap and 3′ Cap).

To construct polysaccharides,

- Select Saccharide from the Display menu to open the Sugar builder window.
- Choose Add menu from the Sugar builder window to open the dialog box for linking monosaccharides (Aldoses, Ketoses, Derivatives, or End groups).
- Each selection opens another dialog box with options for choosing specific saccharide residues in pyranose or furanose forms, anomer (α, β, or acyclic), isomer (D or L), and type of link (linkage type).
- Add (pick) saccharide residues from the list of aldoses (hexoses in aldopyranose form, pentoses in aldofuranose form, and tetraoses in open-chain form), ketoses (hexoses in ketofuranose form, pentoses and tetraose in open-chain form), derivatives (glucosamine, galactosamine, N-acetylnuraminic acid, N-acetyl muramic acid, inositol, 2-deoxyribose, rhamnose, fucose, and apiose), and blocking groups (H, NH_2, $=O$, COO—, methyl, lactyl, O-methyl, N-methyl, O-acetyl, N-acetyl, phosphoric acid, sulfate, N-sulfonic acid) to build polysaccharides.

The Databases menu also permits the mutation of the selected amino acid residue(s) of a protein molecule. Choose Mutate from the Database menu to open a listing of amino acids. Highlighting the candidate amino acid effects the mutation.

To compare structures of two molecules by overlapping,

- Open the first molecule.
- Choose Merge command from the File menu to open the second molecule.
- Set different colors for the two structures (Display→Color after selecting the molecule).
- Select matching residue(s) from each of the two molecules via Select→Residue→Select to open a dialog box.
- Enter residue number under By number/Residue number. (Repeat the process in order to use more than one residue for the overlaying.)
- Choose RMS Fit and Overlay on the Display menu and pick the desired overlaying option (Molecular numbering or Selection order). This places one molecule on top of another (Figure 14.12).

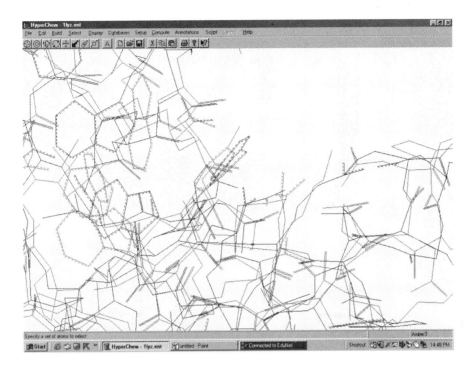

Figure 14.12. Superimposition of molecular structures with HyperChem. Pigeon lysozyme structure (red) derived from homology modeling with Swiss-PDB Viewer is overlapped against chicken lysozyme structure, pdb1lyz.ent (black). Two catalytic residues, Glu35 and Asp 52 (chicken lysozyme), are highlighted (green).

14.4. WORKSHOPS

1. Aldopentose and aldohexose may exist in furanose and pyranose forms. The former favors the furanose form while the latter prefers the pyranose form. Would the energetic differences between the two forms as exemplified by the following two pairs of monosaccharides be sufficient to rationalize the preferences.

(a) β-D-Ribofuranos versus β-D-ribopyranose

(b) α-D-Glucofuranose versus α-D-glucopyranose

2. Both purine and pyrimidine bases tautomerize between enol form and keto form. Calculate the energy of tautomerization for thymidine.

Enol   Keto

3. Arachidonic acid ($5E, 8E, 11E, 14E$-eicosatetraenoic acid) is a precursor for the biosynthesis of prostaglandins. Apply torsion angle rotations involving bond C9–C10 and C10–C11 to search for the prostaglandin-like conformation.

4. The nicotinamide adenosine dinucleotide (NAD^+) exits in *anti-* and *syn-* conformations:

anti   *syn*

Compare their conformational energies as affected by solvation under periodic boundary conditions.

5. Monosaccharide units are linked together by different glycosidic linkages to form oligosaccharides of diverse structures and conformations. Explore this diversity by performing geometry optimization of the following groups of oligosaccharides of D-glucopyranose (Glc*p*):

(a) α,α-Trehalose [α-Glc*p*-(1→1) αGlc*p*], maltose [αGlc*p*-(1→4) αGlc*p*], and isomaltose [αGlc*p*-(1→6) αGlc*p*]

(b) Hexasaccharides of Glc*p* linked by α-1,4 linkages, [αGlc*p*-(1→4) αGlc*p*]$_3$, versus those linked by β-1,4 linkages, [βGlc*p*-(1→4) βGlc*p*]$_3$

6. The double-stranded structure of DNA is stabilized by the hydrogen bonds formed between A and T pairs and between G and C pairs. Model the hydrogen bond interactions between $=O\cdots HN<$ of purine and pyrimidine bases for the AT pair at 2.70 ± 0.05 Å (between N and N or O at positions 1 and 6) and the GC pair at 2.95 ± 0.05 Å (between N and N or O at positions 1, 2 and 6).

(a) dAMP and TMP

(b) dGMP and dCMP

Perform geometry optimization until RMS gradient is equal to or less than 0.100. Formulate your conclusions with reference to their interaction energies.

7. Compare energies and conformation of the following pairs of biomolecules by performing geometry optimization (steepest descent then conjugated gradient until the energy difference is less than 1 kcal/mol) and molecular dynamics (heating to 300 K to be followed by equilibration for 5 ps at 300 K).

(a) Met-enkephalin (TyrGlyGlyPheMet) versus Leu-enkephalin (TyrGlyGly-PheLeu)

(b) Double-stranded hexadeoxyribonucleotides of AT chains versus GC chains

8. Retrieve nucleotide sequences (fasta files) of yeast cytosolic and mitochondrial Gly-tRNA and submit them to RNA folding to obtain their secondary (cloverleaf) structures and thermochemical data of foldings.

9. Retrieve nucleotide sequence (fasta file) and atomic coordinates (pdb file) of yeast Asp-tRNA. Perform folding analysis/molecular modeling to display graphics of the following:

(a) Secondary structure showing the maximum hydrogen bond formations

(b) Tertiary structure with highlighted anticodon

(c) Tertiary structure with molecular surface plot

(d) Tertiary structure with electrostatic potential plot

10. Deduce the probable conformations for the two chains of a hammerhead ribozyme with the following sequences by performing folding analysis, geometry optimization and dynamic simulation (heating to and equilibrate at 300 K for 5 ps).

Chain A: GUGGUCUGAUGAGGCC
Chain B: GGCCGAAACUCGUAAGAGUCACCAC

Compare the modeled structure with that of X-ray crystallographic structure (299D.pdb).

REFERENCES

Allen, M. P., and Tildesley, D. J. (1967) *Computer Simulation of Liquids.* Oxford University Press, New York.

Allinger, N. L., Yuh, Y. H., and Lii, J. H. (1989). *J. Am. Chem. Soc.* **111**:8551–8566.

Altona, C., and Faber, D. H. (1974). *Top. Curr. Chem.* **45**:1–38.

Berkert, U., and Allinger, N. L. (1982). *Molecular Mechanics.* American Chemical Society, Washington, D.C.

Bohne, A., Lang, E., and von der Lieth, C.-W. (1998). *Mol. Model.* **4**:33–43.

Bohne, A., Lang, E., and von der Lieth, C.-W. (1999). *Bioinformatics.* **15**:767–768.

Boyd, D. B., and Lipkowitz, K. B. (1982). *J. Chem. Ed.* **59**:269–274.

Boyd, D. B. (1995) *Rev. Comput. Chem.* **6**:383–417.

Brooks, B. R., Bruccoleri, R. E., Olafson, B. D., States, D. J., Swaminathan, S., and Karplus, M. (1983). *J. Comput. Chem.* **4**:187–217.

Brooks C. I., III, Karplus, M., and Pettitt, B. M. (1988). *Proteins: A Theoretical Prospective of Dynamics, Structure and Thermodynamics.* John Wiley & Sons, New York.

Chang, G., Guida, W. C., and Still, W. C. (1989) *J. Am. Chem. Soc.* **111**:4379–4386.

Clark, M., Cramer R. D., III, and van Opensch, N. (1989) *J. Comput. Chem.* **10**:982–1012.

Columbia University, Department of Chemistry, http://www.cc.columbia.edu/cu/chemistry/mmod/mmod.html/. MACROMODEL, v4.5

Gogonea, V., Urez, D. S., van der Vaart, A., and Merz, K. M. Jr. (2001) *Curr. Opin. Struct. Biol.* **11**:217–223.

Hehre, J. W., Radom, L., Schleyer, P., and Pople, J. (1986). *Ab Initio Molecular Orbital Theory.* John Wiley & Sons, New York.

Herman, J., Berendsen, H. J. C., van Gunsteren, W., and Postma, J. P. M. (1984). *Biopolymers.* **23**:1513–1518.

Höltje, H.-D., and Folkeis, G. (1997) *Molecular Modeling: Basic Principles and Applications.* VCH, New York.

Howard, A. E., and Kollman, P. A. (1988) *J. Med. Chem.* **31:**1669–1675.

Jalaie, M., and Lipkowitz, K. B. (2000) *Rev. Comput. Chem.* **14:**441–486.

Jorgensen, W. L., and Tirado-Rives, J. (1988) *J. Amer. Chem. Soc.* **110:**1657–1666.

Leach, A. R. (1996) *Molecular Modeling Principles and Applications.* Longman, Essex.

Mathew, D. H., Sabina, J., Zuker, M., and Turner, D. H. (1999) *J. Mol. Biol.* **288:**911–940.

McCammon, J. A., and Harvey, S. C. (1987) *Dynamics of Proteins and Nucleic Acids.* Cambridge University Press, New York.

Metropolis, N., Rosenbluth, A. W., Rosenbluth, M. N., Teller, A. H., and Teller, E. (1953) *J. Chem. Phys.* **21:**1087–1092.

Nemathy, G. , Porttle, M. S., and Scheraga, H. (1983) *J. Phys. Chem.* **87:**1882–1887.

Osawa, E., and Lipkowitz, K. B. (1995) *Rev. Comput. Chem.* **6:**355–381.

Schlecht, M. F. (1998) *Molecular Modeling on the PC.* Wiley-VCH, New York.

Warshel, A. (1991) *Computer Modeling of Chemical Reactions in Enzymes and Solutions.* John Wiley & Sons, New York.

Weiner, S. J., Kollmann, P. A., Case, D. A., Singh, U. C., Ghio, C., Alagona, G., Profeta, S. Jr., and Weiner, P., (1984). *J. Am. Chem. Soc.* **106:**765–784.

Weiner, S. J., Kollman, P. A., Nguyen, D. T., and Case, D. A. (1986) *J. Comput. Chem.* **7:**230–252.

Wiberg, K. B. (1965). *J. Am. Chem. Soc.* **87:**1070–1078.

15

MOLECULAR MODELING: PROTEIN MODELING

This chapter focuses on the application of molecular modeling to calculate, manipulate, and predict protein structures and functions. Concepts of structure similarity/overlap, homology modeling, and molecular docking, which are special concerns of protein biochemists, are considered. Approaches to protein modeling by the use of two programs (Swiss-Pdb Viewer and KineMage) and two online servers (B and CE) are described.

15.1. STRUCTURE SIMILARITY AND OVERLAP

15.1.1. General Consideration

With protein structures, we can observe similarities in topology or even in structural details. Comparison of protein structures can reveal distant evolutionary relationships that would not be detected by sequence alignment alone. In devising measures of similarity and difference between two proteins, it is sometimes clearer to note how to proceed in comparing sequences than to do so in comparing structures. If the amino acid sequences of two proteins can be aligned, then we can either count the number of identical residues or use a similarity index between amino acids. Such an index would take the form of a 20×20 matrix, M, such that each entry corresponds to a pair of amino acids and M_{ij} gives a measure of the similarity between any pair of amino acids. The similarity between two sequences is then the sum of values for each pair of aligned amino acids taken from the matrix, plus a correction to account for the gaps found in the sequences at sites of insertions or deletions of amino acids (Needleman and Wunsch, 1970). Indeed, it is by maximizing such a similarity score that the optimal alignment of two sequences is conventionally calculated (Chapter 11).

In three dimensions the problem can be more complex. If two protein structures are very closely related and we align the residues, it is then easy to superpose the corresponding residues and to measure and analyze the nature of the deviations in position of corresponding atoms. Structure tends to change more conservatively than sequence, and it is not uncommon to be able to recognize a relationship between proteins from their structures, when no evidence of homology appears in the sequences (Bränden, 1980; Levitt and Chothia, 1976). Using computer graphics, it is possible to superpose two (or more) structures in one picture, and this can reveal, immediately and obviously, which features are conserved and which are different between two related proteins or in the conformational changes during allosteric transition/upon complexation. It is this facility in which lies the real advantage of computer graphics.

There are two ways to superpose structures. One may display pictures of each structure, and use interactive graphics to provide the facility to rotate and translate one with respect to the other, superposing the two "by eye." The second method is numerical. By selecting corresponding sets of atoms from two structures, a program can calculate the best "least-square fit" of one set of atoms to another. Both methods are useful: Fitting by eye permits rapid experimentation to assess the goodness of fit of different portions of the molecules. Numerical fitting permits quantitative comparisons of the relative goodness of fit of different structures and substructures.

Suppose we are dealing with two or more protein structures that contain regions in which the backbone atoms are almost congruent. We wish to superpose the two structures by moving one with respect to the other so that the corresponding atoms in the well-fitting regions are optimally matched. There are two aspects to this problem:

1. Choosing the corresponding sets of atoms from the two structures.

2. Finding the best match of the corresponding atoms.

15.1.2. Rigid-Body Motions and Least-Squares Fitting

The general numerical approach employs rigid body motions and least-squares fitting. Given two sets of points: x_i, $i = 1, 2, \ldots, N$ and y_i, $i = 1, 2, \ldots, N$ (here x_i and y_i are vectors specifying atomic coordinates), find the best rigid motion of the points $y_i \to Y_i$ such that the sum of the squares of the deviations $\Sigma |x_i - Y_i|^2$ is a minimum.

Two mathematical facts that we apply without proof are:

1. The most general motion of a rigid body is a combination of a rotation and a translation.

2. At the minimum, the mean positions (colloquially, the centers of gravity) of the two sets of points coincide.

It follows that Y_i must be equal to $Ry_i + t$, where R is a rotation matrix and t is a translation vector, and that to minimize

$$\sum |x_i - (Ry_i + t)|^2$$

A program must choose t to move the set y_i so that its center of gravity coincides with that of the x_i, and determine the optimal rotation matrix R. These solutions provide the following:

1. The minimum value of the root-mean-square (RMS) deviation between corresponding atoms:

$$\Delta = [\sum \{x_i - (Ry_i + t)\}^2/N]^{1/2}$$

    This quantity is a useful measure of how similar the structures are.

2. A translation vector t and rotation matrix R: From the rotation matrix, one can derive the angle of rotation θ.

3. Assessment of the fit: By calculating the transformed points $Y_i = Ry_i + t$, one can report the deviation in individual atomic positions $|x_i - Y_i|$. Scanning such a list can reveal that certain portions of the selected regions fit well and that others do not.

Finding which regions fit well and which do not is an important step in the analysis of the structural similarities/differences. The repeated least-squares fitting of different set of atoms called "sieving" procedure is carried out to extract a well-fitting subset.

15.1.3. Identification of Similar Substructure

One difficulty in identifying similar substructures is that it is hard to formulate the problem in such a way to yield a unique answer. For this reason the use of interactive graphics as a guide to formulating the appropriate question is generally adapted. The basic idea is that if some set of corresponding residues from each of two proteins fits well, then any subsets of corresponding residues will also fit well though the converse is not necessarily true. For two related structures, the main question is finding the best *geometric transformation* in order to superimpose their frameworks together as closely as possible and evaluate their degree of similarity. Generally, the superimposition is carried out by finding the rigid-body translation and minimizing the root-mean-square (RMS) distance between corresponding atoms of the two molecules; that is, two structures are superimposed and the square root is calculated from the sum of the squares of the distances between corresponding atoms:

$$\text{RMS (rms)} = \left\{ \sum_i^N (|u_i - v_i|)^2 \right\}^{1/2}$$

where u_i and v_i are corresponding vector distances of the ith atom in the two structures containing N atoms. The result is a measure of how each atom in the structure deviates from each other, and an RMS value of 0–3 Å signifies strong structural similarity.

It is useful to set a threshold for goodness of fit—for example, 0.5 Å. By recording only those pairs of substructures for which RMS deviation, ΔRMS $\leqslant$ 0.5, one has generated a list of pairs of well-fitting substructures. The next step is to work toward larger substructures by merging entries in this list. The procedure can be iterative; given a list of pairs of well-fitting substructures, form the unions of pairs of entries, fit them, and append to the list any new well-fitting substructures discovered. Another approach is to characterize the conformation by the sequence of main-chain conformational angles ϕ and ψ, and extract sets of consecutive residues with similar conformations (Levine et al., 1984).

15.1.4. Structural Relationships Among Related Molecules

Families of related proteins tend to retain similar folding patterns. If one examines sets of related proteins, it is clear that general folding pattern of the structural core is preserved. But there are distortions that increase in magnitude as the amino acid sequences diverge. In proteins, the common core generally contains the major elements of secondary structures and segments flanking them (Chapter 12), including active site peptides. Large structural changes in the regions outside the core make it difficult to measure structural change quantitatively by straightforward application of simple least-squares superposition techniques. To define a useful measure of structural divergence, it is necessary first to extract the core and then carry out the least-squares superposition on the one alone (Chothia and Lesk, 1986).

For each major element of secondary structure of given two related proteins, perform a succession of superposition calculations, which include the main-chain atoms (N, Cα, C, O) of corresponding secondary structural elements plus additional residues extending from either end. Include more and more additional residues now identified as well-fitting contiguous region containing an element of secondary structure plus flanking segments. After finding such pieces corresponding to all common major elements of secondary structure, do a joint superposition of the main chain of all of them.

15.1.5. Homology Modeling

One of the ultimate goals of protein modeling is the prediction of 3D structures of proteins from their amino acid sequences. The prediction of protein structures rely on two approaches that are complementary and can be used in conjunction with each other:

1. Knowledge-based model combining sequence data to other information, such as homology modeling (Hilbert et al., 1993; Chinea et al., 1995).
2. Energy-based calculations through theoretical models and energy minimization, such as, *ab initio* prediction (Bonneau and Baker, 2001).

The most promising methodology relies on modification of a closely related (homologous sequence), functionally analogous molecule whose 3D structure has been elucidated. This is the basis of homology modeling for deriving a putative 3D structure of a protein from a known 3D structure. Residues are changed in the sequence with minimal disturbance to the geometry, and energy minimization optimizes the altered structure.

The understanding is that functionally analogous proteins with homologous sequences will have closely related structures with common tertiary folding patterns. When sequence homology with known protein is high, modeling of an unknown

TABLE 15.1. Some Web Sites for Protein Structure Alignment

Program	URL Address	Reference
CE	http://cl.sdsc.edu	Shindyalov and Bourne (1998)
Dali	http://www2.ebi.ac.uk/deli	Holm and Sander (1993)
K2	http://sullivan.bu.edu/k2	Szustakowski and Weng (2000)
ProSup	http://anna.came.sbg.ac.at/prosup/main.html	Feng and Sippl (1996)
TOP	http://bioinfol.mbfys.lu.se/TOP	Lu (2000)
VAST	http://www.ncbi.nlm.nih.gov:80/structure/VAST/vast.shtml	Gibrat et al. (1996)

structure by comparison can be carried out with reasonable success. However, it is noted that structure homology may remain significant even if sequence homology is low; that is, 3D structure seems better conserved than the residue sequence. It is shown that protein pairs with a sequence homology greater than 50% have 90% or more of the residues within a structurally conserved common fold (Stewart et al., 1987). The homology modeling normally consists of four steps:

1. Start from the known sequences.
2. Assemble fragments/substructures from different, known homologous structures.
3. Carry out limited structural changes from a known neighboring protein.
4. Optimize the structure by energy minimization.

In principle, predicting structure from the sequence by comparison to a known homologous structure is satisfactory when sequence homology is greater than 50% (Chothia and Lesk, 1986). Part of the problem of homology modeling at lower levels of similarity is to correctly align unknown and target proteins. Sequence alignments are more or less straightforward for levels of above 30% pairwise sequence identity. The region between 20% and 30% sequence identity (the twilight zone) is less certain. A means to automatically intrude into the twilight zone by detecting remote homologues (sequence identity <25%) are threading techniques (Bryant and Altschul, 1995). A sequence of unknown structure is threaded into a sequence of known structure, and the fitness of the sequences for that structure is assessed.

Comparing and overlapping two protein structures quantitatively remain an active area of development in structural biochemistry. Methods for protein comparison generally rely on a fast full search of protein structure database. Some of these methods that are available over the Internet are listed in Table 15.1.

15.2. STRUCTURE PREDICTION AND MOLECULAR DOCKING

15.2.1. *Ab initio* Prediction of Protein Structure

It is well established that the sequence of the amino acid constituting a protein is of prime importance in the determination of its 3D structure and functional properties.

Because sequence assignment is much faster and easier than 3D structure determination, the prediction of the 3D structure from the sequence of amino acids has been a great challenge for protein modeling (Bonneau and Baker, 2001). Energy-based structure prediction relies on energy minimization and molecular dynamics. The method is faced with the problem of a large number of possible multiple minima, making the traversal of the conformational space difficult and making the detection of the real energy minimum or native conformation uncertain.

The genetic algorithm (Dandekar and Argos, 1994; Koza, 1993) is applied to the problem of protein structure prediction with a simple force field as the fitness function to generate a set of suboptimal/native-like conformations. Because the number of probable conformations is so large (for the main chain conformations with 2 torsion angles per residue and assuming 5 likely values per torsion angle, a protein of medium size with 100 residues will have $(2 \times 5)^{100} = 10^{100}$ conformations even if we further assume optimal (constant) bond lengths, bond angles, and torsion angles for the side chains), it is computationally impossible to evaluate all the conformations to find the global optimum. Various constraints and approximations have to be introduced. For example, an assumption of constant bond lengths and bond angles by carrying out folding simulation in vacuum simplifies the total energy expression to

$$E = E_{tor} + E_{vdW} + E_{elec} + E_{pe}$$

where E_{tor}, E_{vdW}, E_{elec}, and E_{pe} are torsion angle potential, van der Waals interaction, electrostatic potential, and pseudoentropic term that drive the protein to a globular state ($E_{pe} \approx 4^{\Delta d}$ kcal/mol, where $\Delta d =$ largest distance between any $C\alpha$ atoms in one conformation). The combination of heuristic criteria with force field components may alleviate the inadequacy in the simplified fitness functions. The secondary structure prediction may be performed to reduce the search space. Thus, either idealized torsion angles or boundaries for torsion angles according to the predicted secondary structures can be used to constrain main-chain torsion angles.

It is shown that the incorrect structures have less stabilizing hydrogen bonding, electrostatic, and van der Waals interactions. The incorrect structures also have a larger solvent accessible surface and a greater fraction of hydrophobic side-chain atoms exposed to the solvent.

15.2.2. Molecular Docking

Molecular docking explores the binding modes of two interacting molecules, depending upon their topographic features or energy-based consideration, and aims to fit them into conformations that lead to favorable interactions. Thus, one molecule (e.g., ligand) is brought into the vicinity of another (e.g., receptor) while calculating the interaction energies of the many mutual orientations of the two interacting molecules. In docking, the interacting energy is generally calculated by computing the van der Waals and the Coulombic energy contributions between all atoms of the two molecules.

Ligand–receptor interaction is an important initial step in protein function. The structure of ligand–receptor complex profoundly affects the specificity and efficiency of protein action. The molecular docking performs the computational prediction of the ligand–receptor interaction and the structures of ligand–receptor complexes. In the complex, ligand and receptor molecules are presumed to adopt the energetically most favorable docking structures. Thus, the goal of molecular docking is to search for the structure and stability of the complex with the global minimum energy.

There are two classes of strategies for docking a ligand to a receptor. The first class uses a whole ligand molecule as a starting point and employs a search algorithm to explore the energy profile of the ligand at the binding site, searching for optimal solutions for a specific scoring function. The search algorithms include geometric complementary match, simulated annealing, molecular dynamics, and genetic algorithms. Representative examples are DOCK3.5 (Kuntz et al., 1982), AutoDock (Morris et al., 1996), and GOLD (Jones et al., 1997). The second class starts by placing one or several fragments (substructures) of a ligand into a binding pocket, and then it constructs the rest of the molecule in the site. Representative examples are DOCK4.0 (Ewing and Kuntz, 1997), FlexX (Rarey et al., 1996), LUDI (Böhm, 1992), GROWMOL (Bohacek and McMartin, 1997), and HOOK (Eisen et al., 1994).

In an interactive docking, an initial knowledge of the binding site is normally required. The ligand is interactively placed onto the binding site. Geometric restraints such as distances, angles, and dihedrals between bonded or nonbonded atoms may facilitate the docking process. Two parameters, the equilibrium value of the internal coordinate and the force constant for the harmonic potential, need to be specified. For interatomic distance, the equilibrium restraint may be the initial length of the bond if you would like a particular bond length to remain constant during a simulation. If you want to force a bond coordinate to a new value, the equilibrium internal coordinate is the new value. Normally, the force constants are 7.0 kcal/mol Å^2 for an interatomic distance (larger for nonbonded distances), 12.5 kcal/mo Å^2 for an angle, and 16.0 kcal/mol Å^2 for a dihedral angle. Such constraints also ensure the confinement of the ligand molecule at proximity of the binding site of the receptor during energy minimization. One may freeze most of the receptor molecule while allowing the ligand and contact residues to move in the field of the frozen atoms. Thus only the selected atoms (e.g., ligand molecule and contact residues) move while other (frozen) atoms influence the calculation.

In an automatic docking for example (Jones et al., 1997; Kuntz et al., 1982; Morris et al., 1996), a ligand is allowed to fit into potential binding cleft of the receptor of known crystal structure. Initially, the surface complementarity between the ligand and the receptor is determined by searching for a geometrical fit using the molecular surface as starting images (spheres). From each surface points, a set of spheres that fill all pockets and grooves on the surface of the receptor is generated, and various criteria are introduced to reduce their number to one sphere per atom. Within this approximation of both ligand and receptor shapes by sets of spheres, the search is made to fit the set of ligand spheres within the set of receptor spheres. The matching algorithm collects all fits that are possible by comparing internal distances in both ligand and receptor and lists pairs of ligands and receptor spheres having all internal distances matching within a tolerance value. Ligand atom coordinates are calculated, and the locations of the ligand atoms are optimized to improve the fit.

There are numerous cases of proteins for which structures have been determined in more than one state of ligation. In some cases, the structures undergo little change, except perhaps for specific and localized changes associated with particular functional residues — for example, triose phosphate isomerase. Other cases, such as citrate synthase, show conformational change in which binding of ligand in a site between two domains leads to a closure of the interdomain cleft (Lest and Chothia, 1984). Finally, there are the long-range integrated conformational changes associated with allosteric transitions of proteins (Perutz, 1989).

15.2.3. Solvation

Solvation can have a profound effect on the results of molecular calculation. Solvent can strongly affect the energies of different conformations. It influences the hydrogen bonding pattern, solute surface area, and hydrophilic/hydrophobic group exposures of protein molecules. Solvation is an important, ongoing problem in protein modeling. At present, a protein molecule is placed in an imaginary box of water molecules, and periodic boundary conditions (Jorgensen et al., 1983) are set at constant dielectric of 1.0. The dielectric constant defines the screening effect of solvent molecules on nonbonded (electrostatic) interactions and can vary from 1 (*in vacuo*) to 80 (in water). For the binding site, a constant dielectric of 4.0 is often used where no measured data are available. Alternatively, a distance-dependent dielectric constant is used to mimic the effect of solvent in molecular mechanics calculations in the absence of explicit water molecules.

15.3. APPLICATIONS OF PROTEIN MODELING

Most comprehensive software programs suitable for protein modeling are commercial packages, some of which are listed in Table 15.2. The application of HyperChem in the biomolecular modeling has been described (Chapter 14) and can be extended to the modeling of protein structures. The aspects of protein modeling will be illustrated with two freeware programs and two online servers.

15.3.1. Protein Modeling with Swiss PDB Viewer

Swiss-PDB Viewer (http://www.expasy.ch/spdbv/mainpage.html) is an application program that provides a user interface for visualization and analysis of biomolecules in particular proteins. The program (Spdbv) can be downloaded from http://expasy.ch/spdbv/text/getpc.htm, and the user guide is available from http://www.expasy.ch/spdbv/mainpage.html. Spdbv implements GROMOS96 force field

TABLE 15.2. Some Commercial Suppliers for Modeling Packages

Supplier	URL Address	Modeling Package
Accelrys, Inc.	http://www.accelrys.com	Cerius^2, Insight II, QUANTA
CambridgeSoft Corp.	http://www.camsoft.com	Chem3D
Hypercube, Inc.	http://www.hyper.com	HyperChem
Tripos, Inc.	http://www.tripos.com	Alchemy, Biopolymer, SYBYL
Wavefunction, Inc.	http://www.wavefun.com	Spartan, PC Spartan

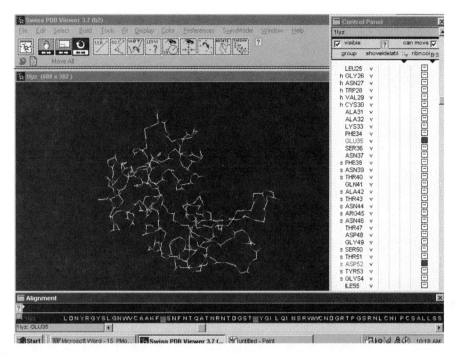

Figure 15.1. Workspace of Swiss-PDB Viewer. The workspace of Swiss-PDB Viewer (SPBBV) consists of main windows (with menu bars, icons, and display window), control panel (with listing of amino acid residues), and align window (with amino acid sequence of the displayed molecule). Lysozyme (1lyz.pdb) with highlighted catalytic residues (Glu35 and Asp52) is displayed.

(Xavier et al., 1998) to compute energy and to execute energy minimization by steep descent and conjugate gradient methods. For standalone modeling, an accompanying Spdbv Loop Database should be downloaded and placed into the "_stuff_" directory. Typically, the workspace consists of a menu bar, tool icons, and three windows; Main (Display) windows, Control panel, and Align window (Figure 15.1). The Control panel and Align window can be turned on and off from the Window menu. The Control panel provides a convenient way to select and manipulate the attributes of individual residues. The first column shows names of amino acid residues/nucleotides with check mark ($\sqrt{}$) toggle between display/hide. The subsequent columns are for the display of side chains, residue labels, dot surfaces (VDW or accessibility), and secondary structures if the check marks ($\sqrt{}$) are present. The last column with small square boxes is used to highlight the residue(s) with color. The Align window permits an easy means of threading sequence in the homology modeling.

The common tools are located at the top of the display window under the menu bar. These include move molecule (translate, enlarge, and rotate) tools and general usage (bond distance, bond angle, torsion angle, label, range, center, fit, mutation, and torsion) tools. Clicking the icon displays an instruction on the text space under the tool icons. For example, upon clicking the label icon, you will be asked to pick an atom. Upon picking an atom, the label (Residue ID and atom class) is displayed. The mutation icon provides a convenient approach to mutate selected residue(s) of the picked atom(s). To perform amino acid mutation of the displayed molecule,

- Color the residue(s) to be mutated from the Control panel.
- Click the mutation icon to open a pop-up list of amino acids.
- Pick the mutating amino acid and hit Enter key to automatically selecting the rotamer.
- Revisualize the molecule (check on the Control panel) if necessary.
- Click the mutation icon again and answer Yes to accept the mutation.
- Repeat the procedures for further mutations.
- Perform energy minimization to release local constraints of the mutant protein.
- Refine the structure by Selecting amino acids making clashes (Select→aa Making clashes) and initiate fixation (Tool→Fix selected sidechains).
- Save (File→Save→Layer) the mutant molecule as mutmolec.pdb.

The Preferences menu provides tools for defining and selecting options such as Loading protein molecule (Cα-trace, backbone, backbone + side chains or ribbon), Labels, Colors, Ribbon, Surface, and Electrostatic potential representations. If hydrogen bonds are to be calculated, the H-Bond detection threshold option should be checked. Hydrogen bonds are detected if an H is within 1.20–2.76 Å of an acceptor in the presence of explicit hydrogen or 2.35–3.20 Å between the prospective donor and acceptor in the absence of explicit hydrogen. If energy minimization is to be performed, the Energy minimization option should be checked. The set preferences can be saved and opened for later operations as filename.prf. The PDB file can be opened and saved from the File menu.

The Edit menu enables an online access to **PROSITE** and **BLAST** searches. The Edit menu also offers utilities for assigning secondary structure types to the selected amino acids. The groups/residues can be selected for visualization and manipulation from the Select menu or the Control panel. The Select features includes group kind (amino acids and nucleotides), group property (basic, acidic, polar or nonpolar), secondary structure, accessible amino acids (dialog box for defining % accessible surface) and neighbors of selected amino acids (dialog box for defining the distance from the selected molecule/residue). The selected residues become highlighted (red) in the Control panel except that the neighbors of selected amino acids appear with side chains on the displayed molecule and as check marks in the side-chain column of the Control panel. After selection, the selected groups/residues can be highlighted from the Control panel by hitting the small square box to open a color palette or saved as newfile.pdb that contains the atomic coordinates of the selected molecule/residues.

The Tools menu provides tools for computing hydrogen bonds, molecular surface, electrostatic potential, threading energy, and force field energy. If the appropriate Show option in the Display menu is checked, hydrogen bonds (with/without distance), surface dots, or electrostatic potential grids (automatic) are displayed on completion of the computation. The threading energies are plotted in the Align window (by clicking the little arrow located at the top of the window).

To perform energy minimization,

- Select Compute Energy (Force field) tool to open a dialog box for specifying angles/bonds to be computed. Be sure to check Show energy report, which is saved in the "temp" directory as molecule.En (*n* according to the order of computations). Each residue needs a topology for the force field to work. The force field calculation of the current version (version 3.7) of Spdbv only treats proteins (i.e., energies of heteroatoms are not calculated).

- Before performing energy minimization, check the Energy Minimization option of the Preference menu.
- On the Energy minimization preferences box, specify the number of steps/cycles and method of minimization (Steepest descent or Conjugate gradients), angles/bonds to be minimized, and conditions for terminating the minimization. Be sure to check the radio button for Lock nonselected residues and the box for Show energy report.
- Choose the residues to be energy-minimized.
- Invoke the Energy Minimization tool of the Tools menu to initiate minimization.
- The results of each cycle are displayed (Figure 15.2) and saved in the "temp" directory as molecule.En (*n* = order of executing minimization).

The structural similarity of two molecules can be compared by superimposition with Spdbv:

- Open the reference molecule and color the whole molecule with reference color (shift-click the square box in the Control panel to open color palette, and select the desired color).

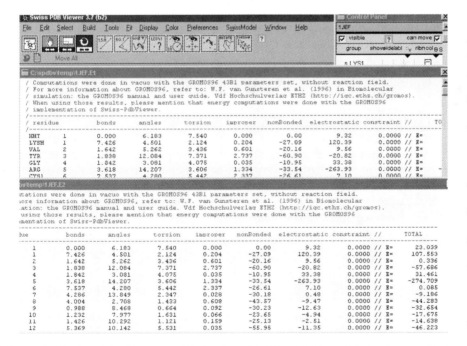

Figure 15.2. Results of energy minimization with SPDBV. The minimization results (per residues) of turkey lysozyme (1JEF.pdb) are shown in two frames (upper frame and continuing lower frame).

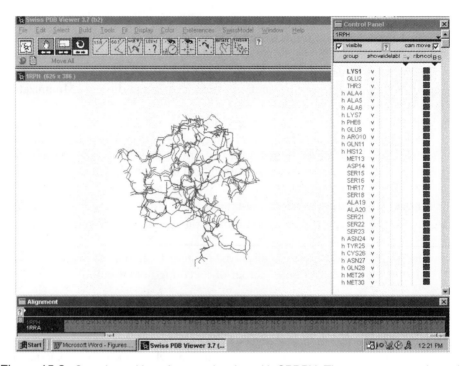

Figure 15.3. Superimposition of two molecules with SPDBV. The structure superimposition of bovine ribonuclease (1RPH.pdb in blue) and rat ribonuclease (1RRA.pdb in red) is shown. The color for the superimposed molecules is assigned from the control panel as illustrated for 1RPH in blue. The corresponding color for color boxes of 1RRA is red. The align window shows sequence alignment of the two molecules.

- Open the trial molecule and color the whole molecule with another color.
- Invoke the Magic Fit tool from the Fit menu. The two molecules become superimposed and their sequences are displayed in the Align window (Figure 15.3).
- As you move the cursor pointing to amino acid residues of the aligned sequences in the Align window, the corresponding residues of the superimposed molecules in the Display window blink from blue to yellow, allowing an easy visualization of the overlap. In addition, the rms distance of the superimposed residues is displayed.
- Activate the trial molecule by clicking its name.
- Select the Generate Structural Alignment tool of the Fit menu to initiate the sequence alignment which can be viewed by clicking the little text icon.
- Save the file as alignefile.txt.
- To improve the superimposition (lowering rms distance), select the Iterative Magic Fit tool from the Fit menu.
- You can save each of the overlapped molecules separately (File→Save→ Layer) as individual.pdb or save the superimposed molecules together (File→Save→Project) as overlap.pdb.

The modeling of a protein structure from its sequence against the known 3D structure of homologous protein(s) in the homology modeling can be attempted with Spdbv as follows:

- Download the trial sequence in fasta format (start with >) and save it as trialseq.txt.
- Open the template structure of the reference molecule (refmol.pdb) and color it with reference color.
- Choose the Load Raw Sequence to Model tool from the SwissModel menu and open trialseq.txt. The trial sequence appears on the Display window as α helix (color it other than the reference color if desired).
- Select the Magic Fit tool from the Fit menu. The α helix changes to the structure overlapping the reference structure.
- Make sure that the option "Update Threading Display Automatically" via SwissModel→Update threading display now is not checked.
- Invoke the Generate Structure Alignment tool of the Fit menu.
- Click the little arrow beside the question mark of the Alignment window to view a plot of threading energies.
- Click smooth to set smooth = 1 and check Update threading display now tool.
- Select Color→By Threading Energy to display threading energy profile of the structure (The mean force potential energy of the polypeptide chain increases with color varying from blue to green, yellow, and then red).
- Activate the trial structure from the Control panel for structural refinement.
- Perform two operations: Select→aa Making Clashes, and Tools→Fix Selected Side Chains.
- Repeat the process to observe a decrease in the number of amino acid residues which make clashes (Figure 15.4).
- Save the trial structure (File→Save→Layer) as trialmol.pdb, which can be submitted to Swiss-Model (http://www.expasy.ch/swissmod/) using Optimise (project) mode for optimization.

The two structures (or part of structures) from two opened pdb files can be merged. The merge can be applied to dock a ligand from one file onto the receptor site of the other. For example,

- Activate the first file (ligand file) and select the ligand molecule.
- Activate the second file (receptor file) and select residues of the receptor site.
- Select Create Merged Layer from the Selection tool of the Edit pull-down menu. The selected structures are merged.
- Saved (File→Save→Layer) as pdb file as shown in Figure 15.5.

PROSITE is a database of biologically significant sites and patterns (amino acid residues) formulated from a known family of proteins as the characteristic familial identifiers or signatures that can be used to detected such family from the query sequences of proteins. If the PROSITE database (prosite.dat downloaded from http://expasy.hcuge.ch/sprot/prosite.html) is installed in the "usrstuff" directory, selecting Search for PROSITE tool from the Edit menu returns a list of ProSite for

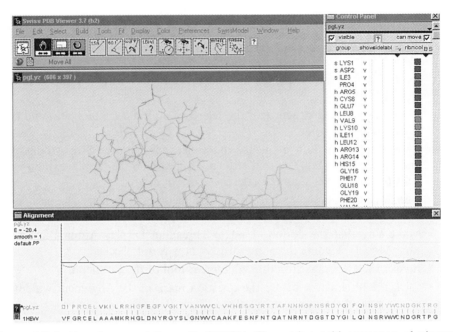

Figure 15.4. Homology modeling with SPDBV. The amino acid sequence of pigeon lysozyme (pgLyz.txt) is modeled against hen's egg-white lysozyme (1HEW.pdb) as the template structure by the automatic threading. The display window shows the modeled pigeon lysozyme (pgLyz.pdb) colored according to threading energy as shown in the align window. The align window illustrated a plot of the mean force potential energy per each residue above the sequence alignment.

the protein molecule. Picking the desired ProSite highlights amino acid residues of the ProSite of the displayed molecule as well as the amino acid residues in the Control panel. This provides a convenient way of preparing pdb files of selected amino acid residues comprising ProSites of proteins (e.g., pdb file of the active site residues).

15.3.2. Representation and Animation with KineMage

KineMage (kinetic images) is an interactive 3D structure illustration software (Richardson and Richardson, 1992; Richardson and Richardson, 1994) that can be downloaded from http://orca.st.usm.edu/~rbateman/kinemage/. It is adapted for the structure representation of biochemical molecules by many biochemistry textbooks. The program consists of two components; PREKIN and MAGE. The PREKIN program interprets/converts pdb file (molecule.pdb) to kinemage file (molecule.kin), which is then displayed and manipulated with the MAGE program. Launch the PREKIN program:

- Click Proceed to input the pdb file and to assign the output file (e.g., molecule.kin). This opens the Starting Ranges dialog box, which offers options (radio buttons) for "Backbone browsing script," "Selection of built-in scripts," "New ranges," "Focus only," and Range list and Reset options.

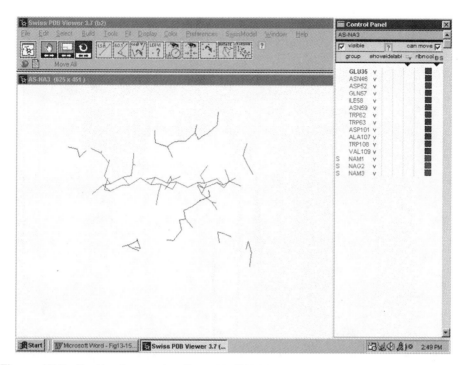

Figure 15.5. Docking by merging files with SPDBV. The docking of trisaccharide (NAM-NAG-NAM) into the active site of lysozyme by merging trisaccharide file (9lyz.pdb) onto lysozyme file (1lyz.pdb). After the merge, only active-site residues (catalytic residues in green and contact residues in blue) and trisaccharide (in red) are selected to display and saved.

- Choose Selection of built-in scripts to open the Built-in Scripts dialog box, which offers various scripts for illustrating macromolecules (Figure 15.6). "CaSS" is the default Browsing backbone script connecting all Cα of the backbone plus disulfides, "mcHb" gives main chain with hydrogen bonds, "aasc" offers Cα backbone with all the amino acid side chains colored by amino acid types, and "lots" includes mc, sc, ca, and heteroatoms of the bound ligands and water. Two scripts for DNA/RNA are "naba" and "separate bases." The "ribbon" representations include (a) a thin ribbon with variable width dependent on the curvature of the backbone (an option for adding an arrow head on β strand) and (b) "ribbon HELIX_SHEET," which offers further options for specifying different widths for α helix, β strand, and β coil and for specifying the number of subunits to be transcribed.

For customerized structural representation (e.g., main chain with side chains of active site residues and bound ligands):

- Select radio button of New Ranges of the Starting Ranges dialog box to open the Range Controls box.
- Enter residue numbers for the start and end residues.

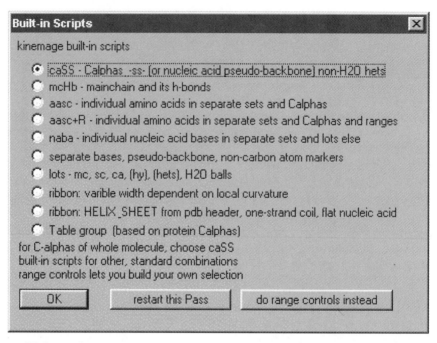

Figure 15.6. Kinemage built-in scripts page of PREKIN program. Build-in scripts program for converting molecule.pdb file into molecule.kin file according to the selection(s) for structural features to be displayed/manipulated. To terminate the selection, click OK button. For more than one selection (structural features), click "restart this Pass" button.

- Check mc (main chain), sc (side chain), or ht (heteroatoms of ligands), respectively, and choose either "OK accept and comes back for more" for multiple selections or "OK accept and end ranges" to terminate the selection.
- Accept defaults from the Focus Options box and specify the subunits to be transcribed in the Run Conditions box.
- Go to File then Quit.

PREKIN compiles molecule.kin file (ascii text file), which can be read and edited with Microsoft Word or Corel WordPerfect. The kinemage file (*.kin) has the following general organization:

```
@text: Description of kinemage to enter into the text window.
@kinemage i: Starting a new kinemage with numbers (where i = 1,2···n)
@caption: List for the caption window
@onewidth: All lines 2 pixels wide
   optional @?viewid {} to @?matrix for displaying different views
                                    derived from *.vw files
@group {identifier} [parameter]: High level display object
@subgroup {identifier} [parameter]: Mid level display object
@vectorlist {identifier} [parameter]: Low  level  display  object,
      [color = ] is the most common
      {atomid} [-/P/L] x,y,z: Individual vectorlist item e.g. coordi-
      nates of atoms
```

The color of structural elements can be changed by changing color description of color = {color} on the appropriate @subgroup or @vectorlist line. The control checkboxes corresponding to @groups are displayed automatically. However, the control checkboxes relating to @subgroup or @vectorlist can be added/deleted by adding/deleting master = {label} on the desired lines where the checkboxes applied. Similarly, the label of the control checkbox can be changed as well. The molecu_i.kin files can be merged in a usual manner. The merged files can remain as independent kinemages (as @kinemage i) or edited to become @group {identifier} within the merged kinemage. The former is displayed under the control of Kinemage menu and is useful if each kinemage consists of different views. The latter offers mechanisms for superimposing structures (if multiple control checkboxes for @group are checked) and animation.

To create animation (displaying structures in successive frames), merge molecu_i.kin files either by text editor or from the Edit menu of MAGE program. Edit the merged file such that the merged kinemages become @groups. Set parameter of @group to animate (adding an animate statement on @group line, i.e., @group {mergedkinemage} animate). For different views of structures, write molecule.vw (from @viewid to @matrix) from the MAGE program and insert molecule.vw files into the molecu_i.kin file. After editing, save the file as newname.kin or molecule.kin.

The following lysozyme.kin file exemplifies the animated structures of lysozyme and its binary complexes showing different colors for the main chain Cα (white), contact side chains (cyan), hidden (off) catalytic side chains (blue), and bound ligand (orange, NAGs) with multiple control checkboxes (main chain, contact, catalytic and NAGs). @groups are derived from merged molecu_i.kin files which can be animated. @subgroups designate different structural components with different colors assigned by @vectorlists. The statement "@vectorlist off {catalytic} color = blue master = {catalytic}" provides a mechanism for highlighting the side chains of two catalytic residues (Glu35 and Asp52) to blue, which is shown initially in cyan via the checkbox.

```
@kinemage 1
@caption
  Lysozyme and its binary complexes.
@onewidth
@zoom 1.00
@zslab 200
@matrix
1.000000  -0.000000  0.000000  0.000000  1.000000  0.000000  0.000000  -
0.000000 1.000000  ·  ·  ·  ·        ·  ·  ·  ·  ·  ·  ·
@group dominant {Lysozyme} animate
@subgroup {mainchain} master = {main chain}
@vectorlist {ca-ca} color = white
{ ca lys a 1 } P 2.420, 10.480, 9.130
{ ca lys a 1 } L 2.420, 10.480, 9.130
{ ca val a 2 } L 2.390, 13.800, 7.270
{ ca phe a 3 } L -1.140, 15.180, 7.230
{ ca gly a 4 } L -2.630, 17.250, 4.400
{ ca arg a 5 } L -4.240, 20.510, 5.660
 ·    ·    ·    ·        ·    ·    ·    ·    ·    ·    ·
@subgroup {sidechain}
```

```
@vectorlist {ss} color = yellow

{ ca cys a 64 } P 4.450, 13.410, 28.940
{ cb cys a 64 } L 4.340, 13.240, 27.440
{ sg cys a 64 } L 5.670, 14.060, 26.450
{ sg cys a 80 } L 7.200, 12.800, 26.570
{ cb cys a 80 } L 7.080, 11.770, 25.070
{ ca cys a 80 } L 5.820, 10.990, 24.930
  .   .   .   .        .   .   .   .   .
@vectorlist {contact} color = cyan master = {contact}
{ ca glu a 35 } P 4.850, 23.610, 15.140
{ cb glu a 35 } L 3.960, 23.120, 16.260
{ cg glu a 35 } L 3.380, 24.160, 17.230
{ cd glu a 35 } L 4.450, 24.740, 18.110
{ oe1 glu a 35 } L 5.560, 24.230, 18.140
{ cd glu a 35 } P 4.450, 24.740, 18.110
{ oe2 glu a 35 } L 4.160, 25.770, 18.750
{ ca asn a 46 } P 15.140, 22.250, 24.040
{ cb asn a 46 } L 13.900, 22.660, 24.830
{ cg asn a 46 } L 12.620, 22.500, 24.030
{ od1 asn a 46 } L 11.720, 21.740, 24.390
{ cg asn a 46 } P 12.620, 22.500, 24.030
{ nd2 asn a 46 } L 12.500, 23.250, 22.940
  .   .   .   .        .   .   .   .   .
@vectorlist off {catalytic} color = blue master = {catalytic}
{ ca glu a 35 } P 4.850, 23.610, 15.140
{ cb glu a 35 } L 3.960, 23.120, 16.260
{ cg glu a 35 } L 3.380, 24.160, 17.230
{ cd glu a 35 } L 4.450, 24.740, 18.110
{ oe1 glu a 35 } L 5.560, 24.230, 18.140
{ cd glu a 35 } P 4.450, 24.740, 18.110
{ oe2 glu a 35 } L 4.160, 25.770, 18.750
{ ca asp a 52 } P 8.750, 18.590, 22.160
{ cb asp a 52 } L 8.590, 19.930, 21.500
{ cg asp a 52 } L 8.940, 21.040, 22.460
{ od1 asp a 52 } L 8.740, 21.050, 23.650
{ cg asp a 52 } P 8.940, 21.040, 22.460
{ od2 asp a 52 } L 9.470, 22.010, 21.910
  .   .   .   .        .   .   .   .   .   .
@group dominant {E-NAG4} animate
@subgroup {mainchain}
@vectorlist {ca-ca} color = white master = {main chain}
  .   .   .   .        .   .   .   .   .   .
@subgroup {NAG4} master = {NAGs}
@vectorlist {sc} color = orange
{ c1 nag b 1 } P 0.740, 27.090, 34.670
{ c1 nag b 1 } 0.740, 27.090, 34.670
{ c2 nag b 1 } 0.280, 26.450, 35.950
{ c3 nag b 1 } -1.070, 26.970, 36.340
{ c4 nag b 1 } -1.150, 28.470, 36.320
{ c5 nag b 1 } -0.520, 29.110, 35.070
  .   .   .   .        .   .   .   .   .   .
```

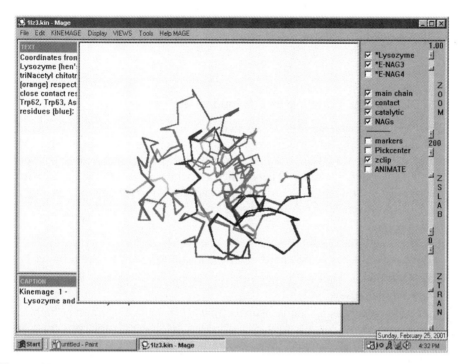

Figure 15.7. Mage desktop window. The desktop window of MAGE program consists of three component windows (graphic window, caption window, and text window) as shown for human lysozyme (1lza.pdb, 1lzb.pdb, and 1lzc.pdb). The displayed structural features can be turned on and off via checkboxes. The check marks ($\sqrt{}$) indicate those structural features that are displayed. For example, the main chain of lysozyme (black), contact residues (light blue), catalytic residues (navy blue), and trisaccharide (NAG$_3$ in red) are checked and displayed. The prefixed asterisk (*) indicates that these structures can be animated by clicking the ANIMATE checkbox successively.

Launch MAGE and click Proceed to display the desktop window, which consists of three component windows: the Caption window, which provides information about the file, the Text window, which describes the features of .kin file; and the Graphic window for interactive viewing of the structure as shown in Figure 15.7. The structural features corresponding to the checkboxes are displayed if they are checked. More than one molecule can be displayed (overlapped) if multiple checkboxes corresponding to the group name are checked. Click the checkbox labeled Animate to display structures (with asterisk in front of the group name) in successive frames (Figure 15.7). To measure geometry, successively click two atoms for the distance and three atoms for the angle. The information appears at the lower left corner of the Graphic window.

To merge *.kin files, select Append File tool of the File menu and open the file to merge. To change default color, invoke Change Color tool of the Edit menu and click the structure element to open Color selection box from which the desired color can be assigned. Different structure views are created/added to the *.kin file by selecting Keep Current View tool of the Edit menu and entering View number and

View ID to the dialog box. This adds @nviewid {viewid}, @nzoom, @nzslb, @ncenter, and @nmatrix (*n* is the view number) lines (ahead of @group) to the .kin file.

15.3.3. Biopolymer Modeling at *B* (Biopolymer) Server

The B server (http://www.scripps.edu/~nwhite/B/indexFrames.html) offers an online biomolecular modeling program that builds biomolecular models, measures their geometry, and implements an AMBER force field in the geometry optimization as well as in molecular dynamics. The Biomer authentification certificate (see http://www.scripps.edu/~nwhite/B/Security/) is required to enable opening/saving files (though it is not needed to run the B program).

- Click the Start B button to open the B window.
- Hit the Grant button to the Biomer certificate to open the file browser box.
- Select Polynucleotide/Polypeptide/Polysaccharide from the Build menu to open a dialog box.
- For polysaccharide, choose connectivity (O1–C[1–6]), anomer (alpha or beta), isomer (L or D), and conformation (define phi and psi angle with omega = 180). Add sugars (alsohexoses or aldopentoses) to build polysaccharide chain.
- For polynucleotide, choose various options such as either DNA or RNA, Form (a, b, c, d, e, t, or z forms), Build (5′ to 3′ or 3′ to 5′), single strand/double strand, and conformation of pentofuranose ring. Add bases (pairs) in the specified direction to build polynucleotide chain.
- For polypeptide, the B program provides options for building various protein conformations including 3–10 helix, alpha helix, alpha helix (L-H), beta sheet (anti-prl), beta sheet (parallel), various beta turns, extended, gamma turns, omega helix, pi helix, polyglycine, and polyproline. Choose the desired conformation and isomer (L or D) and then add amino acids from N-terminus to construct polypeptide chain.
- Cap the termini if desired for polypeptide or polysaccharide. Click Done button to view the molecule on the display window (Figure 15.8).

The Tools menu provides utilities to calculate net mass, net charge, energy and to invert chirality. For energy minimization,

- Select the Molecular mechanics menu and choose Steepest descent as the initial Minimizer that can be changed to Conjugated gradient later.
- Click Minimize to start energy minimization. The gradient and energy are displayed on the lower left corner of the display window.

For dynamic simulation,

- Select the Molecular dynamics menu to open the dialog box. Specify Time, Temperature, and Step size for Heating cycle, Equilibrium period, and Cooling cycle.

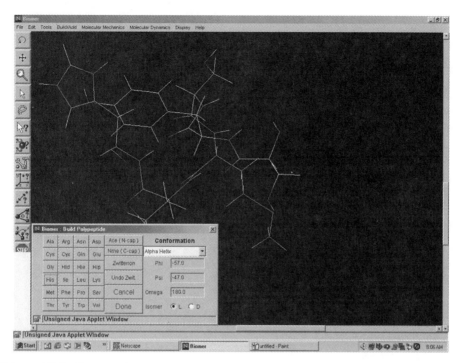

Figure 15.8. Construction of polypeptide chain at B server. The inset shows the build polypeptide dialog box. The chain is built from N-terminus by clicking amino acids from the dialog box successively.

- Click Start trajectory to initiate simulation. The lower left corner shows the progress of the simulation.
- Click the Grant button to the certificate to save the molecule.pdb.

15.3.4. Structure Comparison and Alignment at *CE* (Combinatorial Extension) Server

The 3D protein structure comparison and alignment can be performed online using the combinatorial extension algorithm (Shindyalov and Bourrne, 1998) at CE server (Shindyalov and Bourne, 2001). On the CE home page (http://cl.sdsc.edu/ce.html), click All or Representatives PDB, then enter PDB ID into the Specify protein chain box to search for structural alignments from the CE database. From the hit list, select desired entries from the check boxes for sequence alignment and select Neighbors to display the structure neighbors superimposing with the protein specified by the query PDB ID.

To perform sequence alignment and structure comparison,

- Click Two Chains to upload two PDB IDs (e.g., user's query pdb file against PDB file supplied by the user or retrieved form the CE database) on the CE home page.
- On the query form, enter the template PDB ID (either from the CE database or from user's directory) for Chain 1 and enter/browse the query pdb file for Chain 2.

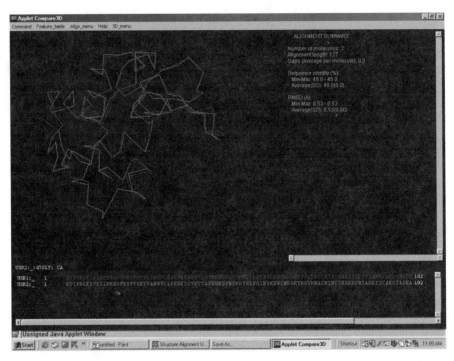

Figure 15.9. Superimposition of 3D protein structures at CE server. The opening graphic window for the superimposition of the modeled (homology modeling with SPDBV) pigeon lysozyme (USR2 in pink) and hen's egg-white lysozyme, 1LYZ.pdb (USR1 in blue). The superimposed structures are displayed with the alignment summary (e.g., sequence identity, rms deviation).

- Click the Calculate Alignment button to receive Structure Alignment result showing sequence alignment of USR1 (template PDB) and USR2 (query PDB), which can be downloaded as a pdb file.
- Click the Press to Start Compare 3D button to open the display window showing the overlapped structures (Figure 15.9). The menu bar consists of Command (Reset, Exit), Feature table (Browse features...), Align_menu (Control penal), and 3D_menu (rotate, translate, zoom, switch C_a/all atoms).
- Select Feature table→Browse features→Structure or Sequence to analyze the superimposed structures. For example, choosing Structure (Phys)-Secondary structure illustrate secondary structural features of the superimposed structures and choosing Structure (Comp)-Difference displays the structural differences between the superimposed structures as shown in Figure 15.10.

To perform sequence alignment/structure comparison of a query pdb file against the CE database,

- Click Uploaded by the user against the PDB to upload the query pdb file.
- Enter the e-mail address and click the Search Database button. The search and analysis results are returned by the e-mail if the submission is followed by

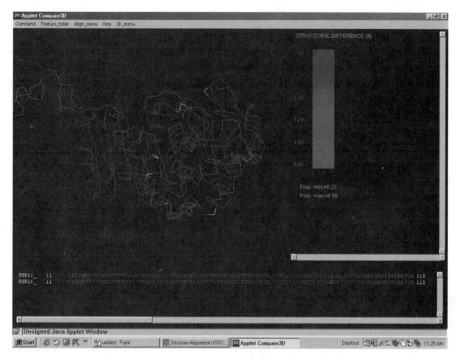

Figure 15.10. Structural comparison of the superimposed structures. Structural differences between the superimposed muscle isozyme (3LDH.pdb) and heart isozyme (5LDH.pdb) of lactate dehydrogenase analyzed by the CE server.

a message: "Your query [ce no.] has been successfully submitted. Results will be sent to [your e-mail address]."

15.4. WORKSHOPS

All software programs discussed in Chapters 14 and 15 whichever are the most appropriate, are to be applied for this workshop.

1. Deduce the probable conformation of porcine glucagon with the following sequence (construct initial structure in α helix from amino acids with HyperChem which lists residue ID in PDB file) by geometry optimization until the difference in energies between the successive steps is less than 1.0 kcal/mol or rms gradient of 0.1 kcal/(mol-Å):

HSQGTFTSDYSKYLDSRRAQDFVQWLMNT

Compare the optimized structure with that of the X-ray crystallographic structure (1GCN.pdb).

2. Retrieve pdb files of lysozyme from different species including chicken (1LYZ.pdb), swan (1GBS.pdb), pheasant (1GHL.pdb), and turkey (1LZY.pdb), and compare their main chain structures. Write a new pdb file for the superimposed structures of chicken and turkey lysozymes.

3. Bovine chymotrypsinogen A is activated by excision of two dipeptides, Ser14-Arg15 and Thr147-Asn148, to form γ-chymotrypsin A and α-chymotrypsin A. The activation of inactive zymogen to active γ-chymotrypsin and α-chymotrypsin can be explained by comparing X-ray crystallographic structures of zymogen versus the active enzyme, which shows the formation of the properly oriented substrate binding pocket (Birktoft, Kraut, and Freer, 1976). The substrate binding pocket of chymotrypsin can be inferred from ProSite around Ser195. Retrieve pdb files of chymotrypsinogen (*Note:* The atomic coordinates of several residues in chymotrypsinogen are not shown in the pdb files due to poor resolution) and γ-chymotrypsin or α-chymotrypsin, and examine the structural change induced by the activation of zymogen in the movement of Gly193. Write the new pdb file illustrating the change. Compare the strain energies and geometries of the catalytic triad — His57, Asp102, and Ser195 — of the optimized chymotrypsinogen and chymotrypsin.

4. The allosteric enzyme, aspartate transcarbamylase, exists in different conformational R-state and T-state. Retrieve the pdb files (2AT1.pdb and 3AT1.pdb) of both states that give the tetrameric structures, each consisting of two larger catalytic (A and C) and smaller regulatory (B and D) subunits. Write the new pdb file for the superimposed catalytic subunit structures and identify major conformational changes accompanied the allosteric transition with reference to

    a. Surface plots

    b. Conformational (Ramachandran) plots

    c. Solvent accessible amino acid residues (e.g., 30% accessible surface)

    d. Amino acid residues interacting (e.g., within 4.0 Å) with the effector, palmitoleic acid (PAM)

    e. Aspartate carbamoyltransferase signature (ProSite) amino acid residues

5. The induced fit model (Koshland, 1958) provides good explanation for enzyme specificity and catalysis. The binding of a substrate induces conformational changes that align the catalytic groups in the correct orientations for catalysis. The 3D structures of *E. coli* dihydrofolate reductase and its binary (with $NADP^+$ or NADPH) as well as ternary (with $NADP^+$ and folate.) complexes have been determined (Bystroff and Kraut, 1991). Retrieve these pdb files (5DFR for apoenzyme, 6DFR and 1DRH for binary complexes, and 7DFR for ternary complex) and examine conformational differences upon complexations. Construct animated kinemage files showing induced conformational changes accompanying formation of the binary and ternary complexes (*Note:* The coordinates for the residues 16–20 are missing for some pdb files, and the coordinates for nicotinamide mononucleotide moiety of $NADP^+$ in 6DRF.pdb are not shown.)

6. Retrieve pdb file of ribonuclease A and use it as the reference coordinates to perform homology modeling of sheep ribonuclease A with the following amino acid sequence:

```
KESAAAKFERQHMDSSTSSASSSNYCNQMMKSRNLTQDRCKPVNTFVHESLADVQAVCSQKNVACKNGQT
NCYQSYSTMSITDCRETGSSKYPNCAYKTTQAEKHIIVACEGNPYVPVHFDASV
```

Apply ProSite to identify pancreatic ribonuclease family signature.

7. The three disulfide bonds linking Cysteines 5 & 55 [5–55], 14 & 38 [14–38], and 30 & 51[30–51] stabilize the tertiary fold (comprising two antiparallel β strands

and two short α helices) of a 58-amino-acid protein, bovine pancreatic trypsin inhibitor (BPTI). The main folding pathway of BPTI has been elucidated (Crighton, 1977). The disulfide bond [30–51] is formed as an initial intermediate. This is followed by the formation of disulfide bonds, [30–51] and [5–55], which adopt a native-like fold to bring Cys14 and Cys38 into correct proximity to form the last disulfide bond and the native structure (1BPI.pdb). Perform mutations by replacing cysteines with serines successively to produce mutants (C5S, C14S, C30S, C38S, C51S, and/or C55S) that mimic the unfolded polypeptide chain and folding intermediates. Use KineMage animation to simulate structural changes accompanied the folding pathway of BPTI at 300 K.

8. The active site of hen's egg-white lysozyme (1LYZ.pdb) is a cleft consisting of six subsites, A to F, which accommodate hexasaccharide with alternative *N*-acetylglucosamine (NAG) and *N*-acetylmuramic acid (NAM). Two catalytic residues, Glu35 and Asp52, are situated on the opposite sides of the active cleft between subsites D and E (Phillips, 1967). Attempt to dock trisaccharide, NAM-NAG-NAM (9LYZ.pdb), to the active site (subsites B–D) of lysozyme interactively by placing C1′ of the reducing NAM at approximately equal distance from the carboxyl oxygens of Glu35 and Asp52 (Tsai, 1997). Perform energy minimization and molecular dynamic with respect to the trisaccharide and the two catalytic residues while freezing the rest of the lysozyme molecule. Deduce the structural changes upon complexation between lysozyme and trisaccharide.

9. Protein engineering has been carried out to redesign substrate specificity of lactate dehydrogenase from *Bacillus stearothermophilus* (Wilks et al., 1988; Wilks et al., 1990):

  a. Mutant (Q102M, K103V, P104S, A235G, A236G) tolerates substrates with large hydrophobic side chains.

  b. Mutant (Q102R) shifts substrate specificity from pyruvate to oxaloacetate.

The amino acid sequences of corresponding regions from dogfish M_4 and *Bacillus stearothermophilus* lactate dehydrogenases are shown below:

```
dogfish    VITAGARQQEGESRLNLV... ... KDVVDSAYEV
B.stear.   VICAGANQKPGETRLDLV... ... VNVRDAAYQI
```

Simulate these mutants for dogfish M_4 lactate dehydrogeanse (6LDH.pdb). After energy minimization, superimpose the wild-type and mutant enzymes and construct their animated kinemage files including views for the complete molecular structures and close-ups of the L-lactate dehydrogenase active site as identified by the ProSite.

10. The two nicotinamide coenzymes, NAD(H) and NADP(H), differ in that the 2′-hydroxy group of the adenylyl moiety of NAD(H) is phosphorylated in NADP(H). Site-directed mutagenesis has been used to identify amino acid residues that are responsible for the coenzyme specificity in oxidoreductases. Glutathione reductase and dihydrolipoamide dehydrogenase catalyze related chemical reactions and bear a striking similarity to one another, both structurally and mechanistically. Both enzymes are dimeric and contain functionally reactive disulfide in addition to two molecules of FAD at the active site. Glutathione reductase catalyzes the electron transfer between NADPH and glutathione, whereas dihydrolipoamide dehydrogenase catalyzes NAD^+-dependent oxidation of dihydrolipoamide. A comparison

of amino acid sequences of the $\beta\alpha\beta$ dinucleotide binding fold of several glutathione reductases and dihydrolipoamide dehydrogenases as shown below suggests candidate residues (bold blue Glu for 2'-hydroxy of NAD(H) and bold red Arg for 2'-phosphate of NADP(H)) for investigating the coenzyme specificity by directed mutagenesis (Bocnegra et al., 1993; Scrutton et al., 1990).

Glutathione reductase:

```
Human          194 GAGYIAVEMAGILSALGSKTSLMIRHDKVLRSFD 227
Drosophila     155 GAGYIGVEMAGILSALGYGT-VMVR-SIVLRGFD 186
Yeast          206 GAGYIGIELAGVFHGLGSETHLVIRGETVLRKFD 239
P. aeruginosa  188 GGGYIAVEFASIFNGLGAETTLLYRRDLFLRGFD 221
E. coli        174 GAGYIAVELAGVINGLGAKTHLFVRKHAPLRSFD 207
```

Dihydrolipoamide dehydrogenase:

```
Human          220 GAGVIGVELGSVWQRLGADVTAVEFLGHVGGVGID 254
Pig            220 GAGVIGVELGSVWARLGADVTAVELLGHVGGIGID 254
Yeast          209 GGGIIGLEMGSVYSRLGSKVTVVEFQPQIGGPM-D 242
P. putida      178 GGGYIGLGLGIAYRKLGQVSTVVEARERILP-TYD 211
E. coli        180 GGGILGLEMGTVYHALGSQIDVVEMFDQVIP-AAD 213
```

Retrieve human erythrocyte glutathione reductase (3GRS.pdb) and perform the triple mutations by replacing Ile217 with Glu (I217E), Arg218 with Phe (R218F), and Arg224 with Gly (R224G). Save the energy-minimized mutant enzyme as 1GRM.pdb. Dock NAD$^+$ and NADP$^+$, respectively, to the mutant enzyme (save the complexes as 1GRM.pdb and 1GRN.pdb, respectively). Evaluate the structural and thermodynamic contributions of these three residues by comparing the mutant enzyme and complexes (1GRM.pdb, 1GRN.pdb and 1GRP.pdb) against the similarly energy-minimized wild-type enzyme and complexes (3GRS.pdb, 1GRA.pdb, and 1GRB.pdb). Construct animated kinemage files to display their structural differences with views of the complete molecular structures and close-ups of FAD, NAD(P) bound to nicotinamide nucleotide-disulfide oxidoreductase active site (identified by ProSite).

REFERENCES

Birktoft, J. J., Kraut, J., and Freer, S. T. (1976) *Biochemistry* **15**:44–81.

Bocnegra, J. A., Scrutton, N. S., and Perham, R. N. (1993) *Biochemistry* **32**:2737–2740.

Bohacek, R. S., and McMartin, C. (1997) *Curr. Opin. Chem. Biol.* **1**:157–161.

Böhm, H. J. (1992) *J. Comput. Aided Mol. Des.* **6**:61–78.

Bonneau, R., and Baker, D. (2001) *Annu. Rev. Biophys. Biomol. Struct.* **30**:173–189.

Brändén, C.-I. (1980) *Q. Rev. Biophys.* **13**:317–338.

Bryant S. H., and Altschul S. F. (1995) *Curr. Opin. Struct. Biol.* **5**:236–244.

Bystroff, C., and Kraut, J. (1991) *Biochemistry* **30**:2227–2239.

Chinea, G., Padron, G., Hooft, R. W. W., Sander, C., and Vriend, G. (1995) *Proteins* **23**:415–428.

Chothia, C., and Lesk, A. M. (1986). *EMBO J.* **5**:823–826.

Crighton, T. E. (1977) *J. Mol. Biol.* **113:**275–293.

Dandekar, T., and Argos, P. (1994) *J. Mol. Biol.* **236:**844–861.

Eisen, M. B., Wiley, D. C., Karplus, M., and Hubbard, R. E. (1994) *Proteins* **19:**199–221.

Ewing, T., and Kuntz, I. D. (1997) *J. Comput. Chem.* **18:**1175–1189.

Feng, Z. K., and Sippl, M. J. (1996) *Fold Des.* **1:**123–132.

Gibrat, J.-F., Madej, T., and Bryant, S. H. (1996) *Curr. Opin. Struct. Biol.* **6:**377–385.

Hilbert, M., Bohm, G., and Jaenicke, R. (1993) *Proteins* **17:**138–151.

Holm, L., and Sander, C. (1993) *J. Mol. Biol.* **233:**123–138.

Jones, G., Willett, P., Glen, R. C., Leach, A. R., and Taylor, R. (1997) *J. Mol. Biol.*, **267:**727–748.

Jorgensen, W. L., Chandrasekhas, J., Madura, J. D., Impey, R. W., and Klein, M. L. (1983). *J. Chem. Phys.* **79:**926–935.

Koshland, D. E. Jr. (1958) *Proc. Natl. Acad. Sci. U.S.A.* **44:**98–104.

Koza, J. (1993) *Genetic Programming.* MIT Press, Cambridge.

Kuntz, I. D., Blaney, J. M., Oatley, S. J., Langridge, R., and Ferrin, T. E. (1982) *J. Mol. Biol.* **161:**269–288.

Lest, A. M., and Chothia, C. (1984) *J. Mol. Biol.* **174:**175–191.

Levine, M., Stuart, D., and Williams J. (1984) *Acta Crystallogr.* **A40:**600–610.

Levitt, M., and Chothia, C. (1976) *Nature* **261:**552–558.

Lu, G. G. (2000) *J. Appl. Crystallogr.* **33:**176–183.

Morris, G. M., Goodsell, D. S., Huey, R., and Olson, A. J. (1996) *J. Comput. Aided Mol. Des.* **10:**293–304.

Needleman, S. B., and Wunsch, C. D. (1970) *J. Mol. Biol.* **48:**443–453.

Perutz, M. F. (1989) *Mechanisms of Cooperativity and Allosteric Regulation in Proteins.* Cambridge University Press, Cambridge.

Phillips, D. C. (1967) *Proc. Natl Acad. Sci., U.S.A.* **57:**484–495.

Rarey, M., Kramer, B., Lenguer, T., and Klebe, G. (1996) *J. Mol. Biol.* **261:**470–489.

Richardson, D. C., and Richardson, J. S. (1992) *Protein Sci.* **1:**3–9.

Richardson, D. C., and Richardson, J. S. (1994) *Trends Biochem. Sci.* **19:**135–138.

Scrutton, N. S., Berry, A., and Perham, R. N. (1990) *Nature* **343:**38–43.

Shindyalov, I. N., and Bourne, P. E. (1998) *Protein Eng.* **9:**739–747.

Shindyalov, I. N., and Bourne, P. E. (2001) *Nucleic Acids Res.* **29:**228–229.

Stewart, D. E., Weiner, P. K., and Wampler, J. E. (1987) *J. Mol. Graph.* **5:**133–140.

Szustakowski, J. D., and Weng, Z. P. (2000) *Proteins* **38:**428–440.

Tsai, C. S. (1997) *Int. J. Biochem. Cell Biol.* **29:**325–334.

Tulinsky, A., Park, C. H., and Skzypezak-Jankun, E. (1988) *J. Mol. Biol.* **202:**885–901.

Wilks, H. M., Hart, K. W., Feeney, R., Dunn, C. R., Muirhead, H., Chia, W. N., Barstow, D. A., Atkinson, T., Clarke, A.R., and Holbrook, J. J. (1988) *Science* **242:**1541–1544.

Wilks, H. M., Halsall, D. J., Atkinson, T., Chia, W. N., Clarke, A. R., and Holbrook, J. J. (1990) *Biochemistry* **29:**8587–8591.

Xavier, D., Mark, A. E. and van Gunsteren, W. F. (1998) *J. Comput. Chem.* **19:**535–547.

Appendix 1

LIST OF
SOFTWARE PROGRAMS

Chapter	Software	Source
2	Microsoft Access	http://www.microsoft.com
	Microsoft Excel	http://www.microsoft.com
	SPSS	http://www.spsscience.com
	SyStat	http://www.spsscience.com
4	ACD/3D Viewer	http://www.acdlabs.com/downloar/download.cgi
	ChemOffice	http://www.chemsoft.com
	Cn3D	http://www.ncbi.nlm.nih.gov/Structure/CN3D/cn3d.html
	ISIS Draw	http://www.mdli.com/download/isisdraw.html
	KineMage	http://orca.st.usm.edu/~rbateman/kinemage/
	RasMol	http://www.umass.edu/microbio/rasmol/index2.htm
	WebLab Viewer Lite	http://www.accelrys.com/viewer/viewerlite/index.html
5	ChemFinder	http://www.chemfinder.com
	PeakFit	http://www.spsscience.com
6	DynaFit	http://www.biokin.com/
7	Leonora	http://ir2lcb.cnrs-mrs.fr/~athel/leonora0.htm
	KinTekSim	http://www.kintek-corp.com/kinteksim.htm
8	Gepasi	http://www.gepasi.org
9	BioEdit	http://www.mbio.ncsu.edu/RnaseP/info/program/BIOEDIT/bioedit.html
11	TGREASE	ftp://ftp.virginia.edu/pub/fasta/
12	PROSITE database	http://www.expasy.ch/databases/prostie/
	PROSITE database	ftp://ftp.ebi.ac.uk/pub/databases/profiles/
	PROSITE database	ftp://ftp.isrec.isb-sib.ch/sib-isrec/profiles/
	WPDB	http://www.sdsc.edu/pb/wpdb

Chapter	Software	Source
13	PHYLIP	http://evolution.genetics.washington.edu/phylip.html
	TreeView	http://taxanomy.zoology.gla.ac.uk/rod/treeview.html
14	Chem3D	http://www.camsoft.com
	HyperChem	http://www.hyper.com
	PC Spartan	http://www.wavefun.com
15	KineMage	http://orca.st.usm.edu/~rbateman/kinemage/
	Swiss-Pdb Viewer	http://expasy.ch/spdbv/text/getpc.htm

Appendix 2

LIST OF
WORLD WIDE WEB SERVERS

Chapter	Web Server	Uniform Resource Locator
2	Javastat	http://members.aol.com/johnp71/javastat.html
3	AltaVista	http://www.altavista.digital.com
	Bioinformatics sites	http://biochem.kaist.ac.kr/bioinformatics.html
	BSM, University College London (UCL)	http://www.biochem.ucl.ac.ul/bsm/dbbrowser/
	Computation Genome Group, Sanger Centre	http://genomic.sanger.ac.uk/
	DBcat	http://www.infobiogen.fr/services/dbcat/
	DBGet DB link of GenomeNet, Japan	http://www.genome.ad.jp/dbget/
	DNA Database of Japan (DDBJ)	http://www.ddbj.nig.ac.jp/
	EasySearcher 2	http://www.easysearcher.com/ez2.html
	Entrez browser of NCBI	http://www.ncbi.nlm.nih.gov/Entrez/
	Eur Bioinfo Institute (EBI)	http://www.ebi.ac.uk/
	Eur Mol Biol Lab (EMBL)	http://www.embl-heidelberg.de/
	Excite	http://www.excite.com
	Expert Protein Analysis System (ExPASy)	http://expasy.hcuge.ch/
	GenBank	http://www.ncbi.nlm.nih.gov/Web/Genbank/
	Google	http://www.google.com
	Harvard genome research DB and servers	http://golgi.harvard.edu

Chapter	Web Server	Uniform Resource Locator
	HotBot	http://www.hotbot.com
	Human genome project information	http://www.ornl.gov/TechResources/Human_Genome
	INFOBIOGEN Catalog of DB	http://www.infobiogen.fr/services/dbcat/
	InfoSeek	http://www.infoseek.com
	Internet Explore	http://www.microsoft.com/
	IUBio archive	http://iubio.bio.indiana.edu/soft/molbio/Listing.html
	Johns Hopkins Univ. OWL Web server	http://www.gdb.org.Dan/proteins/owl.html
	Links to other bio-Web servers	http://www.gdb.org/biolinks.html
	Lycos	http://www.lycos.com
	Munich Info Center for Protein Seq (MIPS)	http://www.mips.biochem.mpg.de/
	Natl Biotech Info Facility	http://www.nbif.org/data/data.html
	Natl Center for Biotech Information (NCBI)	http://www.ncbi.nlm.nih.gov/
	Netscape Navigator	http://home.netscape.com/
	Pedro's list for molecular biologists	http://www.public.iastate.edu/~pedro/
	PDB at Resear Collabor for Struct Bioinfo (RCSB)	http://www.rcsb.org/pbd/
	Protein Information Resource (PIR)	http://pir.georgetown.edu
	Survey of Molecular Biology DB and servers	http://www.ai.sri.com/people/mimbd/
	SWISS-PROT, Protein Seqence DB	http://expasy/hcuge.ch/sprot/
	TIGR Database (TDB)	http://www.tigr.org/tdb/
	WebCrawler	http://www.webcrawler.com
	Yahoo	http://www.yahoo.com
4	Chemical MIME .	http://www.ch.ic.ac.uk/chemime/
	PDB at RCSB	http://www.rcsb.org/pdb/
	ReadSeq	http://dot.imgen.bcm.tmc.edu:9331/seq-util/Options/readseq.html
	SMILES tutorial	http://www.daylight.com/dayhtml/smiles/
	TOPS cartoons	http://www3.ebi.ac.uk/tops/
	AAindex: Parameters for amino acids	http://www.genome.ad.jp/dbget-bin/
	BioMagResBank (BMRB)	http://www.bmrb.wisc.edu/pages/homeinfo.html
	EBI:Sequences of proteins/ncleic acids	http://www.ebi.ac.uk/
	Entrez: Sequences of proteins/nucleic acids	http://www.ncbi.nlm.nih.gov/Entrez/
	European Large Subunit rRNA Database	http://rrna.uia.ac.be/lsu/index.html
	European Small Subunit rRNA Database	http://rrna.uia.ac.be/ssu/index.html
	GlycoSuiteDB	http://www.glycosuite.com/

Chapter	Web Server	Uniform Resource Locator
	Histone Database	http://genome.nhgri.nih.gov/histones/
	IUBMB	http://www.chem.qmw.ac.uk/iubmb/
	Klotho: General metabolites	http://ibc.wustl.edu/klotho/
	Lipid Bank: Comprehensive lipid info	http://lipid.bio.m.u-tokyo.ac.jp/
	LIPID: Membrane lipid structures	http://www.biochem.missouri.edu/LIPIDS/membrane_lipid.html
	Merck manual	http://www.merck.com/pubs/mmanual/
	Monosaccharide database	http://www.cermav.cnrs.fr/databank/mono/
	Mptopo: Membrane protein topology	http://blanco.biomol.uci.edu/mptopo/
	PDB: 3D structures of biomacromolecules	http://www.rcsb.org/pdb/
	RNA modification database	http://medlib.med.utah.edu/RNAmods
	RNA Structure database	http://rnabase.org
	Spectral Database Systems (SDBS)	http://www.aist.go.jp/RIODB/SDBS/menu-e.html
	tRNA Sequence Database	http://www.uni-bayreuth.de/departments/biochemie/trna/
6	Biomolecular Interaction Network DB (BIND)	http://www.binddb.org/
	Cell Signaling Network Database (CSNDB)	http://geo.nihs.go.jp/csndb/
	GPCRDB	http://www.gpcr.org/7tm/
	NucleaRDB	http://www.receptors.org/NR/
	Receptor database	http://impact.nihs.go.jp/RDB.html
	ReliBase	http://relibae.ebi.ac.uk/
7	2-Oxoacid dehydrogenase complex	http://qcg.tran.wau.nl/local/pdhc.htm
	Aldehyde dehydrogenase	http://www.ucshc.edu/alcdbase/aldhcov.html
	Aminoacyl-tRNA synthetases	http://rose.man.poznan.pl/aars/index.html
	Brenda: General enzyme DB	http://www.brenda.uni-koeln.de/
	CAZy: Carbohydrate active enzymes	http://afmb.cnrs-mrs.fr/~pedro//CAZY/db.html
	EMP: Summary of enzy literatures	http://wit.mcs.anl.gov/EMP/
	Enzyme Commission, EC	http://www.chem.qmw.ac.uk/iubmb/enzyme/
	ENZYME database	http://www.expasy.ch/enzyme/
	Enzyme Structures Database	http://www.biochem.ucl.ac.uk/bsm/enzyme/index.html
	Esther: Esterases	http://www.ensam.inra.fr/cholinesterase/
	G6P dehydrogenase	http://www.nal.usda.gov/fnic/foodcomp/
	Leonora: Enzyme kinetics	http://ir2lcb.cnrs-mrs.fr/~athel/leonora0.htm
	LIGAND	http://www.genome.ad.jp/dbget/ligand.html
	MDB: Metalloenzymes	http://metallo.scripps.edu/
	Merops: Peptidases	http://www.bi.bbsrc.ac.uk/Merops/Merops.htm
	PKR: Protein kinase	http://pkr.sdsc.edu
	PlantsP: Plant protein kinases & phosphatase	http://PlantP.sdsc.edu

Chapter	Web Server	Uniform Resource Locator
	Promise: Prosthetic group/Metal enzymes	http://bmbsgi11.leads.ac.uk/promise/
	Proteases	http://delphi.phys.univ-tours.fr/Prolysis
	REBASE: Restriction enzymes	http://rebase.neh.com/rebase/rebase.html
	Ribonuclease P Database	http://www.mbio.ncsu.edu/RnaseP/home.html
8	Biocatalysis/ BiodegradatDatabase (UM-BBD)	http://www.labmed.umn.edu/umbbd/index.html
	Boehringer Mannheim	http://expasy.houge.ch/cgi-bin/search-biochem-index
	EcoCyc	http://ecocyc.panbio.com/ecocyc/
	KEGG, home	http://www.genome.ad.jp/kegg/
	KEGG Pathway site	http://www.genome.ad.jp/dbget/
	PathDB	http://www.ncgr.org/software/pathdb/
	Soybase	http://cgsc.biology.yale.edu/
	TRANSFAC	http://transfac.gbf.de/TRANSFAC/cl/cl.html.
9	American Type Culture Collection	http://www.atcc.org/
	BLAST at DDBJ	http://spiral.genes.nig.ac.jp/homology/top-e.html
	BLAST at EBI	http://www2.ebi.ac.uk/blastall/
	BLAST at NCBI	http://www.ncbi.nlm.nih.gov/BLAST/
	Codon Usage DB	http://www3.ncbi.nlm.nih.gov/htbin-post/Taxonomy/wprintge?/ mode = t
	DDBJ	http://www.ddbj.nig.ac.jp/.
	EMBL	http://www.ebi.ac.uk/
	Entrez	http://www.ncbi.nlm.nih.gov
	European Bioinformatics Institute (EBI)	http://www.ebi.ac.uk/embl/Access/index.html
	FASTA at EBI	http://www2.ebi.aci.uk/fasta3/)
	FASTA at PIR	http://www-nbrf.geogetown.edu/pirwww/search/fasta.html
	GenBank	http://www.ncbi.nlm.nih.gov/Genbank/
	Genome Information Broker, GIB	http://mol.genes.nig.ac.jp/gib/
	Genome OnLine Database	http://igweb.integratedgenomics.com/GOLD/
	Keynet	http://www.ba.cnr.it/keynet.html
	Molecular Probe Database	ftp://ftp.biotech.ist.unige.it/pub/MPDB
	Primer3, Whitehead Institute/MIT	http://www-genome.wi.mit.edu/cgi-bin/primer/primer3_www.cgi
	REBASE	http://rebase.neb.com/rebase/rebase.html
	Riken Gene Bank	http://www.rtc.riken.go.jp/
	Web Primer	http://genome-www2.stanford.edu/cgi-bin?SGD/web-pirmer
	Webcutter	http://www.firstmarket.com/cutter/cut2.html
10	Aat	http://genome.cs.mtu.edu/aat.html
	AceDB	http://www.sanger.ac.uk/Software/Acedb/
	AtDB	http://genome-www.stanford.edu/Arabidopsis
	BCM GeneFinder	http://dot.imgen.bcm.tmc.edu:9331/gene-finder/gfb.html
	Bibliographies, computl. gene recognition	http://linkage.rockefeller.edu/wli/gene/right.html
	Celera	http://www.celera.com

Chapter	Web Server	Uniform Resource Locator
	Cellular Response Database	http://LH15.umbc.edu/crd
	CGG GeneFinder	http://genomic.sanger.ac.uk/gf/gfb.html
	ClustalW at EBI	http://www.ebi.ac.uk/clustalW/
	Codon usage database	http://biochem.otago.ac.nz:800/Transterm/homepage.html
	CropNet	http://synteny.nott.ac.uk
	dbEST	http://www.ncbi.nlm.nih.gov/dbEST/index.html
	DBGet	http://www.genome.ad.jp/dbget/
	DNA Data Bank of Japan	http://www.ddbj.nig.ac.jp
	EBI	http://www.ebi.ac.uk/
	EcoGene	http://bmb.med.miami.edu/EcoGene/EcoWeb
	EID	http://mcb.harvard.edu/gilbert/EID/
	EMBL Nucleotide Sequence Database	http://www.ebi.ac.uk/embl.html
	EMGlib	http://pbil.univ-lyon1.fr/emglib/emglib.html
	Entrez	http://www.ncbi.nlm.nih.gov
	EPD	http://epd.isb-sib.ch/seq_download.html
	Eukaryotic polymerase II promoter server	http://www.epd.isb-sib.ch
	ExInt database	http://intron.bic.nus.edu.sg/exint/extint.html
	ExPASy, tools	http://www.expasy.ch/tools/
	FlyBase	http://www.fruitfly.org
	GDB	http://www.gdb.org
	GenBank,	http://www.ncbi.nlm.nih.gov/GenBabk/
	GeneCards	http://bioinformatics.weizmann.ac.il/cards/
	GeneID	http://www1.imim.es/software/geneid/index.html
	GeneMark at EBI	http://www2.ebi.ac.uk/genemark/
	GeneQuiz of EBI	http://columbs.ebi.ac.uk:8765/
	GeneStudio	http://studio.nig.ac.jp/
	Genie	http://www.fruitfly.org/seq_tools/genie.html
	GenLang	http://www.cbil.upenn.edu/genlang/genlang_home.html
	Genome Information Broker	http://gib.genes.ac.jp/gib_top.html
	GenScan	http://genes.mit.edu/GENSCAN.html
	Globin Gene Server	http://globin.csc.psu.edu
	Grail	http://compbio.ornl.gov/Grail-1.3/Toolbar.html
	Grail/Grail 2	http://compbio.ornl.gov/gallery.html
	HuGeMap	http://www.infobiogen.fr/services/Hugemap
	Human Developmental Anatomy	http://www.ana.ed.ac.uk/anatomy/database/
	INE	http://rgp.dna.affrc.go.jp/gict/INE.html
	Intron server	http://nutmeg.bio.indiana.edu/intron/index.html
	Keynet	http://www.ba.cnr.it/keynet.html
	Merck Gene Index	http://www.merck.com/mrl/merck_gene_index.2.html
	MitBase	http://www3.ebi.ac.uk/Research/Mitbase/
	Mzef	http://www.cshl.org/genefinder/
	NCBI	http://www/ncbi.nlm.nih.gov/
	NRSub	http://pbil.univ-lyon1.fr/nrsub/nrsub.html
	ORF Finder	http://www.ncbi.nlm.nih.gov/gorf/gorf.html
	PEDANT	http://pedant.mips.biochem.mpg.de/
	Procrustes	http://www.hto.usc.edu/software/procrustes
	RepeatMasker	http://www.genome.washington.edu/analysistools/repeatmask.htm

Chapter	Web Server	Uniform Resource Locator
	Result of ClustalW at EBI	http://www.ebi.ac.uk/servicestmp/.html
	RsGDB	http://utmmg.med.uth.tmc.edu/sphaeroides
	Sanger Centre, home	http://genomic.sanger.ac.uk/
	Sanger Centre, Nucleotide Sequence Analysis	http://genomic.sanger.ac.uk/gf/gfb.html
	Sanger Centre, Genomic database	http://genomic.sanger.ac.uk/inf/infodb.shtml
	SEQANALREF	http://expasy.hcuge.ch
	SDG	http://genome-www.stanford.edu/Saccharomyces
	TIGR	http://www.tigr.org/tdb/tdb.html
	UniGene	http://www/ncbi.nlm.nih.gov/UniGene/
	UTRdb	http://bigrea.area.ba.cnr.it:8000/srs6/
	Uwisc	http://www.genetics.wisc.edu/
	WebGene, home	http://www.itba.mi.cnr.it/webgene/
	WebGene, Genebuilder	http://125.itba.mi.cnr.it/~webgene/genebuilder.html
	WebGene, GeneView	http://125.itba.mit.cnr.it/~webgene/wwwgene.html
	WebGene, ORFGene	http://125.itba.mi.cnr.it/~webgene/wwworfgene2.html
	WebGene, Splicing signals prediction	http://www.itba.mi.cnr.it/~webgene/wwwspliceview.html
	WebGeneMark	http://genemark.biology.gatech.edu/GeneMark/ webgenemark.html
	ZmDB	http://zmdb.iastate.edu/
11	AAindex database	http://www.genome.ad.jp/dbget/aaindex.html
	ABIM	http://www.up.univ-mrs.fr/~wabim/d_abim/compo-p.html
	BCM, Baylor College of Medicine	http://dot.imgen.bcm.tme.edu.9331/multi-align/multi-align.html
	BLOCKS	http://www.blocks.fhcrc.org/
	ClustalW at EBI	http://www2.ebi.ac.uk/clustalw/
	DDBJ	http://srs.ddbj.nig.ac.jp/index-e.html
	EBI	http://www2.ebi.ac.uk/clastalw/
	Entrez	http://www.ncbi.nlm.nih.gov/Entrez
	ExPASy, home	http://www.expasy.ch/
	ExPASy Proteomics tools	http://www.expasy.ch/tools/
	Homology Search System of DDBJ	http://spiral.genes.nig.ac.jp/homology/tope.html
	IBC of Washington University at St. Louis	http://www.ibc.wustl.edu/msa/clustal.html
	IDENTIFY	http://dna.Stanford.edu/identify/
	MIPS	http://www.mips.biochem.mpg.de/
	OWL	http://www.bioinf.man.ac.uk/dbbrowser/OWL/
	PANAL, University of Minnesota	http://mgd.ahc.umn.edu/panal/run_panal.html
	PIR	http://pir.georgetown.edu
	PredictProtein of Columbia University	http://cubic.bioc.columbia.edu/predictprotein/
	PRINTS	http://www.bioinf.man.ac.uk/dbbrowser/PRINTS/
	PROPSEARCH	http://www.embl-heidelberg.de/prs.html
	ProtScale	http://www.expasy.ch/cgi-bin/protscale.pl
	PROSITE	http://www.expasy.ch/sprot/prosite.html

Chapter	Web server	Uniform resource locator
	Proteome Analysis Database	http://ebi.ac.uk/proteome/
	SRS of EBI	http://srs.ebi.ac.uk/
	SWISS-PROT	http://expasy.hcuge.ch/sprot/sprot-top.html
	TMAP	http://www.embl-heidelberg.de/tmap/tmap_info.html
	Tmpred	http://www.ch.embnet.org/software/TMPRED_form.html
12	3D_PSSM	http://www.bmm.icnet.uk/~3dpssm/
	AAindex	http://www.genome.ad.jp/aaindex/
	ASTRAL	http://astral.stanford.edu/
	BCM, Baylor College of Medicine	http://www.hgsc.bcm.tmc.edu/search_launcher/
	BLOCKS	http://www.blocks.fhcrc.org/
	BMERC	http://bmerc-www.bu.edu/psa/index.html
	CATH	http://www.biochem.ucl.ac.uk/bsm/cath/
	CBS, Center for Biol Sequence Analysis	http://www.cbs.dtu.dk/
	CBS, NetGly	http://www.cbs.dtu.dk/services/NetOGly-2.0/
	CBS, NetPhos	http://www.cbs.dtu.dk/services/NetPhos/
	CBS, SignalP	http://www.cbs.dtu.dk/services/SignalP/
	DOE-MBI	http://fold.doe-mbi.ucla.edu/Login/
	Enzyme Structure Database, Enzyme DB	http://www.biochem.ucl.ac.uk/bsm/enzyme/index.html
	ExPASy Proteomics tools	http://www.expasy.ch/tools/
	FUGUE	http://www-cryst.bioc.cam.ac.uk/~fugue/prfsearch.html
	G To P site	http://spock.genes.nig.ac.jp/~genome/gtop.html
	JOY	http://www-cryst.bioc.cam.ac.uk/~joy/
	Jpred	http://jura.ebi.ac.uk:8888/index.html
	LIBRA I	http://www.ddbj.nig.ac.jp/E-mail/libra/LIBRA_I.html
	Mitochondrial targeting sequence prediction	http://www.mips.biochem.mg.de/cgi-bin/proj/medgen/mitofilter
	ModBase	http://guitar.rockefeller.edu/modbase/
	MMDB	http://www.ncbi.nlm.nih.gov/Entrez/structure.html
	nnPredict	http://www.cmpharm.ucsf.edu/~nomi/nnpredict.html
	NPS@	http://npsa-pbil.ibcp.fr
	Nucleic Acid Database (NDB)	http://ndbserver.rutgers.edu/NDB/ndb.html
	Protein Data Bank, PDB	http://www.rcsb.org/pdb/
	PDBSum	http://www.biochem.ucl.ac.uk/bsm/pdbsum/
	PEST sequence search	http://www.icnet.uk/LRITu/projects/pest/
	Pfam	http://www.sanger.ac.uk/
	PPSearch	http://www2.ebi.ac.uk/ppsearch/
	Predator	http://www.embl-heidelberg.de/cgi/predator_serv.pl
	PredictProtein	http://cubic.bioc.columbia.edu/predictprotein/
	ProDom	http://www.toulouse.inra.fr/prodom.html
	ProfileScan	http://www.isrec.isb-sib.ch/software/PFSCAN_form.html
	PROSITEscan	http://www.expasy.ch/tools/scnpsite.html
	Protein Data Bank at RCSB	http://www.rcsb.org/pdb/
	PSIpred	http://insulin.brunel.ac.uk/psipred/
	Relibase	http://relibase.ebi.ac.uk/
	SCOP	http://scop.mrc-lmb.cam.ac.uk/scop/

Chapter	Web Server	Uniform Resource Locator
	SSThread	http://www.ddbj.nig.ac.jp/E-mail/ssthread/www_service.html
	SSThread of DDBJ	http://www.ddbj.nig.ac.jp/E-mail/ssthread/www_service.html
	TargetP	http://www.cbs.dtu.dk/services/TargetP/
	ClustalW at EMI	http://www.ebi.ac.uk/
13	BCM, Baylor College of Medicine	http://dot.imgen.bcm.tme.edu.9331/multi-align/multi-align.html
	DDBJ	http://www.ddbj.nig.ac.jp/
	EBI	http://www2.ebi.ac.uk/clastalw/
	List of phylogenetic analysis software	http://evolution.gentics.washington.edu/philip/software.html
	ReadSeq	http://dot.imgen.bcm.tmc.edu:9331/seq-util/Options/readseq.html
	WebPHYLIP	http://sdmc.krdl.org.sg:8080/~lxzhang/phylip/
14	AMBER server	http://www.amber.ucsf.edu/ amber/amber.html
	B server	http://www.scripps.edu/~nwhite/B/indexFrames.html
	Cambridge Crystallography Data Centre, CCDC	http://www.ccdc.cam.ac.uk/
	DNA mfold	http://bioinfo.math.rpi.edu/~mfold/dna/form1.cgi
	Protein Data Bank, PDB	http://www.rcsb.org/pdb/
	RNA mfold	http://bioinfo.math.rpi.edu/~mfold/rna/form1.cgi
	SWEET	http://dkfz-heidelberg.de/spec/sweet2/doc/mainframe.html
	SWEET template	http://www.dkfz-heidleberg.de/spec/sweet2/doc/input/liblist.html
	Swiss-Pdb Viewer	http://www.expasy.ch/spdbv/mainpage.html
15	Accelrys, Inc.	http://www.accelrys.com
	B server	http://www.scripps.edu/~nwhite/B/indexFrames.html
	Biomer authentification certificate	http://www.scripps.edu/~nwhite/B/Security/
	CambridgeSoft Corp.	http://www.camsoft.com
	CE server	http://cl.sdsc.edu/ce.html
	Dali	http://www2.ebi.ac.uk/deli
	Hypercube, Inc.	http://www.hyper.com
	K2	http://sullivan.bu.edu/kenobi
	ProSup	http://anna.camesbg.ac.at/prosup/main.html
	Swiss-Model	http://www.expasy.ch/swissmod/
	Swiss-Pdb Viewer	http://www.expasy.ch/spdbv/mainpage.html
	TOP	http://bioinfo1.mbfys.lu.se/TOP
	Tripos, Inc.	http://www.tripos.com
	VAST	http://www.ncbi.nlm.nih.gov:80/Structure/VAST/vast.shtml
	Wavefunction, Inc.	http://www.wavefun.com

Appendix 3

ABBREVIATIONS

ACR	ancient conserved region
AI	artificial intelligence
ANOVA	analysis of variance
BLAST	basic local alignment search tool
bp	basepairs
cAMP	$3',5'$-cyclic adenosine monophosphate
CD	circular dichroism
cDNA	complementary deoxyribonucleic acid(s)
CDS (cds)	coding sequence
CGG	Computational Genomics Group
DB (db)	database(s)
DDBJ	DNA Database of Japan
DF	degree of freedom
DNA	deoxyribonucleic acid(s)
EBI	European Bioinformatics Institute
EC	Enzyme Commission
EF	electrofocusing
(E)FF	(empirical) force field
EMBL (embl)	European Molecular Biology Laboratory
EST	expressed sequence tag
ExPASy	Expert Protein Analysis System
FAD	flavine adenine dinucleotide
FT	fourier transform
FTP (ftp)	file transfer protocol
GIF (gif)	graphical interchange format
GTO	Gaussian-type orbital

HMM	hidden Markow model
HPLC	high performance liquid chromatography
HSP	high scoring pair
HTML	hypertext markup language
HTTP (http)	hypertext transfer protocol
IC	International Nucleotide Sequence Database Collaboration
ID	identification
INDEL	insertion and deletion
IP	Internet protocol
IR	infrared
ISP	Internet service provider
IUBMB	International Union of Biochemistry and Molecular Biology
JIPID	Japanese International Protein Sequence Database
JPEG (jpg)	joint photographic experts group
LAN	local area network
MD	mutation data
MolD	molecular dynamics
MIME	multipurpose Internet mail extensions
MIPS	Munich Information Center for Protein Sequences
MM	molecular mechanics
MMDB	Molecular Modeling Database
mRNA	messenger ribonucleic acid(s)
MS	mass spectrum
NAD(P)	nicotinamide adenine dinucleotide (phosphate)
NBRF	National Biomedical Research Foundation
NCBI	National Center for Biotechnology Information
NDB	Nucleic Acid Database
NJ	neighbor joining method
NMR	nuclear magnetic resonance
ORD	optical rotatory dispersion
ORF	open reading frame
OUT	operational taxonomic unit
PAGE	polyacrylamide gel electrophoresis
PAM	point accepted mutation
PC	personal computer
PCR	polymerase chain reaction
PDB (PDB)	Protein Data Bank
PIR	Protein Information Resource
PKC	protein kinase C
PPP	point to point protocol
PS (ps)	post script
PWM	position weight matrix
QM	quantum mechanics
RCSB	Research Collaboratory for Structural Bioinformatics
RMS (rms)	root mean square
RNA	ribonucleic acid(s)
rRNA	ribosomal ribonucleic acid(s)
SCF	self-consistent field
SDS	sodium dodecyl sulfate
SEM	standard error of the mean
SQL	structured query language
SRS	sequence retrieval system
SS	sum of squares

STO	Slater-type orbital
TCP	transmission control protocol
TOPS	protein topology cartoons
TrEMBL	translated EMBL
tRNA	transfer (soluble) ribonucleic acid(s)
UPGMA	unweighted pair group method using arithmetic means
URL	uniform resource locator
UTR	untranslated region
UV	ultraviolet
WWW (www)	World Wide Web

Note: Abbreviations used for Web servers, symbols for amino acids, nucleotides (bases) and sugars are not listed.

INDEX